Progress in SEPARATION AND PURIFICATION

Volume 1

An Interscience Series

Progress in
SEPARATION AND
PURIFICATION

VOLUME 1

Edited by

EDMOND S. PERRY

Research Laboratories
Eastman Kodak Company
Rochester, New York

INTERSCIENCE PUBLISHERS
a division of John Wiley & Sons, New York · London · Sydney

Library of Congress Catalog Card Number 67-29539
Printed in the United States of America

Preface

Through the ages man has been aware of the importance of separating the valuable from the less valuable in nature's mixtures. This trait has persisted; it is difficult to single out an area of science of a science-based industry where separations and purifications do not play an important role. Yet, in spite of this heritage, only within recent times has the science of separations been recognized in its individuality and has finally been accorded independent identity. The beginning of this era is difficult to pinpoint, but it is approximately the time of the second world war. Certainly, the science and technology generated in the struggle to separate the fissionable isotopes, to isolate and purify marketable quantities of antibiotics, the discovery of gas-phase chromatography, and the industries based on solid-state phenomena were important factors in leading the way for the science of separations and purifications as we know it today.

In the intervening twenty-five to thirty years, there has occurred an extensive development in this field. The literature has grown immensely; it is still expanding at a rapid rate for both the science and the technology of the subject, and a forum is needed for specialists as well as those new to the field or those working in related areas to keep up with the development. *Progress in Separation and Purification* is devoted to this purpose. Its broadest objective is to provide its readers with a high degree of current awareness on the progress being made in the large and complex field. We hope that the Series will help the practitioner to keep abreast of the ever growing literature by providing him with authoritative summaries on significant new developments and critical evaluations of new methods, apparatus, and techniques. The organization and condensation of the literature which is now dispersed throughout the chemical, biological, and nuclear sciences will also help to bring to the science and technology of separations its rightful status.

Fast and expeditious reporting are particular objectives of this Series. We plan to publish volumes at intervals commensurate with the procurement of articles. Manuscripts will be processed as received, and a volume will be issued when sufficient material has been assembled. The choice of subject and the manner in which ideas

and opinions are expressed are essentially left to the discretion of
the authors. Our only request of them is to render a service which
will be of value to the reader and to the science and technology
of separation and purification.

The nine articles in this first volume fall within the spirit of this
liberal policy. Each author is an acknowledged leader in the field
and has written on a subject with which he has had intimate experience. The random order of appearance of the articles in the volume
has helped to expedite publication. We shall be happy to receive
suggestions and recommendations for improving the service this Series
purports to provide, and we invite inquiries for publication in the
Series from authors.

My sincere gratitude goes to the authors of this first volume who
were willing to embark with me on this new venture. Special thanks
are due Dr. Arnold Weissberger, who suggested this undertaking and
has provided advice and counsel in getting it under way.

Edmond S. Perry

January, 1968

Contributors to Volume 1

H. J. BIXLER, *Vice President, Research, Amicon Corporation, Lexington, Massachusetts*

J. FELDMAN, *Research Department, U.S. Industrial Chemicals Co., Division of National Distillers and Chemical Corporation, Cincinnati, Ohio*

O. D. FRAMPTON, *Research Department, U.S. Industrial Chemicals Co., Division of National Distillers and Chemical Corporation, Cincinnati, Ohio*

T. H. GOUW, *Chevron Research Company, Richmond, California*

K. KAMMERMEYER, *University of Iowa, Iowa City, Iowa*

R. LEMLICH, *Department of Chemical Engineering, University of Cincinnati, Cincinnati, Ohio*

A. S. MICHAELS, *President, Amicon Corporation, Lexington, Massachusetts*

W. G. PFANN, *Bell Telephone Laboratories, Inc., Murray Hill, New Jersey*

C. J. VAN OSS, *Serum and Plasma Departments, Milwaukee Blood Center, Inc., and Associate Professor of Biology, Marquette University, Milwaukee, Wisconsin*

M. VERZELE, *Laboratory of Organic Chemistry, State University of Ghent, Belgium*

Contents

Principles of Foam Fractionation*

ROBERT LEMLICH

Department of Chemical Engineering
University of Cincinnati
Cincinnati, Ohio

* The present work is supported in part by F.W.P.C.A. Research Grants WP–00161 and WP–00814 from the U.S. Department of Interior.

I. INTRODUCTION

Foam fractionation is a method of partially separating the components of a solution by virtue of differences in their surface activities. The basic idea behind the operation can be explained in the following way.

Consider an open vessel containing a liquid solution. Whatever might be the composition in the bulk of the liquid, the composition of the surface will usually be different. Thus if the surface molecules could be skimmed off, a partial separation of components could be effected.

Obviously, such a procedure would be very awkward. A means for generating greater surface is required. This can be accomplished by deliberately sparging with air or some other gas. The gas is introduced near the bottom of the liquid, and bubbles rise to the top. If the system foams, the foam will overflow, carrying off selectively adsorbed solute on the surface of the bubbles.

A schematic diagram representing a simple batch operation is shown in Figure 1a. Simple continuous operation is shown in Figure 1b. A special case which is discussed later is shown in Figure 1c. Foam fractionation can also occur "incidentally" as when foam is produced by shaking or by the release of dissolved gas. The difference in composition between beer foam and liquid is an example of the latter.

Quite a number of different solutions have now been successfully foam fractionated. These of course include situations in which the solute to be separated is itself surface active. The large-scale removal of detergents from sewage and the laboratory separation of proteins are examples. However foam fractionation has also been used for the separation of non-surface-active solutes. This has been accomplished by the addition of a suitable surfactant which either combines with the solute in question or simply adsorbs it at the bubble surface. In either case the solute is then carried off in the foam. A good example here is the removal of trace radioactive cations. An ion or other solute which is not itself surface active, but which is surface

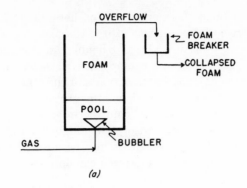

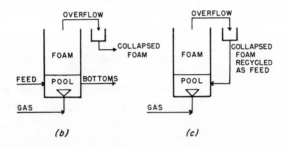

Fig. 1. Foam fractionation in the simple mode: (*a*) batchwise operation, (*b*) continuous flow operation, (*c*) continuous operation with recycle of collapsed overflow as feed.

adsorbed through the addition of a surfactant, is termed a "colligend" (1). For an extensive compilation of various solutions which have been foam fractionated by one means or another, the reader is referred to the reviews by Schoen (2), Rubin and Gaden (3), Cassidy (4), and Shedlovsky (5). The present chapter deals primarily with the principles of foam fractionation, some of which have only recently come to light.

Foam fractionation is actually just one member of a group of related separation techniques which are based on selective adsorption at the surface of rising bubbles. Recently, the generic name "adsorptive bubble separation methods" was proposed for the group, with "adsubble methods" as a convenient contraction (6).

The sizes of the particles separated in the various adsubble methods range from molecular (and sometimes colloidal) in foam fractionation to macroscopic in the familiar froth flotation (ore flotation). Between these two extremes, and to some extent overlapping them and each other, can be found the methods of ion flotation (1,7), microflotation (8,9), and precipitate flotation (10). Nonfoaming adsubble methods include solvent sublation (1) and bubble fractionation (11). A recent paper (11a) proposes a comprehensive scheme of nomenclature.

Needless to say, foam fractionation and the other adsubble methods should never be confused with gas desorption. The latter involves absorption of volatile material into the interior of the bubbles. The adsubble methods involve adsorption of nonvolatile material at the surface of the bubbles. Furthermore, the adsubble methods can enrich as well as strip. Gas desorption can only strip.

II. ADSORPTION

A. Equilibrium

The equilibrium adsorption of dissolved material at a gas–liquid interface is given by Gibbs (12) as eq. (1) where γ is the surface tension, a_i is the activity of the ith component, $\mathbf{R}$ is the gas constant, T is the absolute temperature, and Γ_i is the surface concentration (sometimes termed "surface excess") of adsorbed component i. Typical units for Γ_i are gram moles per square centimeter.

$$d\gamma = -\mathbf{R}T\Sigma\Gamma_i d \ln a_i \tag{1}$$

Unfortunately, this equation is often difficult to apply in practice. Generally speaking, it requires knowledge of activity coefficients so as to obtain activities, and precise measurement of surface tension changes which are sometimes very small. As a result, it is usually practical to apply it only in certain special cases. Foremost among these is the case of a single nonionic surface-active solute in otherwise pure solvent at sufficiently low concentration that the activity coefficient is constant. A nonionic detergent in water below the critical micelle concentration (c.m.c.) is an example. For this case, eq. (1) simplifies to eq. (2),

$$\Gamma = -\frac{1}{\mathbf{R}T}\frac{d\gamma}{d \ln C} \tag{2}$$

where the concentration of surfactant is Γ at the surface and C in the bulk. Equation (2) has been well confirmed by experiment.

With a simple uni-univalent ionic surfactant, eq. (1) simplifies into eq. (3) below the critical micelle concentration.

$$\Gamma = \frac{-1}{2\mathbf{R}T}\frac{d\gamma}{d \ln C} \tag{3}$$

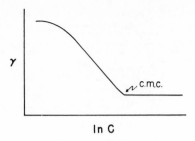

Fig. 2. Typical curve for surface tension versus logarithm of concentration in the liquid.

However, if a considerable excess of one of the ions should be present in solution (due for example to the presence of a salt) the factor of 2 will not appear because C_i for that ion will be constant. Thus eq. (2) will apply instead. Even in the absence of excess ions, experimental results obtained by foaming have not always confirmed the factor of 2 in eq. (3). The reasons for this are not yet clear.

Figure 2 shows the typical shape of a plot of γ versus ln C for a surfactant in water. In accord with eq. (2) or (3), the linear portion of the curve at concentrations below the c.m.c. corresponds to constant Γ. This represents saturation of the surface. Above the c.m.c. eqs. (2) and (3) do not apply. The micelles constitute another species. Nevertheless, the bulk of available evidence indicates that essentially the same constant Γ applies above the c.m.c. as well (at least for foam fractionation), and the addition of further surfactant goes primarily to the formation of micelles. The entire region of constant Γ is generally believed to represent saturation or complete coverage of the surface with a monolayer.

Figure 3 shows a typical curve for Γ versus concentration. According to simple adsorption theory, the lower portion of the curve is linear and can be expressed by eq. (4).

$$\Gamma = KC \tag{4}$$

The upper portion levels off at saturation. For un-ionized material, the entire curve can be approximated by a Langmuir isotherm (13)

$$\Gamma = KC/(1 + K'C) \tag{5}$$

where K and K' are constant. Thus K/K' is Γ at saturation. On the basis of the size of the molecules and their close packing at the surface,

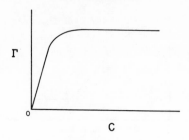

Fig. 3. Typical curve showing the effect of concentration in the liquid on concentration at the surface.

for many surfactants Γ is approximately 3×10^{-10} g-mole/cm² at saturation (14).

The general shape of the curve in Figure 3 often applies not only to the adsorption of a surfactant at the surface but also to the adsorption of a colligend on that surfactant layer. Of course, the particular curves will differ one from the other. Also, the adsorption of colligend will be affected by the surfactant concentration (14), especially if micelles are formed since these will compete for adsorption against the surfactant at the surface (1,15). Below the c.m.c. the adsorption of colligend ions as counterions is theoretically little affected by surfactant concentration provided of course that sufficient surfactant is present to saturate the surface with a surface-active monolayer. If the surfactant present is insufficient for this, then K for the colligend counterion is no longer constant but theoretically becomes directly proportional to Γ for the surfactant. However, foaming is difficult at such low surfactant concentrations.

The surface adsorption of a colligend through chelation or the formation of some other surface-active compound with the surfactant can also be affected by surfactant concentration. It has been suggested that this occurs partly through competition for surface adsorption between the compound on the one hand and the unreacted surfactant on the other (16).

B. Effect of Micelles on Adsorption

It is possible to estimate the effect of micelles on the adsorption of a colligend on a surface-active monolayer under equilibrium conditions, assuming a linear isotherm and assuming the surfactant molecules constituting the monolayer are E times as effective in adsorbing the colligend as are the surfactant molecules which constitute the micelles. This is accomplished in the following way:

Let the concentration of unadsorbed colligend counterions be C_1 in the absence of micelles but at sufficient surfactant concentration to assure a surfactant monolayer. Let C_2 be the concentration of unadsorbed colligend counterions in the presence of micelles. Then by eq. (4),

$$\Gamma_1 = KC_1 \tag{6}$$

and

$$\Gamma_2 = KC_2 \tag{7}$$

where Γ_1 and Γ_2 are the concentration of adsorbed colligend counterion at the surface in the absence and in the presence of micelles, respectively. In the latter case the ratio of colligend adsorbed at the surface to surfactant adsorbed at the surface is KC_2/Γ_s, where Γ_s is the concentration of surfactant at the surface. Assuming all surfactant added above the c.m.c. goes to the formation of micelles, the quantity of colligend counterion adsorbed by the micelles per unit volume of solution is then $(C_s - C_{sc})KC_2/(\Gamma_s E)$ where C_s is the surfactant concentration in the solution, and C_{sc} is the c.m.c. Striking a mass balance for the colligend counterion in a unit volume of pool liquid yields eq. (8).

$$\frac{(C_s - C_{sc})KC_2}{\Gamma_s E} + C_2 = C_1 \tag{8}$$

The mass balance represented by eq. (8) neglects the adsorption at the surface of the bubbles in the liquid pool since the holdup at the surface is generally small compared to the material in the pool. (If otherwise, eq. (8) can be modified to incorporate the holdup with little difficulty.) Combining eqs. (6)–(8) yields

$$\Gamma_2 = \frac{\Gamma_1}{1 + (C_s - C_{sc})K/(\Gamma_s E)} \tag{9}$$

Equation (9) gives the effect of the micelles on Γ for the colligend counterion. The effect on the apparent equilibrium constant can be obtained by defining $K = K_1 = \Gamma_1/C_1$ and $K_2 = \Gamma_2/C_1$, and combining these definitions with eq. (9). The result is

$$\frac{1}{K_2} = \frac{1}{K_1} + \frac{(C_s - C_{sc})}{\Gamma_s E}$$

C. Adsorption Measured through Foam Fractionation

For a surfactant, or for a colligend adsorbed on the surfactant layer, Γ can be found by foam fractionating in the simple mode (17). In the foam, the component in question is partly in the interstitial liquid and partly adsorbed at the bubble surfaces. For spherical bubbles of various diameter d_i, each present in proportion to the number n_i, the rate of surface overflow in the foam is $6G/d$ where G is the volumetric gas flow rate and d is determined according to eq. (10).

$$d = \frac{\Sigma n_i d_i^3}{\Sigma n_i d_i^2} \tag{10}$$

Thus the rate of overflow of the component in the adsorbed state is $6G\Gamma/d$.

If there is no bubble coalescence in the foam, the concentration in the interstitial liquid will be essentially that in the liquid pool, C_W. Thus, from the foregoing,

$$C_Q Q = C_W Q + 6G\Gamma/d \tag{11}$$

where Q is the volumetric rate of foam overflow on a gas-free (collapsed) basis, and C_Q is the concentration in the collapsed foam. Rearranging yields eq. (12).

$$\Gamma = \frac{(C_Q - C_W)Qd}{6G} \tag{12}$$

Thus, from measurements of Q, d, G, C_Q, and C_W, Γ can be found.

Figure 4 shows some apparatus for making such measurements which was developed by Brunner and Lemlich (17). It is a foam fractionation column operating in the simple mode according to Figure 1c, with return of collapsed foam to the bottom in order to maintain steady-state conditions in a convenient manner. The feed inlet is located near the bubbler to counteract local depletion of adsorbable material in the pool. However, if the surfactant concentration in the return stream should be above the c.m.c. and the micelles should be too slow to dissociate upon subsequent dilution in the liquid pool, then any adverse adsorptive effect by such micelles can be avoided by not returning the collapsed foam and feeding the pool with fresh feed instead. Alternatively, but somewhat less satisfactorily, a short batchwise run can be conducted with a liquid pool large enough to maintain a sufficiently constant concentration throughout the run.

Prehumidified nitrogen is used in order to avoid any spurious evaporative effects. A very low height of foam is employed to reduce foam residence time so as to minimize coalescence within the foam. A substantial gas rate also helps, partly by reducing foam residence time and partly by producing a wetter foam (meaning a foam of

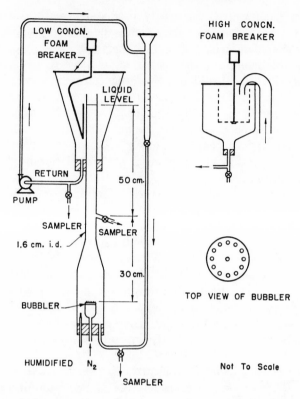

Fig. 4. Standard separator. [From *Ind. Eng. Chem. Fundamentals,* **2**, No. 4, 297 (1963).]

higher liquid content) which is less likely to coalesce. Such internal coalescence not only alters bubble sizes but releases adsorbed material into the interstitial liquid, some of which drains back down to the pool. This reduces the applicability of eqs. (11) and (12). It may also set up an undesirable local concentration gradient below the pool surface. The sum of these errors inclines toward a falsely high Γ. If necessary, in order to minimize any vertical concentration gradient in the liquid pool, the column can be widened to encourage pool mixing. Then with sufficient pool height to assure a close approach to equilibrium between bubble surface and pool liquid, and with negligible coalescence within the rising foam, the apparatus operates essentially as a one-theoretical-stage separator.

At low concentrations of surfactant, the foam is especially unstable and so is allowed to overflow freely from the column into the inverted

Erlenmeyer flask where it is easily broken by gently stirring with a bent rod. Any bends or constrictions would have promoted excessive internal coalescence by straining the foam, thus rupturing bubbles. For high concentrations of surfactant, this is less serious and gentle bends may be tolerated. However sharp constrictions are not employed until the foam has passed over its highest point so that internal coalescence induced by the constrictions will not cause liquid to drain back through the rising foam. The stable foam overflow produced by a high concentration of surfactant is conveniently broken by being whirled through a spinning perforated can inside a large inverted plastic bottle.

Bubble diameters are measured either optically or photographically. This is convenient to do through the glass walls of the column. However there are certain sources of error which should be recognized. The foam bubbles are somewhat distorted at the walls. There is also the question of whether the local distribution of bubble sizes at a wall is the same as that in the interior of the foam. Finally, even if it is, observations at the wall (or at any plane, even one cut through frozen foam) are not representative of the bubble distribution as a whole because the plane tends to miss the smaller bubbles and hit the larger ones. These points are considered in detail by de Vries (18).

Fortunately, much of the difficulty can be obviated by selecting a sparger (bubbler) which gives bubbles of fairly uniform size. To this end, a fritted glass bubbler may not be suitable. Rather, a group of carefully matched capillary tubes or a good spinneret should be employed. Except for its low capacity, a single short capillary tube is sometimes best of all.

Bubble diameters can be optically or photographically measured in the liquid pool. Optical distortion due to curvature of the glass can be avoided by immersing the lower part of the column in an external pool of liquid with flat glass walls.

Γ can also be measured by operating continuously in the simple mode as shown in Figure 1b. Experimentally, this is somewhat more cumbersome than the recyclic operation of Figure 1c. However, recycle of micelles is avoided and eq. (16), which is derived later, can be applied to find Γ. Furthermore, if internal coalescence is a problem, the very similar eq. (24) can be applied with the quantity in parentheses approximated by the constant 6.3. This is discussed later.

The reader is referred to the author's publication for further details (17). For a discussion of chemical equilibrium between surface and bulk, the reader is referred to Sebba (1).

D. Illustrative Problem 1

Statement

A continuous foam fractionation column is operating in the simple mode with 4 cm³/sec of prehumidified nitrogen. Collapsed foam, which is collected overhead at the rate of 0.2 cm³/sec contains 5×10^{-7} g-mole/cm³ of surfactant and 5×10^{-12} g-mole/cm³ of a colligend counterion. The average bubble size according to eq. (10) is 0.1 cm. Liquid withdrawn from the pool contains 2×10^{-7} g-mole/cm³ of surfactant and 2×10^{-13} g-mole/cm³ of the colligend. If internal coalescence is negligible, estimate the surface concentration for the surfactant and for the colligend.

Solution

Applying eq. (12) to the surfactant,

$$\Gamma = \frac{[(5 \times 10^{-7}) - (2 \times 10^{-7})] \times 0.2 \times 0.1}{6 \times 4}$$

$\Gamma = 2.5 \times 10^{-10}$ g-mole surfactant per cm²

Now applying eq. (12) to the colligend,

$$\Gamma = \frac{[(5 \times 10^{-12}) - (2 \times 10^{-13})] \times 0.2 \times 0.1}{6 \times 4}$$

$\Gamma = 4 \times 10^{-15}$ g-mole colligend per cm²

E. Illustrative Problem 2

Statement

For the liquid of the previous problem with the same concentration of colligend, if the critical micelle concentration of the surfactant is 4×10^{-7} g-mole/cm³, estimate the equilibrium surface concentration of adsorbed colligend when the concentration of the surfactant in the liquid is 6×10^{-7} g-mole/cm³. Assume the surfactant molecules that constitute the micelles are equally effective in adsorbing colligend as are the surfactant molecules that constitute the surface monolayer.

Solution

For this system, $E = 1$. From eq. (6) and the results of the previous problem,

$$K = \Gamma_1/C_1 = (4 \times 10^{-15})/(2 \times 10^{-13}) = 0.02 \text{ cm}$$

Substituting in eq. (9),

$$\Gamma_2 = \cfrac{4 \times 10^{-15}}{1 + \cfrac{[(6 \times 10^{-7}) - (4 \times 10^{-7})] \times 0.02}{2.5 \times 10^{-10}}}$$

$$\Gamma_2 = 2.35 \times 10^{-16} \text{ g-mole/cm}^2$$

III. MODES OF COLUMN OPERATION

A foam fractionation column can be operated in several modes. These include the simple mode which has already been discussed, the stripping mode, the enriching mode, and the combined mode. For comparison, the four modes are shown for continuous operation in Figure 5.

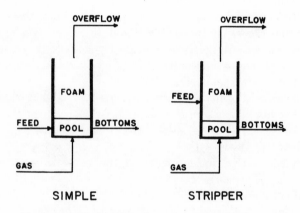

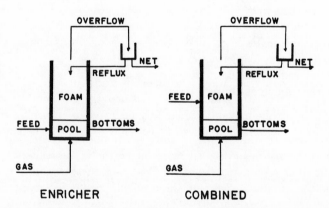

Fig. 5. Continuous foam fractionation columns showing four modes of operation.

In the stripping mode, feed enters the foam some distance above the liquid pool and trickles down countercurrently through the rising foam. This tends to replace interstitial liquid of pool composition with interstitial liquid of feed composition. The overall result is an improvement in the degree of stripping. Quantitatively, this is expressed as an increase in the decontamination factor C_F/C_W or a decrease in the stripping ratio C_W/C_F, where C_W is the concentration in the pool or bottoms and C_F is the concentration in the feed.

It is interesting to note that stripping operation can be so successful in removing surfactant that the amount which remains may be insufficient to properly remove the colligend if any. In such event some additional surfactant can be added to the pool.

The enriching mode involves the use of reflux. A portion of the collapsed overhead is fed back to the top of the column. From there it trickles back down through the rising foam. Since the reflux is much richer than the rising interstitial liquid, the mass transfer resulting from this countercurrent action can greatly cascade the enrichment and yield a much higher concentration in the overhead product than would be obtainable without reflux (19).

If D is the rate at which net overhead product is withdrawn, RD is the rate of reflux flow, where R is the reflux ratio. R can vary from zero (no reflux—simple mode) to infinity (total reflux). By simple material balance, $Q = (R + 1)D$.

C_D is defined as the concentration in the net overhead product. Thus, $C_D = C_Q$. The enriching ratio C_D/C_F is a useful measure of the degree of enrichment.

Figure 5 shows the use of external reflux. However this is not the only source of reflux. Bubble coalescence in the rising foam releases adsorbed material plus some liquid, both of which drain back down through the foam. This constitutes internal reflux. Indeed, some internal coalescence, and hence some internal reflux, is nearly always present in foam fractionation.

Internal reflux can be deliberately increased by widening the column, either along its entire length or just for some distance at the top. This slows the linear rate of foam rise and so permits more thorough drainage of interstitial liquid. This drainage does not itself represent true (useful) reflux. However, it does increase the concentration in the overhead product by reducing the liquid content. It also thins the films separating bubbles. Widening increases their residence time in the column. Both of these effects make for more frequent film rupture in the column. The resulting coalescence releases enriched liquid which constitutes true reflux.

For columns of large diameter, an expanded section at the top may not prove satisfactory if some foam forms pockets in the expanded section thus forcing the rest of the foam to pass through without slowing down sufficiently. In such a situation, Haas and Johnson (20) recommend a horizontal drainage section instead.

Of course lengthening the column is another means for encouraging internal coalescence and reflux by increasing the residence time. The greater length also furnishes increased opportunity for countercurrent contact between the internal streams, thus increasing the overall enrichment.

In batch operation, a long column with internal coalescence can be employed to effect the separation of a complex mixture into fractions according to differences in surface activity. This is accomplished by first running the column for a while with a gas rate just sufficient to drive the foam up nearly to the top of the column. Under these conditions the rate of foam generation just balances the rate of internal coalescence, and the column operates under total internal reflux.

The gas rate is then raised very slowly, and successive fractions of foam are collected separately. The essentially complete removal of a component is sometimes signaled by a sudden drop in foam height within the column. In any case, the gas rate is continually raised until no more foam comes off. The last bit of foam, which is too unstable to reach the top, can be collected from one or more ports lower down the column. The final overall result is a separation into fractions, the sharpness of which depends in part on the composition of the original charge, the height of the column, and the control of the gas rate.

The enriching mode and the stripping mode can be combined to furnish the advantages of both. The resulting combined mode is shown in Figure 5 for continuous operation.

IV. TYPICAL COLUMN PERFORMANCE

Figure 6 shows some apparatus employed by Harper and Lemlich (21) to study typical overall column performance. The column was of glass, 174 cm tall, with an internal diameter of 4.8 cm. The liquid pool comprised approximately 110 cm of this height. Foam filled the remaining 64 cm and overflowed into a rotary foam breaker. Reflux was returned through a multichannel pump. Net overhead product and bottoms were withdrawn through other channels of the pump.

Feed consisted of a $2.2 \times 10^{-4} M$ aqueous solution of the commercial

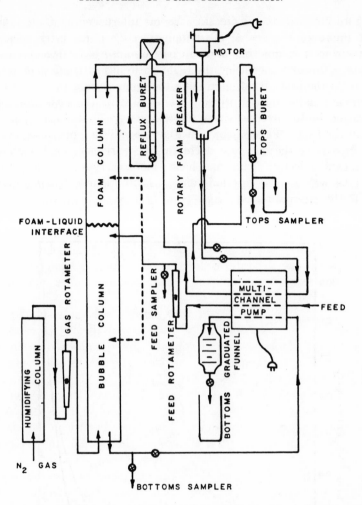

Fig. 6. Apparatus for studying overall column performance in various modes of operation. [From *Ind. Eng. Chem., Process Design Develop.,* **4,** No. 1, 14 (1965).]

surfactant Triton X-100 which has the nominal formula C_8H_{17}—C_6H_4—$(OCH_2CH_2)_{9.7}OH$. It was fed through one channel of the pump at the rate of 32 cm³/min to one of three feed locations.

A factorial group of nine runs was conducted with the feed entering the liquid pool 5 cm below the foam–pool interface. Results are shown in Figures 7 and 8. (The original reference shows the points (0,0) and (1,1) as lower limits for Figures 7 and 8, respectively. However this is not necessarily the case.)

Figure 7 reveals that increasing the gas rate decreases C_W/C_F, that is, it improves the degree of stripping. Higher gas rates generate more surface per unit time and so carry up adsorbed surfactant more rapidly. However increasing the reflux ratio has little effect on stripping. On the other hand, increasing the reflux ratio greatly increases enrichment, as is to be expected. At total reflux the surfactant concentration in the overhead was approximately 70 times as great as that in the feed. This is shown in Figure 8. This figure also reveals that increasing the gas rate decreases the enrichment. A high gas rate means a wetter foam, and hence more carryover of interstitial liquid as well as less internal reflux. R in Figures 7 and 8 refers only to the external reflux.

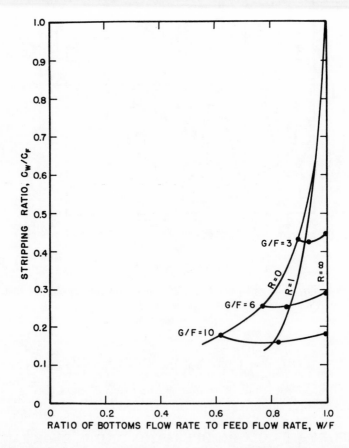

Fig. 7. Typical stripping performance from an enriching column separating a surfactant from water. Decreasing ordinate represents improvement in the degree of stripping.

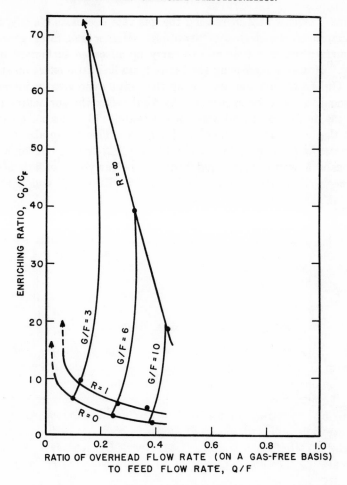

Fig. 8. Typical enriching performance for an enriching column separating a surfactant from water. Increasing ordinate represents improvement in the degree of enrichment.

Lowering the feed inlet to a point approximately midway up the liquid pool was of only slight effect. However raising it to a level 16 cm above the foam–pool interface improved the stripping considerably. This follows from the counter-current action mentioned earlier. The subject of counter-current action is further discussed later.

V. THEORY OF COLUMN OPERATION AND PERFORMANCE

It is possible to make certain quantitative predictions regarding the performance of foam fractionation columns.

A. Simple Mode

As shown earlier, eq. (11) applies to operation in the simple mode. Rearranging that equation gives

$$C_Q = C_W + \frac{6G\Gamma_W}{Qd} \tag{13}$$

where the subscript in Γ_W is added to emphasize that Γ_W is in equilibrium with C_W. By overall mass balance for continuous operation,

$$F = Q + W \tag{14}$$

and
$$C_F F = C_Q Q + C_W W \tag{15}$$

Combining eqs. (13)–(15) yields eq. (16) for continuous operation in the simple mode (22).

$$C_W = C_F - \frac{6G\Gamma_W}{Fd} \tag{16}$$

B. Stripping Mode

For a long stripping column, that is, one in which the feed enters the foam at a sufficient height above the pool, the downflowing interstitial liquid will closely approach equilibrium with the rising interstitial liquid. The concentration pinch will occur at the feed level. The rising interstitial liquid at the top of the column will have a concentration of C_F (if internal coalescence above the feed inlet is negligible). Replacing C_W in eq. (11) with C_F, replacing Γ with Γ_F, and replacing the constant 6 with 6.59 (for reasons stated later) gives

$$C_Q = C_F + \frac{6.59G\Gamma_F}{Qd} \tag{17}$$

where Γ_F is the surface concentration in equilibrium with C_F. A height of several feet is often sufficient to assure a close pinch. Combining eqs. (14), (15), and (17) yields eq. (18) for a long stripping column (22).

$$C_W = C_F - \frac{6.59G\Gamma_F}{Wd} \tag{18}$$

Since W must be less than F, and Γ_F cannot be less than Γ_W, it is clear that C_W from eq. (18) must be lower than C_W from eq. (16). With Γ a known function of C, the comparison permits one to estimate how much further the bottoms can be purified by stripping operation as compared to simple operation.

The small change in the constant was made for the following reasons. The original constant 6 is based on spherical bubbles. In a well-drained foam the bubbles press against each other and are therefore more polyhedral. This is especially so with bubbles of uniform size. Within the foam, the dodecahedral shape predominates (23–25). (This may not seem so under casual observation through a glass retaining wall; however, it must be borne in mind that at the glass the bubbles are connected to a continuous wall and not just to each other.)   For a regular dodecahedron the ratio of surface to volume is $6.59/d$. Accordingly, the constant in eqs. (17) and (18) is more properly 6.59.

However, no such change is made for the simple mode. The constant remains 6 in eqs. (11)–(13) and (16). This is because here the additional surface adsorption, which occurs after the spherical bubbles leave the pool and distort to form polyhedral bubbles in the foam, occurs at the expense of material already in the interstitial liquid. In the simple mode this liquid is not replaced by any fresh incoming stream. Accordingly, the incremental adsorption resulting from the distortion contributes no net increase to the upflow of material. Therefore the constant 6 remains unchanged for the simple mode.

The preceding equations, as well as those to follow, must not be used beyond the range of reason. For example, indiscriminate substitution of numbers could yield negative values for C_W. This of course would be absurd.

C. Illustrative Problem 3

Statement

A foam fractionation column operating continuously in the simple mode gives a decontamination factor of 4.0 for the removal of a particular detergent from water under a certain set of conditions wherein the foam overflow rate (on a gas-free basis) is 0.05 times the feed rate. Under similar conditions, what should the decontamination factor theoretically be for highly efficient stripping?

Solution

Rearranging eq. (16),

$$1 - \frac{C_W}{C_F} = \frac{6G\Gamma_W}{FC_Fd}$$

Rearranging eq. (18),

$$1 - \frac{C_W}{C_F} = \frac{6.59 G \Gamma_F}{W C_F d}$$

Assuming Γ is constant (meaning the adsorbed detergent forms a saturated monolayer), and then dividing the second equation by the first while noting that $W = 0.95F$ here, yields upon substitution of the given decontamination factor.

$$\frac{1 - (C_W/C_F)_{\text{strip}}}{1 - \frac{1}{4}} = \frac{6.59}{6 \times 0.95}$$

Solving, $(C_F/C_W)_{\text{strip}} = 7.5$ which is the theoretical decontamination factor for stripping operation. Of course it must be borne in mind that in changing from the simple mode to the stripping mode, "similar conditions" may be difficult to achieve in practice. In particular, the lower C_W for stripping may make for a higher surface tension and hence a larger d.

D. Illustrative Problem 4

Statement

The system of the previous problem also contains a colligend counterion which is present at a trace level of concentration. For this colligend the decontamination factor is 1.8 for continuous operation in the simple mode. Under similar conditions, what should it theoretically be for highly efficient stripping?

Solution

Since the concentration is very low, the adsorption isotherm should be linear. Therefore, from eq. (4), $\Gamma_W = KC_W$ and $\Gamma_F = KC_F$.

Rearranging eq. (16) and substituting for Γ_W,

$$\frac{C_F}{C_W} - 1 = \frac{6GK}{Fd}$$

Rearranging eq. (18) and substituting for Γ_F,

$$1 - \frac{C_W}{C_F} = \frac{6.59 GK}{Wd}$$

Noting that $W = 0.95F$ here, substituting the given decontamination factor for the simple mode, and dividing the second equation by the first, yields

$$\frac{1 - (C_W/C_F)_{\text{strip}}}{1.8 - 1} = \frac{6.59}{6 \times 0.95}$$

Solving the above equation we obtain $(C_F/C_W)_{strip} = 13.3$ which is the theoretical decontamination factor for stripping operation.

E. Enriching Mode

Figure 9 shows the flow of material for that portion of an enriching column which is situated above the liquid pool. This portion is, of course, the foam. Q' is the rate of the upward flowing interstitial liquid leaving the pool. S is the surface-to-volume ratio for a foam bubble, namely $6/d$ for a sphere or $6.59/d$ for a regular dodecahedron. C_L is the concentration in the downflowing interstitial liquid just before it enters the liquid pool, and C_D is the concentration in the collapsed foam product overhead.

A material balance (17) over the envelope shown in Figure 9 yields eq. (19).

$$\frac{C_D}{C_W} = \frac{(R+1)GS\Gamma_W}{QC_W} + \frac{C_L}{C_W} + \frac{(R+1)Q'(1 - C_L/C_W)}{Q} \qquad (19)$$

As was the case for the stripping column, if the enriching column is sufficiently long the downflowing interstitial liquid may closely approach equilibrium with the rising interstitial liquid (17). The pinch

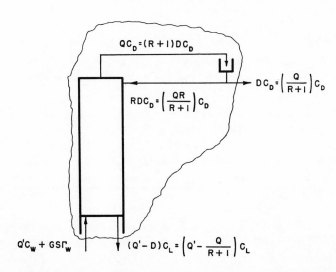

Fig. 9. Material balance around an enriching column. [From *Ind. Eng. Chem. Fundamentals*, **2**, No. 4, 298 (1963).]

will occur just above the liquid pool. Then $C_L = C_W$ and eq. (19) reduces to eq. (20).

$$C_D = C_W + GS\Gamma_W/D \qquad (20)$$

Careful consideration shows that here the incremental adsorption due to the distortion of bubbles from spheres to dodecahedra can only be partly compensated by the downflowing liquid (reflux). Said analysis shows that effectively $S = [6.59 - 0.59/(R + 1)](1/d)$. Substituting in eq. (20) yields eq. (21) for long enriching columns (22).

$$C_D = C_W + \left(6.59 - \frac{0.59}{R + 1}\right)\frac{G\Gamma_W}{Dd} \qquad (21)$$

The range of the parenthetical coefficient is only 6 to 6.59 over the entire possible R range of 0 to ∞.

It is worth noting that eq. (13) is a special case of eq. (21). Also, since $D = Q/(R + 1)$, it is evident that the use of reflux can greatly increase the degree of enrichment.

If the pool is fed continuously at rate F and bled continuously at rate W, an overall balance states that

$$F = D + W \qquad (22)$$

and

$$C_F F = C_D D + C_W W \qquad (23)$$

Combining eqs. (21)–(23) gives eq. (24).

$$C_W = C_F - \left(6.59 - \frac{0.59}{R + 1}\right)\frac{G\Gamma_W}{Fd} \qquad (24)$$

Comparison with eq. (16) which was derived from the simple mode confirms that, unlike the situation for enrichment, the use of reflux does little to improve the degree of stripping. However, for a given C_W and C_F the use of reflux does reduce D/F. This can be of value.

The various equations developed for the several modes apply to the surfactant as well as to any colligend present. As mentioned earlier, Γ for a surfactant can usually be taken as fixed.

Under some conditions, C_D may exceed the critical micelle concentration. Thus micelles may be present in the reflux. These of course would not adversely affect Γ or C_D for the surfactant. Interestingly enough, for a sufficiently long column, theoretically the micelles should not affect C_W or C_D for any colligend either. In fact, for such a column there should be no overall effect at all. To be sure, the micelles may shift the equilibrium adversely, but by the analysis leading to eq. (21) the overall performance of a sufficiently long enricher should not be affected by the shift. In such an enricher, any micelles in the downflowing interstitial liquid are all transferred to the rising interstitial liquid by the time the downflowing liquid reaches the pool. The transferred micelles are then carried up and out the top where a portion of them are continuously bled off in the net overhead product.

The upflowing interstitial stream is larger than the downflowing interstitial stream. This means that any pinch must occur at the start of the upflowing interstitial stream. This can be shown in several ways, including an analogy with an equivalent proof based on the second law of thermodynamics for a long counterflow water-to-water heat exchanger (26). The upflowing stream originates from the pool, and so at its start has a concentration of C_W in the interstitial liquid and Γ_W at its surface.

Of course, the immunity to the effect of micelles does not extend to short columns. For such columns any adverse shift in equilibrium must be balanced off against the beneficial cascading effect of reflux. However, even in such cases, the net effect of using reflux may still be very favorable to the enrichment. Naturally, for the surfactant itself, no adverse effect would be expected at all. Short columns will be considered in some detail later.

F. Combined Mode

For a long combined column; that is, a column consisting of a sufficiently long stripping section and a sufficiently long enriching section, the concentration pinch between the counterflowing streams will occur at the level of the feed. So, by means of an analysis similar to that which preceded, eq. (25) is obtained for a long column operating continuously in the combined mode (22)

$$C_D = C_F + \frac{6.59G\Gamma_F}{Dd} \tag{25}$$

Equation (17) is a special case of eq. (25). Combining eq. (25) with an overall balance gives

$$C_W = C_F - \frac{6.59G\Gamma_F}{Wd} \tag{26}$$

Equation (26) is identical with eq. (18).

G. Illustrative Problem 5

Statement

A continuous foam fractionation column is to operate in the combined mode to remove and concentrate a surfactant, which has a surface concentration of 3×10^{-10} g-mole/cm², from a feed containing 5×10^{-7} g-mole/cm³ of the surfactant. The ratio of gas rate to total collapsed overflow rate is to be 30. With a properly averaged bubble diameter of 0.2 cm, what is the richest overhead that can be expected for a reflux ratio of 10?

Solution

The richest overhead would be obtained with a very long column. Substituting $Q = (R + 1)D$ into eq. (25) gives

$$C_D = C_F + \frac{6.59(R + 1)G\Gamma_F}{Qd}$$

G/Q is given as 30. Substituting numerical values,

$$C_D = 5 \times 10^{-7} + [6.59(10 + 1)30 \times 3 \times 10^{-10}]/0.2$$

$C_D = 3.76 \times 10^{-6}$ g-mole/cm³ maximum.

H. Finite Columns

At this point, the question naturally arises as to how long a column must be in order to qualify as "sufficiently" long. Unfortunately, no general answer can be given short of infinity. Too many factors are involved. At modest reflux ratios, a height of several feet may give a reasonably close approach. At higher reflux ratios, a much greater height may be needed. The author's experience with aqueous solutions of the commercial anionic surfactant Aresket-300 showed roughly a 90% approach to theoretical C_D/C_W at reflux ratios in the neighborhood of 2 with an enriching column 100 cm tall (17). Roughly comparable results were obtained with Triton X-100 (27).

1. OPERATING LINES AND EQUILIBRIUM CURVES

The foregoing equations for long columns can be viewed as limiting equations. Actual performance for short columns can be examined from the familiar points of view which have been established for contacting operations in general. These include the concepts of the theoretical (perfect) stage and the transfer unit (28).

For stripping assemblies, an equilibrium curve of Γ versus C has been used in conjunction with an operating line of slope $\Delta\Gamma/\Delta C = F/GS$ (20,29). This implies $GS\Delta\Gamma = F\Delta C$. Such an equality neglects changes in the solute content of the upflowing interstitial liquid. For some situations this approximation is quite satisfactory. However, for other stripping conditions and for refluxing columns (enriching or combined), the approximation may lead to significant error. When the surface is saturated with the substance in question (as for example a surfactant) the method breaks down completely. For these reasons the present author prefers a different approach which takes into account the material in the upflowing liquid. This approach is outlined next.

Consider the counterflowing streams at some level within a continuous foam fractionation column which may be a stripper, an enricher, or a combination of both. Let the upflowing stream rate be U, and let the concentration in the upflowing interstitial liquid be C. The upflowing rate of the pertinent component is then $UC + GS\Gamma$. Define an effective stream concentration $\bar{C}$ such that $U\bar{C} = UC + GS\Gamma$. Then,

$$\bar{C} = C + GS\Gamma/U \qquad (27)$$

The upflowing interstitial liquid at any level in the column is assumed to be in equilibrium with the upflowing surface at that level. (While this assumption may not be exact, it is clearly more realistic than ignoring the liquid upflow entirely!) Γ can now be related to C from a known equilibrium relationship such as Figure 3. Then with S as, say $6.59/d$, $\bar{C}$ is computed for various values of C at the G/U ratio involved. $\bar{C}$ is then plotted against C to give a new equilibrium diagram.

For the important special case of the linear isotherm, eq. (4) applies. The new equilibrium curve is therefore a straight line through the origin, namely,

$$\bar{C} = (1 + GSK/U)C \qquad (28)$$

At the other extreme, if Γ is constant, the new equilibrium curve is a straight line but with an intercept of $GS\Gamma/U$ and a slope of unity.

On the same coordinates of $\bar{C}$ versus C, the operating line is now plotted. By simple material balance it will have a slope of $\Delta\bar{C}/\Delta C = L/U$ where L is the interstitial liquid downflow rate and C now refers to the downflow. Actual placement of the line depends, of course, on entrance and exit conditions.

The flow rates of the counterflowing streams can be estimated directly from the inlet and outlet stream rates. Thus for a stripper, $L = F$ and $U = Q = F - W$. For an enricher, $L = RD$ and

$U = Q = (R + 1)D$. However L and U so determined are really only the *apparent* downflow and the *apparent* upflow. The true downflow and the true upflow should include the internal recycle of liquid. This recycle is in addition to any from internal coalescence. For example, the true upflow in an enricher is really Q' which, in the absence of internal coalescence, is constant up to the reflux inlet level. At that level part of Q' "turns around" to join RD and both descend through the column.

The internal recycle results from the fact that the interstitial channels are not of fixed dimensions but can expand or contract in a complicated way to accommodate interstitial flow. The amount of recycle can be estimated for any given situation by means of the theory of Leonard and Lemlich for interstitial drainage and overflow which is discussed later. However the computation is somewhat complicated. So for simplicity, L and U will be taken at their apparent values. This implies that mass transfer takes place between the apparent streams rather than the true streams. Within a consistent set of calculations, the results are satisfactory for many purposes.

After the operating line is placed, theoretical stages can be stepped off between it and the equilibrium curve. Figure 10 illustrates the calculation with a stripper where eq. (4) applies, and Figure 11 illustrates a combined column where Γ is constant.

2. TRANSFER UNITS

As an alternative to theoretical stages, the number of transfer units can be determined for the foam. Based on the upflowing stream, this number N_U is given by

$$N_U = \int_{\bar{C}_{W*}}^{C_Q} \frac{d\bar{C}}{\bar{C}^* - \bar{C}} \tag{29}$$

where $\bar{C}$ is related to C by the operating line, and $\bar{C}^*$ is related to C by the equilibrium curve. The lower limit of the integral is the concentration in the upflowing stream just above the pool. Equating it to $\bar{C}_W^*$ is predicated on the pool behaving as a theoretical stage.

Based on the downflowing stream, the number of transfer units N_L is given for the foam by

$$N_L = \int_{C_L}^{C_{top}} \frac{dC}{C - C^*} \tag{30}$$

where C is related to $\bar{C}$ by the operating line, and C^* is related to $\bar{C}$ by the equilibrium curve. C_L is the concentration in the downflowing

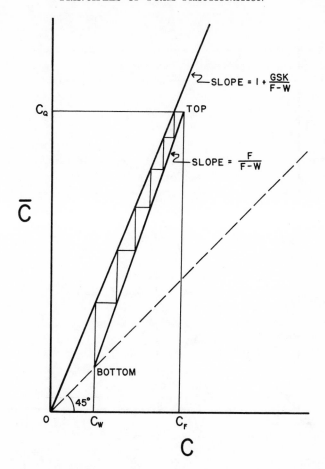

Fig. 10. Graphical stagewise calculation (via apparent streams) for a stripper. The equilibrium curve is based on a linear isotherm $\Gamma = KC$.

stream just before it enters the pool. Taking the pool as a theoretical stage, C_L can be found from C_W by material balance around the pool. Of course, if one wishes to lump together the separation achieved by the pool and the transfer units in the foam, a lower integration limit of C_W can be used in both eqs. (29) and (30), provided the feed inlet is not to the liquid pool.

When the equilibrium curve can be expressed by a simple equation, the number of transfer units can be found by substitution followed by direct integration, and the number of theoretical stages can be found by the calculus of finite differences (30). For example, with

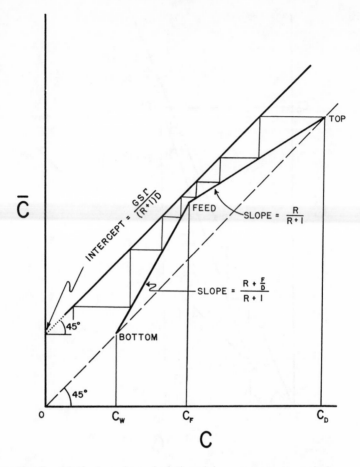

Fig. 11. Graphical stagewise calculation (via apparent streams) for a stripper-enricher combination with feed entering at the match point. The equilibrium curve is based on a constant surface concentration Γ.

a continuous stripping column a material balance yields eqs. (31)–(33), and eq. (28) gives eqs. (34) and (35).

$$F = Q + W \tag{31}$$

$$C = (Q/F)\bar{C} + (W/F)C_W \tag{32}$$

$$C_Q = (F/Q)C_F - (W/Q)C_W \tag{33}$$

$$\bar{C}^* = (1 + GSK/Q)C \tag{34}$$

$$\bar{C}_W^* = (1 + GSK/Q)C_W \tag{35}$$

Combining eqs. (29) and (31)–(35) yields eq. (36) for a continuous stripper operating with a system for which equilibrium adsorption is governed by the linear isotherm $\Gamma = KC$.

$$N_U = \frac{F}{GSK - W} \ln \frac{F(GSK - W)(C_F/C_W) + FW}{GSK(GSK + F - W)} \qquad (36)$$

3. EFFECT OF MICELLES ON COLUMN PERFORMANCE

If micelles are a factor, they can be taken into account. In a re-fluxing column their effect on the adsorption equilibrium of a colligend should be greatest at the top of the column because that is where their concentration is the greatest. Further down the column their effect diminishes as their concentration decreases. For a sufficiently long column their effect vanishes, as discussed earlier.

Equation (9) cannot be used directly to estimate the effect of micelles introduced by reflux into the foam. Equation (9) is derived from eq. (8), and the latter is based on a material balance within the pool where the ratio of surface to volume is small. In the foam this ratio is large and so eq. (8) would first have to be modified to include surface adsorption by adding a term of the order $GSKC_1/U$ to the right-hand side and a term of the order $GSKC_2/U$ to the left-hand side. These changes would in turn modify eq. (9). Also, the variation in C_s along the column must be included. This can be accomplished by stagewise or transfer unit calculation for the surfactant to give values for C_s at each stage or transfer unit, and then substituting these values into the modified eq. (9) to give Γ_2 for the colligend at each stage or transfer unit. Thus $\bar{C}$ at each stage or transfer unit can be found from eq. (27) with $\Gamma = \Gamma_2$. Stagewise or transfer unit calculations can then be carried out for the colligend. Trial and error may be necessary to obtain a match.

4. COLUMN EFFICIENCY

As in the case with other contacting operations, the determination of the number of theoretical stages or transfer units is often only part of the picture. The efficiency of the actual stages, or the actual column height equivalent to a theoretical stage (abbreviated HETS) is frequently required too. Alternatively, the column height equivalent to a transfer unit (abbreviated HTU) may be needed.

These quantities are difficult to predict and, generally speaking, must be determined by experiment. They depend on the intimacy of contact between the countercurrent streams. Fortunately, true foam is by its nature a good packing. This is especially true for relatively dry foam, that is, foam of relatively low liquid content. The interstitial suction opposes channeling and tends to distribute the downcoming liquid evenly over the cross section of the column, thus assuring good contact. HTU values as low as 1 cm have been reported (20).

A single axially located feed inlet generally gives a reasonably uniform distribution of feed in a foam column which does not exceed 2 in. in diameter. In wider columns a feed distributor is advisable, especially for wetter foam. In a 24-in. diameter column, a distributor which supplied 37 streams on 3¾-in. triangular spacing reportedly proved satisfactory (20). The streams impinged on baffles and then dripped into the foam with minimal kinetic energy.

As foam becomes wetter, the interstitial suction decreases and channeling becomes more likely. When the volumetric fraction of the foam occupied by the liquid reaches the void fraction for close packed spheres (which is 0.26 for uniform spheres), the bubbles become free to move. Further increase in the liquid fraction, especially if much beyond 0.30, makes for excessive channeling and poor contact (20,31).

Stages have also been investigated. Alternately placed mixing and settling stages have been used. These were made by constructing a column of alternately narrow and wide diameter (14). However, insofar as the present author is aware, it has not been established that such a configuration is superior to a uniform column of the wider diameter.

Sieve plates, with foam rising up the erstwhile downcomers and liquid dripping down through the holes, gave indifferent results (32). So did screen plates. Bubble cap plates passing foam through the slots yielded efficiencies of up to 30% (14). Individual bubbled pools in countercurrent array have been suggested (29).

It appears from the available evidence that submerging the bubbler 1 ft below the pool surface provides more than enough depth to assure a close approach to equilibrium. Thus the pool at the bottom of a foam fractionation column can generally be considered to be one theoretical stage.

I. Illustrative Problem 6

Statement

An experimental foam fractionation column is operated as a continuous stripper with no bottom stream. In other words, after passing down and up the column, the entire feed exists as the overflow foam. The feed is aqueous and contains surfactant plus 10^{-14} g-mole/cm^3 of an undesirable ion. When the volumetric gas rate is 20 times

as great as the volumetric feed rate, the bubble surface is 100 cm²
per cm³ of gas and the liquid pool contains 10^{-15} g-moles/cm³ of
the ion. The linear equilibrium adsorption constant for the ion is
10^{-4} cm. A very small aqueous stream of concentrated surfactant
is added to the pool in order to maintain sufficient surfactant, but
without exceeding the critical micelle concentration. Estimate the
concentration in the pool if the height of the foam is doubled.

Solution

This problem can be attacked directly with eq. (36). However
it is more instructive to show the use of the prior equations.

In the present problem, $L = F = U = Q$ and $G/F = 20$. From
eq. (28) or (34),

$$\bar{C}^* = (1 + GSK/Q)C \tag{34}$$

By substitution,

$$\bar{C}^* = (1 + 20 \times 100 \times 10^{-4})C = 1.2C$$

and

$$\bar{C}_W^* = 1.2C_W = 1.2 \times 10^{-15}$$

Since $W = 0$, by simple material balance

$$\bar{C} = C$$

and

$$C_Q = C_F = 10^{-14}$$

By substitution

$$\bar{C}^* = 1.2\bar{C}$$

Substitution in eq. (29),

$$N_U = \int_{1.2 \times 10^{-15}}^{10^{-14}} \frac{d\bar{C}}{0.2\bar{C}} = 10.6$$

Let the new situation with the height doubled be denoted by a prime.
Thus $N_U' = 2 \times 10.6 = 21.2$ since doubling the foam height of a
stripper (with negligible internal coalescence) should double the
number of transfer units in the foam. Substituting in eq. (29) for the
new situation

$$21.2 = \int_{1.2C_{W'}}^{10^{-14}} \frac{d\bar{C}}{0.2\bar{C}} = 5 \ln \frac{10^{-14}}{1.2C_{W'}}$$

Solving, $C_{W'} = 1.2 \times 10^{-16}$ g-mole/cm³ in the pool.

VI. BUBBLE FRACTIONATION

Oddly enough, there are some situations where the pool can constitute *more* than one theoretical stage! If the pool itself is tall and

narrow, down-flowing liquid arriving at the pool surface may not mix thoroughly with the rest of the liquid in the pool. As a result, a vertical concentration gradient will set up in the pool. This gradient will further enrich the surface of the rising bubbles before they have even left the pool. Under these conditions, the pool will act as more than a single theoretical stage. In other words, the pool will exhibit a stage efficiency of over 100%.

This happy phenomenon is less likely with operation that is truly in the simple mode, that is, without internal coalescence. Without a rich downflow, little beneficial gradient will form in the pool. With the long stripper, enricher, and combined column, the downflow to the pool surface is there—but, as discussed earlier, it does not arrive rich. Thus only with the short counterflow columns, especially the enricher, can a large gradient form.*

An interesting extreme case of this phenomenon occurs with total internal reflux which may even be virtually instantaneous. This last situation is tantamount to having no foam formation at all. When the bubbles reach the pool surface the gas in them merely passes off, leaving the adsorbed material behind. The concentration gradient so developed can then be put to use since it represents a partial separation of the components in the liquid. The author has named this technique "bubble fractionation." Its potential application is to surface-active systems which, because of low concentration, or for some other reason, foam very slightly or not at all. Batch operation is shown in Figure 12a.

If a bit of foam does form but is ignored in that rich *liquid* is withdrawn from the pool, the technique is still bubble fractionation. However if the foam is withdrawn, then the technique is of course foam fractionation, but with bubble fractionation assisting if there is a vertical concentration gradient in the pool. Indeed, deliberately using bubble fractionation may sufficiently enrich the upper layers of the liquid pool to permit foam fractionation of an otherwise unfoamable system. Thus bubble fractionation can act as a booster to foam fractionation operating above it.

The separation attainable in bubble fractionation is enhanced by increasing the height of the pool and the generation of bubbles. Certain soluble dyes have been effective in giving a separation visible to the eye. In batch operation with several parts per million of crystal violet chloride dissolved in a column of water several centimeters in diameter and several feet tall, the concentration at the top reached

---

* However, a small stripping gradient is always possible.

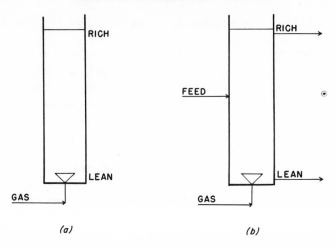

Fig. 12. Bubble fractionation: (*a*) batchwise operation, (*b*) continuous flow operation.

as high as 10 times as great as that at the bottom under favorable conditions. Impurities and additives can increase or decrease the separation.

Continuous bubble fractionation is shown in Figure 12*b*. Rich liquid is withdrawn from the top and lean liquid from the bottom. Feed enters between the two, preferably at or near the level of matching concentration.

Further discussion of bubble fractionation is beyond the scope of the present paper. For additional information, including a theoretical approach, the reader is referred to work on the subject by the author and his co-workers (11,33,34).

VII. FOAM DRAINAGE AND OVERFLOW

A. Theory

So far, little has been said about quantitatively estimating the magnitude of the foam overflow or the magnitude of the true interstitial liquid flow. These are clearly important in developing a comprehensive picture of foam fractionation. Accordingly, a brief outline will now be presented of the theory for foam drainage and overflow developed by Leonard and Lemlich (22). Results of this theory agree fairly well with available experimental data (27,32,35). For further details of theory or experiment, the reader is referred to the original papers.

The bubbles in a foam press against each other, more or less flattening the faces or films in between. These films intersect three at a time to form channels or capillaries which are often called Plateau borders (abbreviated here as PB). With bubbles of uniform size, this means a dihedral angle of 120° between intersecting films, which in turn implies that the bubbles are nearly regular dodecahedra (23–25). This can be shown from geometrical considerations.

The regular dodecahedron was therefore selected as the model for the shape of a typical bubble within the foam. The typical bubble is thus considered as having 12 congruent regular pentagonal faces. Of course it is recognized that in reality bubbles will deviate from this ideal, so alternatively the model can be viewed as a kind of average of shapes. At this point it is again worth mentioning that the view through a glass retaining wall can be misleading to the casual observer because local packing distorts the bubble shapes from what they are in the interior of the foam.

Foam drainage occurs primarily through the interconnecting network of capillaries rather than from film to film (25,36). The problem is therefore to predict the flow through the capillaries.

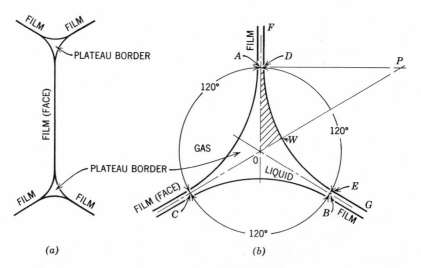

Fig. 13. Capillaries (Plateau borders) in cross section: (a) two capillaries with the film between them, (b) a magnified view of one capillary, illustrating the sixfold symmetry. Capillary wall DWE has the radius r_0 from point P. [From Am. Inst. Chem. Engrs. J., **11**, 19 (1965).]

Figure 13a shows two capillaries in cross section. The curvature of their boundaries exerts a suction on the films. Figure 13b is a magnified view of a single capillary. The capillary boundary is pictured as being of radius r_0. The cross section exhibits sixfold symmetry. Thus, flow through only one such symmetrical section need be examined.

Flow through the capillaries is incompressible and laminar. Accordingly, for steady rectilinear flow through a vertical capillary, the equation of motion in cylindrical form (37) for a Newtonian liquid with constant density ρ and viscosity μ is

$$-\frac{\rho}{\mu}\left(g - \frac{1}{\rho}\frac{\partial p}{\partial Z}\right) = \frac{1}{r}\frac{\partial}{\partial r}\left(r\frac{\partial v_z}{\partial r}\right) + \frac{1}{r^2}\frac{\partial^2 v_z}{\partial \theta^2} \tag{37}$$

where g is the acceleration of gravity, r the radial coordinate, θ the angular coordinate, Z the axial coordinate (in this case vertically downward), v_z the linear axial velicity, and $\partial p/\partial Z$ the axial pressure gradient.

The boundary conditions at the two straight borders of the symmetrical section are simply those of mirror symmetry. However, at the curved wall the situation is more complicated. With some foams, such as those of protein origin, the curved wall may be essentially rigid due to the formation of the well-known protein skin. Under such conditions, v_z at the wall is zero. However, this rigidity is not generally the case. With other foams, such as detergent foams, the wall can move. Of course there is still some resistance. The measure of this drag resistance is the surface viscosity μ_s which typically has the units of dyne sec/cm. This can be compared with dyne sec/cm² for the conventional viscosity μ which is its three-dimensional analog. Assuming μ_s to be Newtonian and constant, a momentum balance at the capillary wall yields eq. (38) as the boundary condition at the wall.

$$\frac{\partial^2 v_z}{\partial \theta^2} = -\frac{\mu r_0^2}{\mu_s}\frac{\partial v_z}{\partial r} \tag{38}$$

The dimensionless quantity $M = \mu r_0/\mu_s$ is a measure of the relative fluidity of the wall. Motion is viewed relative to the center of the junction of the capillary and the flat film for reasons discussed in the original papers (22,32) and subsequently (38). This location is shown as point A in Figure 13b. Points B and C represent similar locations.

Equation (37) was solved with the three boundary conditions by means of finite differences at 53 nodal points. This was accomplished with a digital computer to solve the resulting matric equation via Crout's reduction (39). Velocity contours were obtained. These are shown in dimensionless form in Figure 14 for a rigid capillary wall of $M = 0$, in Figure 15 for a moderately fluid wall of $M = 2.48$, and in Figure 16 for a very fluid wall of $M = 61.9$.

However, the capillaries are generally not vertical. For a capillary inclined at some angle α to the horizontal, the axial component of gravity is $g \sin \alpha$. The axial pressure gradient is also that for the vertical multiplied by $\sin \alpha$. Therefore the axial velocity is now $v_z \sin \alpha$ where v_z is for a similar capillary which is vertical. The downward vertical component of the axial velocity is accordingly $v_z \sin^2 \alpha$. This is all relative to a "moving observer," designated MO, moving upward with the foam bubbles. Accordingly, the vertical component is now

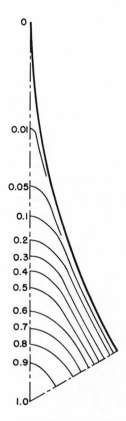

Fig. 14. Dimensionless velocity contours within a capillary with a rigid wall, i.e., with $M = 0$. The film thickness shown here is negligible.

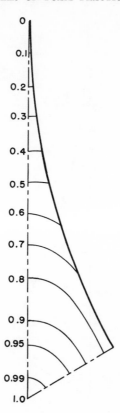

Fig. 15. Dimensionless velocity contours within a capillary with a moderately fluid wall of $M = 2.48$. The film thickness shown here is negligible.

written $v_{z,\text{MO}} \sin^2 \alpha$. Relative to a "stationary observer," SO, standing on the surface of the Earth, the foam bubbles move upward at a linear velocity of $v_{f,\text{SO}}$. Subtracting yields $v_{f,\text{SO}} - v_{z,\text{MO}} \sin^2 \alpha$ as the net upward velocity at a point in a capillary, relative to a stationary observer. Multiplying this net velocity by the differential horizontal cross section $dA_{\text{PB}h}$ and integrating over the entire horizontal cross section of all the capillaries at some arbitrary level in the foam column will give the net upward flow of interstitial liquid, Q_{SO}.

Said integration requires knowledge of the number of capillaries at the column cross section as well as the distribution of their inclination α. This was determined in the following way.

From geometrical considerations it can be shown that the probability of a single random capillary being oriented at α is $\cos \alpha \, d\alpha$. For

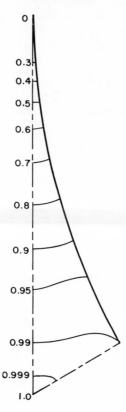

Fig. 16. Dimensionless velocity contours within a capillary with a very fluid wall of $M = 61.9$. The film thickness shown here is negligible.

a capillary that *is* inclined at α, but otherwise randomly located in the column, the probability of intersection by a particular horizontal plane is obviously $P \sin \alpha$ where P is the capillary packing factor defined as capillary length divided by foam column height. Therefore the compound probability of a capillary being at α and also being at a particular horizontal plane (cross section of the column) is the product or $P \sin \alpha \cos \alpha \, d\alpha$. The last expression becomes the number of such capillaries when there are many capillaries in the column, as is actually the case of course. This assumes the capillaries are random.

Equation (39) indicates this aforementioned integration for Q_{SO}.

$$Q_{\mathrm{SO}} = P \int_0^{\pi/2} \int_{A_{\mathrm{PBh}}} (v_{f,\mathrm{SO}} - v_{z,\mathrm{MO}} \sin^2 \alpha) \sin \alpha \cos \alpha \, d\alpha \, dA_{\mathrm{PBh}} \quad (39)$$

It is convenient to replace $A_{PBh} \sin \alpha$ with A_{PB} which is the capillary cross section perpendicular to the capillary axis. P is related to bubble diameter and column cross section by geometry. The bulk foam is presumed to move in plug flow. This is generally the case in a column of uniform cross section because of the structural rigidity of the bulk foam.

Equation (39) was recast into dimensionless form and the double integration carried out numerically with a digital computer. The results are expressed as dimensionless curves in the original paper. These curves permit estimation of the drainage rate, upflow rate, and bulk foam density. All that are required are physical properties and conditions of operation. No empirical constants are required. The theory is more rigorous and complete than the theories of other workers. These earlier theories suffer from three vitiating assumptions, namely that the interstitial flow takes place through channels which are vertical with rigid walls and of either substantially circular (40,41) or parallel plane (12,42–44) cross section.

The results for column overflow can be simplified to give eq. (40) for dry foam, that is, for foam of low liquid content.

$$Q = \frac{G^2\mu}{Ag\rho d_0{}^2}\, \phi\!\left(\frac{\mu^3 G}{\mu_s{}^2 g\rho A}\right) \qquad (40)$$

As before, Q is the volumetric rate of foam overflow on a gas-free basis. A is the cross sectional area of the column. For bubbles of nonuniform size, the appropriate average bubble diameter d_0 is given by eq. (41).

$$d_0 = \left(\frac{\Sigma n_i d_i{}^3}{\Sigma n_i d_i}\right)^{1/2} \qquad (41)$$

Function ϕ was evaluated theoretically according to several alternatives regarding the detailed relationship between interstitial downflow and bulk foam upflow (22,27). Two of the evaluations are shown as curves in Figure 17. The difference between the curves is small over much of the range (35). However, the asymptotes do differ appreciably. Additional work is in progress to examine this difference further. In the meantime the author recommends that the solid curve be used for eq. (40).

For moderately wet foam, the theoretical correction necessary for the higher liquid content can be conveniently approximated by simply multiplying Q from eq. (40) by the quantity $(1 + 3Q/G)$. This gives a corrected Q.

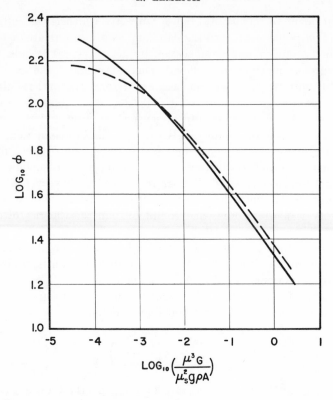

Fig. 17. Theoretical evaluation of ϕ. The solid curve (27) has an asymptote of $\log_{10} \phi = 2.41$. The broken curve (22) has an asymptote of $\log_{10} \phi = 2.27$. Use of the solid curve is recommended (35).

When internal coalescence is present, d_0 must be taken at the top of the column.* This is important. A small, but significant, degree of coalescence can easily go unnoticed to the eye, especially with bubbles that are nonuniform to begin with. For very severe internal coalescence, extremely wet foam, highly nonuniform bubble sizes, or very nonuniform or tortuous column cross section, the theoretical model breaks down.

B. Surface Viscosity and Other Properties

With dilute solutions, μ and ρ can usually be conveniently taken as those of the solvent which is generally water. However, the surface

---

* But d is another matter. An average between top and bottom has been suggested (27). However, on theoretical grounds, it is more correct to take d at the pool surface for eqs. (21) and (24), and at the feed level for eqs. (17), (18), (25), and (26).

viscosity μ_s is a separate property.   It can be measured in several ways.   Unfortunately, there is a lack of consensus on the best way for doing so.   Different methods give quite different results.   This is due partly to differences in technique and the presence of impurities, and partly to the fundamental nature of surface viscosity.   It is not simple.   Surface flow can involve not only shear, but dilation and elasticity (45).   Temperature can have a marked effect too, especially if "melting" of a rigid surface layer occurs.

Fortunately, ϕ is not a highly sensitive function of μ_s.   In fact, for high values of μ_s as are common with denatured protein solutions, ϕ theoretically approaches an asymptote which is independent of μ_s. Joly (46) has compiled measurements of μ_s for various aqueous systems.   Typical values range from under 10^{-4} dyne sec/cm for some detergent solutions to over 10^{-1} dyne sec/cm for some protein solutions.   However, for foam drainage and overflow, and therefore for estimating ϕ, a range of 10^{-4} to 10^{-3} dyne sec/cm may be more realistic.

The present author has found the method of Mysels (36) to be useful as a simple means for roughly measuring μ_s in foam.   In this method, μ_s is obtained from the motion of the black spots which form in a well-drained film.   Such films can be obtained in a foam column by simply shutting off the gas flow and feed flow, thus allowing the standing foam to drain thoroughly.   As such drainage proceeds, various colors appear in the bubble faces due to the diffraction of light.   If a film does not break prematurely, its color eventually turns silver-white with a thickness t_w on the order of 10^{-5} cm.   Black spots form in this film.   The thickness t_b of these spots is on the order of 10^{-6} cm, and their blackness stems from the diffractive cancellation of light.   Being thinner than the surrounding white film, the black spots are buoyant and therefore rise up through (within) the film. This is shown schematically in Figure 18.   Refined methods for mea-

Fig. 18.   Typical pattern of black in white film showing the rising black spots.

suring t_w and t_b exist (36,47). However, in view of other uncertainties, undue effort in this connection is not warranted here.

The inclination angle β of the film from the horizontal, the radius r_b of a more or less circular black spot, and the linear velocity u within the film for this same black spot are all estimated optically. The results are then substituted in eq. (42) which is essentially the surface analog of Lamb's relationship (48) for a buoyant, infinitely long, horizontal cylinder rising at terminal velocity through a homogeneous viscous fluid.

$$\mu_s = \frac{-r_b{}^2\rho g(t_w - t_b)\sin\beta}{8u}\left(0.0772 + \ln\frac{ur_b\rho t_w}{8\mu_s}\right) - \frac{\mu t_w}{2} \qquad (42)$$

The term $\mu t_w/2$ is generally negligible. Solution for μ_s is by trial.

Since the method is only very approximate while the measurements are so easy to obtain, the determination should be repeated many times and the results for μ_s averaged so as to achieve greater reliability. In the work of Leonard and Lemlich (32), 16 such determinations for aqueous Triton X-100 at 25°C yielded a mean μ_s of 1.01×10^{-4} dyne sec/cm with a standard deviation of 0.47×10^{-4} dyne sec/cm for the individual measurements. By statistical t-test this corresponds to a 95% confidence interval of $\pm 0.25 \times 10^{-4}$ dyne sec/cm on the mean.

There is another way to estimate μ_s. One can run a foam column with the system in question, measure (or otherwise evaluate) Q, G, A, μ, g, ρ, and d_0, and substitute in eq. (40) using the wetness correction factor $(1 + 3Q/G)$ if required. Then from ϕ in Figure 17, μ_s can be found. Naturally, μ_s determined in this manner will suffer from whatever shortcomings exist in the theory. On the other hand, for subsequent use in estimating Q under changed conditions, μ_s determined in such a manner offers the prospect of canceling these shortcomings as well as cancelling the experimental error.

C. Implications of the Theory

Besides providing a theoretical estimate of Q by eq. (40) for any mode of column operation, the drainage theory also provides means for estimating the bulk foam density and the interstitial liquid upflow and downflow rates at any level in the column, barring internal coalescence. The quantitative considerations are beyond the scope of the present chapter and the interested reader must be referred to the original paper for further information (22).

The present discussion will be restricted to certain general results of the theory. Perhaps the most significant of these relate to $\mathfrak{D}$, the volumetric density of the bulk foam, which is expressed as interstitial liquid volume divided by bulk foam volume.

According to the drainage theory, $\mathfrak{D}$ within any section of the column should be uniform along the column provided there is no internal coalescence and provided any reflux or feed entering the foam is well distributed over the column cross section. For example, in a combined column, $\mathfrak{D}$ should be uniform from the surface of the liquid pool up to the feed inlet, and then uniform at another value from the feed inlet up to the reflux inlet, and then uniform again at still another value from the reflux inlet up to the top of the column. This means that the HTU or the HETS in any section of a column is theoretically uniform along that section because the foam is uniform. It also means that increasing the height of a column which is truly operating in the simple mode should be of no value since, odd as it may appear at first glance, there is theoretically no change in $\mathfrak{D}$ with height. Of course, if significant internal coalescence is present (as it often is), the situation would be quite different. Finally, the theory shows that $\mathfrak{D}$ at the top of the column is not equal to $Q/(Q + G)$ as one might have thought offhand. Rather, $\mathfrak{D}$ is somewhat greater than $Q/(Q + G)$. This inequality is related to the true interstitial downflow, upflow, and recycle mentioned earlier.

D. Illustrative Problem 7

Statement

A foam fractionation column of 7-cm diameter is to operate at 25°C with the surfactant Triton X-100 in water. If the gas rate is 90 cm³/sec and the average bubble diameter according to eq. (41) is 0.12 cm at the top of the column, estimate the rate of foam overflow.

Solution

Taking the viscosity and density of the solution as equal to those for water at 25°C, $\mu = 8.94 \times 10^{-3}$ dyne sec/cm² and $\rho = 1.00$ g/cm³. $A = (\pi/4) \times 7^2 = 38.5$ cm². Substituting in eq. (40),

$$Q = \frac{90^2 \times 8.94 \times 10^{-3}}{38.5 \times 980 \times 1.00 \times 0.12^2} \times$$

$$\phi \left(\frac{(8.94 \times 10^{-3})^3 \times 90}{(1.01 \times 10^{-4})^2 \times 980 \times 1.00 \times 38.5} \right)$$

$$Q = 0.133 \times \phi(0.167)$$

$$\log_{10} 0.167 = -0.78$$

From the solid curve of Figure 17, $\log_{10} \phi = 1.54$. Therefore, $\phi = 34.7$. Substituting,

$$Q = 0.133 \times 34.7 = 4.62$$

Applying the wetness factor of $(1 + 3Q/G)$,

$$Q = \left(1 + \frac{(3 \times 4.62)}{90}\right) 4.62 = 5.34$$

Rounding off, $Q = 5.3$ cm³/sec of foam overflow on a gas-free basis. Neglecting the small density of gas, this represents 5.3 g/sec of foam overflow.

E. Illustrative Problem 8

Statement

Estimate the approximate range of relative change in the overflow rate of a fairly dry foam from a foam fractionation column if the gas rate is doubled but the bubble sizes remain unchanged.

Solution

At high ϕ, such as might be the case with the comparatively high μ_s of a protein foam, ϕ approaches a constant asymptote. This is shown in Figure 17. Thus from eq. (40), Q is directly (linearly) proportional to G^2. Therefore, in this case, doubling G should quadruple Q. Since the foam is fairly dry, to a first approximation the wetness correction is neglected.

At lower ϕ, such as might be the case with lower μ_s in a detergent foam, the solid curve in Figure 17 becomes approximately linear with a slope of about -0.27. This means ϕ is directly proportional to $G^{-0.27}$. Thus from eq. (40), Q is directly proportional to $G^{1.73}$. Therefore doubling G multiplies Q by $2^{1.73}$ or about 3.3. Again the wetness correction is neglected.

Thus, for a fairly dry stable foam, the effect of doubling G should be to multiply Q by a factor ranging from about 3.3 to 4.0 For a wet foam or for a coalescing foam, the factor can be greater.

F. Experimental Confirmation

Experimental confirmation of the theory has been obtained in various ways. Figure 19 shows one of several columns which were used for this purpose (27). The column was of glass with an internal diameter of 4.6 cm. It was surrounded with a circulating water jacket to maintain the temperature at 25°C. Pairs of calibrated electrodes

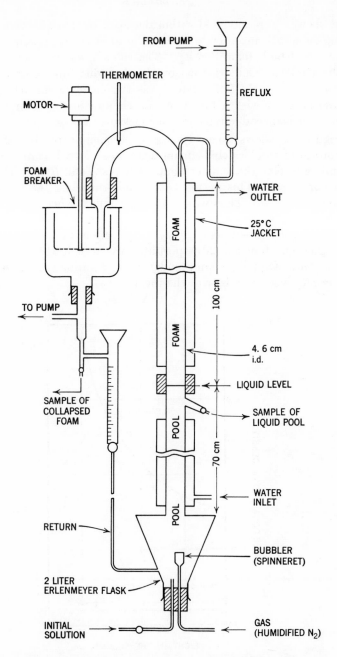

Fig. 19. Foam fractionation column equipped for reflux and return. Seven pairs of electrodes (not shown) are located within the column to measure the bulk foam density at various levels.

(not shown) were mounted within the foam column at seven locations along its length in order to measure local electrical conductivity and hence local bulk foam density. The column was run both with and without reflux. Collapsed net overhead product was returned as feed to the liquid pool. The system was the nonionic surfactant Triton X-100 in water with a bit of sodium chloride added to provide sufficient electrical conductivity for easy measurement.

Figure 20 shows some typical experimental results (27). The reciprocal of the local bulk foam density is plotted against its corresponding level in the foam column. Curves A and B are each straight and horizontal. This means $1/\mathfrak{D}$ is uniform along the column, and therefore so is $\mathfrak{D}$. Thus with or without reflux, when internal coalescence is negligible the uniformity in $\mathfrak{D}$ which is predicted by theory is borne out by experiment. In this connection it is of interest to note that Haas and Johnson (20) report little variation in HTU with column length for a stripping column under good operating conditions. Such uniformity also accords with the theory.

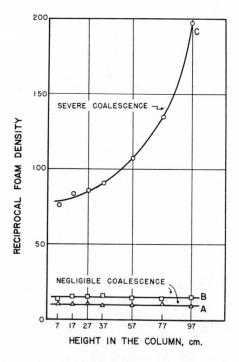

Fig. 20. Typical measurements of bulk foam density at various levels within the vertical foam column.

Of course, if internal coalescence is not negligible, $\mathfrak{D}$ will decrease up the column, which means of course that $1/\mathfrak{D}$ will increase. Curve C illustrates this for a case of severe internal coalescence.

Experimental results (27,32,38,49) also verify the prediction of Q by eq. (40). Data were assembled (35) from 46 runs at 25°C with three different columns. The system was aqueous Triton X-100 at pool concentrations ranging from 2×10^{-7} to 20×10^{-7} g-moles/cm³. R ranged from 0 to 13, A ranged from 13.1 to 16.6 cm², d_0 ranged from 0.05 to 0.40 cm, G ranged from 1.3 to 10.4 cm³/sec and Q ranged from 0.001 to 0.405 cm³/sec. Stripping operation was employed in some runs by feeding a solution of 5×10^{-7} g-moles/cm³ into the foam. For all runs, a comparison was then made between Q measured experimentally and Q predicted by eq. (40) utilizing the solid curve of Figure 17 and the wetness correction factor $(1 + 3Q/G)$. The surface viscosity was taken to be 1.01×10^{-4} dyne sec/cm, as determined separately from the motion of the black spots.

Figure 21 shows the results of the test. Theoretical Q/A is plotted against experimental Q/A. The diagonal is the line of theoretical perfection. Deviation from the line is quite modest, especially in view

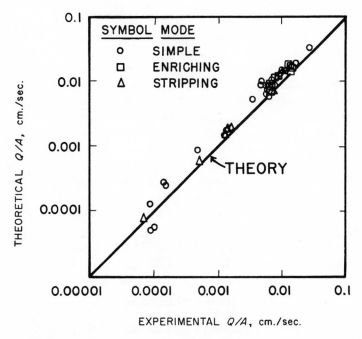

Fig. 21. Test of theory for predicting the foam overflow rate.

of the wide range of variables involved. Over more than a 400-fold range of foam overflow rate, the average deviation of theory from experiment is only 26%. This is deemed to be fairly good agreement for such a wide range.

Of course, the agreement could be improved even further by using μ_s determined from some of the operating data for the column, as described earlier. Such determinations in a pool-fed column showed that μ_s increased somewhat as the interstitial concentration was raised either by the use of reflux or by the use of a higher pool concentration. However, such values of μ_s were not employed in the test. For the sake of impartiality, only the one aforementioned value determined from the independent method of black spots was used.

The drainage theory can also be used to relate the bulk foam density of a stationary foam to the rate of a continuous liquid feed draining down through the foam keeping it wet. This applies at steady state and in the absence of appreciable coalescence. The theory predicts that $\mathfrak{D}$ will be uniform along the column. Of course, if the feed is stopped, allowing the foam to drain in unsteady state, $\mathfrak{D}$ will begin to vary along the column and with time.

The procedure for relating $\mathfrak{D}$ to L at steady state, or vice versa, does not involve ϕ. Also, it does not involve any empirical constants. It is summarized by an information flow diagram presented elsewhere (35). Comparison of theory against experiment yields favorable agreement (35).

From experimental data for aqueous Triton X-100 taken with the column shown in Figure 19, true L and true U were obtained (38) from the drainage theory. L and U so determined include the internal recycle mentioned earlier. Mass transfer coefficients were then obtained for the transfer of surfactant between the two true interstitial streams under conditions of reflux. Of course the bubble surfaces were saturated.

Values for the said mass transfer coefficient k_L based quite arbitrarily on the capillary wall area a_0 are represented elsewhere (27). Those values are easily converted to transfer units since it can be readily shown that $N_U = k_L a_0 / U$ for this case. Dividing the column height of 100 cm by N_U gives the height of a transfer unit. Values so obtained for HTU ranged from 4 to 21 cm for that series of experiments.

VIII. COALESCENCE

As indicated earlier, bubble coalescence within the rising foam can be important. This internal coalescence stems from two sources.

One is the diffusion of gas from small bubbles to large bubbles. While this is not usually significant with respect to foam fractionation, with a sufficiently high column residence time it can be. The second source is the rupture of films which separates bubbles. This can easily be very significant.

A. Gas Diffusion

When an interface between two fluids is curved, a pressure difference exists across it. This pressure difference Δp is related to γ and the radii of curvature r_1 and r_2 by the equation of Laplace and Young, namely,

$$\Delta p = \gamma(1/r_1 + 1/r_2) \tag{43}$$

Thus, for example, the suction of the capillaries which was referred to earlier can be estimated by setting $r_1 = r_0$ and $r_2 = \infty$. This gives suction $\Delta p = \gamma/r_0$.

For the present discussion, eq. (43) is applied directly to a bubble For a spherical bubble of radius $\bar{R}$ in air, $r_1 = r_2 = \bar{R}$. Also, there are two nearly equal surfaces, an inner and an outer. So for this case, eq. (43) yields $\Delta p = 4\gamma/\bar{R}$ where Δp is the pressure excess within the bubble with respect to the atmosphere around it. For a spherical bubble submerged within a liquid, the pressure excess with respect to the adjacent liquid is $\Delta p = 2\gamma/\bar{R}$ since there is only one bubble surface. Thus the pressure difference between two submerged spherical bubbles of unequal size is $\Delta p = 2\gamma(1/\bar{R} - 1/\bar{R}')$ where $\bar{R}'$ is the radius of the larger bubble and $\bar{R}$ is the radius of the smaller. This pressure difference causes gas to diffuse from the smaller bubble, through the liquid, into the larger bubble. This difference makes the larger bubble grow still larger and the smaller bubble grow still smaller, thus increasing the driving force and hence the rate even further. Given sufficient time, the smaller bubble will disappear altogether.

Since it is impossible to have perfectly identical bubbles in a dispersion it is evident that some change must always occur in the distribution of bubble sizes. The rate of this change depends not only on the deviation in bubble sizes but also on the resistance to gas diffusion from bubble to bubble. De Vries (18) has derived and experimentally verified certain relationships which describe how the distribution of bubble sizes changes with time under the influence of gas diffusion. According to his work, if the number of bubbles of radius $\bar{R}$ is $F(\bar{R})d\bar{R}$, and if eq. (44) describes the frequency distribution $F(\bar{R})$ in terms of $\bar{R}$ and a distribution parameter b, then eq. (45) gives the effect of interbubble gas diffusion on b as it changes from b_0 to b_τ over time interval τ.

$$F(\bar{R}) = 6b\bar{R}/(1 + b\bar{R}^2)^4 \tag{44}$$

$$b_\tau = b_0/(1 + k\tau)^2 \tag{45}$$

The quantity k treated as a constant is expressed by eq. (46), where $\mathbf{R}$ is the gas constant, T the absolute temperature, $\bar{D}$ the diffusivity of gas in liquid, $\bar{S}$ the solubility (per unit pressure) of gas in liquid, p_a the atmospheric pressure, and t the average distance between bubbles.

$$k = \frac{4\mathbf{R}T\bar{D}\bar{S}\gamma b}{p_a t} \tag{46}$$

While, strictly speaking, the foregoing relationships are for spherical bubbles, they should nevertheless apply in a general sort of way to polyhedral foams as well.

B. Film Rupture

The persistence of foam results from the persistence of the films separating the bubbles. The latter characteristic is a manifestation of the film elasticity which is defined by Gibbs (12) as $2s\,d\gamma/ds$ where s is the surface of the film.

Physically, this elasticity stems from a depletion of surfactant at the film surface when the surface is stretched. This depletion raises the surface tension which results in the exertion of a restoring force which opposes rupture.

Stretching depletes the adsorbed surfactant for two reasons. Firstly, there is the momentary depletion due to the inability of surfactant molecules to diffuse instantaneously from the interior of the film to the surface. This is known as the Marangoni effect. Secondly, in thin films the limited supply of surfactant in the interior may be insufficient to completely recoat the surface regardless of diffusion rate. This is termed the Gibbs effect. The relative importance of the two effects is still a matter of controversy. A third effect, which can be important with some ionic surfactants in thin films, is the electrostatic repulsion between the charged surfaces. This opposes film thinning and rupture.

It is very difficult to predict rupture. Kitchener (45) summarizes some of the approaches which have been employed. They yield activation energies which are directly proportional to the square of the film thickness. Various suggestions have been put forth as to the causative agents for rupture. These include thermal fluctuation, pressure fluctuation, spontaneous nucleation of vapor, external vibration, cosmic radiation, and local internal stress resulting from readjustment of the packing as the bubble sizes change due to interbubble gas diffusion.

C. External Coalescence

Generally speaking, it is desirable to collapse the foam overflow. If external reflux is to be employed, this is a necessity. However, even otherwise, it is frequently desirable since uncollapsed foam occupies a great deal of volume.

Some foams collapse readily upon standing. Others require special measures. Slowly rotating perforated centrifuges have been used successfully in a number of the investigations already cited. Discharging from the foam breaker on to a Teflon sheet instead of glass was found to be helpful in reducing any foam which remained uncollapsed (20).

An ingenious method, which has been applied in a large unit for foaming pollutants from sewage, is to run collapsed foam on to uncollapsed foam thereby mechanically breaking the latter (50). Of course, for very persistent foams this method may be difficult to apply successfully.

Other methods for collapsing foam include sonic and ultrasonic vibration, thermal means, and chemical additives. The last is not suitable for operation with external reflux, or where the additive would constitute an undesirable contaminant in the collapsed foam. For laboratory purposes the choice of method is largely one of convenience. With larger units the relative costs may prove decisive.

IX. GAS EMULSIONS

If the gas rate to a foam fractionation column is increased, the rate of bubble formation will be increased. This in turn will increase the rate at which surface, and therefore adsorbed material, flows off overhead. However, the liquid content of the foam will also be increased, so the concentration of the collapsed overhead will be lower. Furthermore, the wetter the foam, the poorer the contact is likely to be between countercurrent streams if such streams are involved. These disadvantages of excessive liquid content were pointed out earlier.

If the gas rate is increased still further, the foam bubbles separate. There are no longer films between the bubbles (which have become quite spherical). The rising gas–liquid dispersion becomes a gas emulsion (51). It is no longer a true foam.

The great propensity toward channeling in gas emulsions makes them poorly suited to stripping or refluxing operation. However, for simple operation, in situations where very high adsorptive throughput is required per unit cross sectional area of the column, the generation of a gas emulsion may be desirable. Of course the resulting additional external drainage load must not be overlooked.

Separation by means of gas emulsions are discussed by Wace and Banfield (14) who present relationships for throughput. They find maximum throughput at a $\mathfrak{D}$ of 0.56. Above that $\mathfrak{D}$ the rising dispersion becomes indistinguishable from the highly aerated pool. For bubbles of 0.1-cm diameter, a $\mathfrak{D}$ of 0.56 reportedly corresponds to a superficial gas velocity G/A of about 2.2 cm/sec. They also report that throughput is not a sensitive function of bubble diameter, and a broad maximum in throughput exists at a diameter of 0.08 cm.

X. CLOSURE

Foam fractionation is not a new technique. It has been a laboratory curiosity for a number of years. However, only recently has the technique been examined from a fundamental point of view. This fundamental approach has been stimulated in part by a greater recognition of the potentialities of the technique in large scale separations. Foam fractionation shows particular promise as an economical means for removing substances present at low concentrations from large volumes of liquid.

The material which has been presented in this chapter summarizes the principles of foam fractionation as they have been developed to date. Some of these ideas can be extended to the other adsubble methods. For example, where a foam is involved, the fundamentals of drainage apply; where a partition process is involved, the concept of the transfer unit applies. It is the author's conviction that through the application of principles, more rational designs of processes can be prepared and more fruitful results obtained.

XI. SYMBOLS

A	Horizontal cross sectional area of the column, cm^2
A_{PB}	Cross sectional area of a capillary taken perpendicular to its axis, cm^2
A_{PBh}	Horizontal cross sectional area of a capillary, cm^2
a_i	Activity of component i
a_0	Wall area of the capillaries, cm^2
b	Bubble distribution parameter, cm^{-2}

b_0 Bubble distribution parameter at the start of interbubble gas diffusion, cm^{-2}

b_τ Bubble distribution parameter after time interval τ, cm^{-2}

C Concentration in the liquid, g-moles/cm^3

$\bar{C}$ Effective upflow concentration, g-moles/cm^3 on a gas-free basis

$\bar{C}^*$ Effective upflow concentration in equilibrium with C, g-moles/cm^3 on a gas-free basis

C^* Downflow concentration in equilibrium with $\bar{C}$, g-moles/cm^3

D Volumetric rate of collapsed net overflow product, cm^3/sec

$\bar{D}$ Diffusivity of gas in liquid, cm^2/sec

$\mathfrak{D}$ Volumetric foam density (volumetric fraction of liquid in the foam)

d Bubble diameter or surface average bubble diameter, cm

d_i Individual bubble diameter, cm

d_0 Edge average bubble diameter, cm

E Relative adsorptive effectiveness of surfactant in the surface versus surfactant in the micelles

F Volumetric feed rate, cm^3/sec

$F(\bar{R})$ Frequency distribution function for $\bar{R}$, cm^{-1}

G Volumetric gas rate, cm^3/sec

g Acceleration of gravity, cm/sec^2

K Equilibrium constant for adsorption, cm

K' Auxiliary constant for Langmuir isotherm, cm^3/g-mole

k Redistribution parameter for interbubble gas diffusion, sec^{-1}

k_L Mass transfer coefficient, cm/sec

L Volumetric rate of interstitial liquid downflow, cm^3/sec

M Dimensionless viscosity modulus $\mu r_0/\mu_s$

N_L Number of transfer units based on the downflowing stream

N_U Number of transfer units based on the upflowing stream

n_i Number of bubbles with diameter d_i

P Capillary packing factor (total length of capillaries per unit height of foam)

Q Volumetric rate of foam overflow on a gas-free basis, cm^3/sec

Q' Volumetric rate of upflow on a gas-free basis just above the liquid pool, cm^3/sec

Q_{so} Volumetric rate of net upflow of interstitial liquid relative to a stationary observer, cm^3/sec

p Pressure, dyne/cm^2

p_a Atmospheric pressure, dyne/cm^2

$\mathbf{R}$ Gas constant, erg/g-moles-°K

R Reflux ratio

$\bar{R}$ Bubble radius, cm

$\bar{R}'$	Radius of the larger bubble, cm
r	Radial coordinate, cm
r_b	Radius of approximately circular black spot, cm
r_0	Radius of curvature of the capillary wall, cm
r_1	Radius of surface curvature, cm
r_2	Radius of surface curvature taken in a plane perpendicular to that for r_1, cm
s	Film surface, cm^2
S	Ratio of surface to volume, cm^{-1}
$\bar{S}$	Solubility of gas in liquid per unit pressure, g-moles/erg
T	Absolute temperature, °K
t	Film thickness or distance separating bubbles, cm
t_b	Film thickness of black spot, cm
t_w	Thickness of silver-white film, cm
U	Volumetric rate of interstitial liquid upflow, cm^3/sec
u	Linear velocity of a black spot moving within a silver-white film, cm/sec
v_z	Linear velocity downward at a point within a vertical capillary, cm/sec
$v_{z,\mathrm{MO}}$	v_z but emphasizing that motion is relative to the rising foam bubbles, cm/sec
$v_{f,\mathrm{SO}}$	Linear upward velocity of the foam bubbles relative to a stationary observer, cm/sec
W	Volumetric rate of bottoms takeoff, cm^3/sec
Z	Axial coordinate within a capillary, cm
α	Angle which a capillary makes with a horizontal plane
β	Angle which a bubble face makes with a horizontal plane
Γ	Concentration at the surface, g-moles/cm^2
γ	Surface tension, dyne/cm
Δ	Difference
μ	Viscosity, dyne sec/cm^2
μ_s	Surface viscosity, dyne sec/cm
π	3.14159 . . .
ρ	Liquid density, g/cm^3
τ	Time interval, sec
ϕ	Dimensionless function

Subscripts for C, $\bar{C}$, $\bar{C}^*$, K, and Γ only

D	Collapsed net overhead product
F	Feed
i	Particular component

L	Downflow just above the pool
Q	Foam overflow on a gas-free basis
s	Surfactant
sc	Critical micelle
top	At the top of the downflowing stream
W	Pool or bottoms
1	In the absence of micelles
2	In the presence of micelles

References

1. F. Sebba, *Ion Flotation*, Elsevier, New York, 1962.
2. H. M. Schoen, *Ann. N.Y. Acad. Sci.*, **137**, 148 (1966).
3. E. Rubin and E. L. Gaden, Jr., "Foam Separation," in *New Chemical Engineering Separation Techniques*, H. M. Schoen, Ed., Interscience, New York, 1962, Chap. 5.
4. H. G. Cassidy, *Fundamentals of Chromatography (Technique of Organic Chemistry*, Vol. 10, A. Weissberger, Ed.), Interscience, New York, 1957.
5. L. Shedlovsky, *Ann. N.Y. Acad. Sci.*, **49**, 279 (1948).
6. R. Lemlich, *Chem. Eng.*, **73**, No. 21, 7 (1966).
7. F. Sebba, *Nature*, **184**, 1062 (1959).
8. A. Dognon and H. Dumontet, *Compt. Rend. Soc. Biol.*, **135**, 884 (1941).
9. A. J. Rubin, E. A. Cassell, O. Henderson, J. D. Johnson, and J. C. Lamb, *Biotechnol. Bioeng.*, **8**, 135 (1966).
10. R. E. Baarson and C. L. Ray, "Precipitate Flotation, a New Metal Extraction and Concentration Technique," presented at American Institute of Mining, Metallurgical and Petroleum Engineers Symposium, Dallas, Texas, 1963.
11. D. C. Dorman and R. Lemlich, *Nature*, **207**, 145 (1965).
11a. B. L. Karger, R. B. Grieves, R. Lemlich, A. J. Rubin, and F. Sebba, *Separation Science*, **2**, No. 3, 401 (1967).
12. J. W. Gibbs, *Collected Works*, Longmans Green, New York, 1928.
13. J. T. Davies and E. K. Rideal, *Interfacial Phenomena*, 2nd ed., Academic Press, New York, 1963.
14. P. F. Wace and D. L. Banfield, *Chem. Process Eng.*, **47**, No. 10, 70 (1966).
15. P. F. Wace and R. Lemlich, discussion, *Am. Inst. Chem. Engrs.—Inst. Chem. Engrs. Symp. Ser. (Inst. Chem. Engrs., London)* **9**, 85 (1965).
16. R. W. Schnepf, E. L. Gaden, Jr., E. Y. Mirocznik, and E. Schonfeld, *Chem. Eng. Progr.*, **55**, No. 5, 42 (1959).
17. C. A. Brunner and R. Lemlich, *Ind. Eng. Chem. Fundamentals*, **2**, 297 (1963).
18. A. J. de Vries, *Foam Stability*, Rubber-Stichting, Delft, 1957.
19. R. Lemlich and E. Lavi, *Science*, **134**, No. 3473, 191 (1961).
20. P. A. Haas and H. F. Johnson, *Am. Inst. Chem. Engrs. J.*, **11**, 319 (1965).
21. D. O. Harper and R. Lemlich, *Ind. Eng. Chem. Process Design Develop.* **4**, 13 (1965).
22. R. A. Leonard and R. Lemlich, *Am. Inst. Chem. Engrs. J.*, **11**, 18 (1965).
23. A. W. Adamson, *Physical Chemistry of Surfaces*, 1st ed., Interscience, New York, 1960.

24. J. J. Bikerman, *Foams: Theory and Industrial Applications*, Reinhold, New York, 1953.
25. E. Manegold, *Schaum*, Strassenbau, Chemie und Technik Verlag, Heidelberg, 1953.
26. R. Lemlich, *J. Chem. Eng. Educ. (Univ. of Cincinnati)*, **3**, No. 2, 53 (1965).
27. S. Fanlo and R. Lemlich, *Am. Inst. Chem. Engrs.—Inst. Chem. Engrs. Symp. Ser. (Inst. Chem. Engrs., London)*, **9**, 75 (1965).
28. R. H. Perry, C. H. Chilton, S. D. Kirkpatrick, *Chemical Engineers Handbook*, 4th ed., McGraw-Hill, New York, 1963.
29. D. L. Banfield, I. H. Newson, and P. J. Alder, *Am. Inst. Chem. Engrs.—Inst. Chem. Engrs. Joint Symp.*, Preprint 1.1, Inst. Chem. Engrs., London, 1965.
30. E. J. Henley and H. K. Staffin, *Stagewise Process Design*, Wiley, New York, 1963.
31. I. H. Newson, *J. Appl. Chem.*, **16**, 43 (1966).
32. R. A. Leonard and R. Lemlich, *Am. Inst. Chem. Engrs. J.*, **11**, 25 (1965).
33. R. Lemlich, *Am. Inst. Chem. Engrs. J.*, **12**, 802 (1966).
34. D. O. Harper and R. Lemlich, *Am. Inst. Chem. Engrs. J.*, **12**, 1220 (1966).
35. F. S. Shih and R. Lemlich, *Am. Inst. Chem. Engrs. J.*, **13**, 751 (1967).
36. K. J. Mysels, K. Shinoda, and S. Frankel, *Soap Films, Studies of Their Thinning*, Pergamon, New York, 1959.
37. R. B. Bird, W. E. Stewart, and E. N. Lightfoot, *Transport Phenomena*, Wiley, 1960.
38. S. Fanlo, Ph.D. Dissertation, University of Cincinnati, 1964.
39. F. B. Hildebrand, *Introduction to Numerical Analysis*, McGraw-Hill, New York, 1956.
40. P. A. Haas and H. F. Johnson, "A Model and Experimental Results for Drainage of Solution between Foam Bubbles," presented at American Chemical Society, 148th National Meeting, Chicago, 1964.
41. G. D. Miles, L. Shedlovsky, and J. Ross, *J. Phys. Chem.*, **49**, 93 (1945).
42. A. P. Brady and S. Ross, *J. Am. Chem. Soc.*, **66**, 1348 (1944).
43. W. M. Jacobi, K. E. Woodcock, and C. S. Grove, Jr., *Ind. Eng. Chem.* **48**, 2046 (1956).
44. S. Ross, *J. Phys. Chem.*, **47**, 266 (1943).
45. J. A. Kitchener, "Foam and Free Liquid Films," in *Recent Progress in Surface Science*, Vol. 1, J. F. Danielli, K. G. A. Pankhurst, and A. C. Riddiford, Eds., Academic Press, New York, 1964, Chap. 2.
46. M. Joly, "Surface Viscosity," in *Recent Progress in Surface Science*, Vol. 1, J. F. Danielli, K. G. A. Pankhurst, and A. C. Riddiford, Eds., Academic Press, New York, 1964, Chap. 1.
47. K. J. Mysels, *J. Phys. Chem.*, **68**, 3441 (1964).
48. H. Lamb, *Hydrodynamics*, 6th ed., Dover, New York, 1945.
49. R. A. Leonard, Ph.D. Dissertation, University of Cincinnati, 1964.
50. C. A. Brunner and D. G. Stephan, *Ind. Eng. Chem.*, **57**, No. 5, 40 (1965).
51. J. J. Bikerman, *Ind. Eng. Chem.*, **57**, No. 1, 56 (1965).

Normal Freezing

T. H. GOUW

*Chevron Research Company,
Richmond, California*

I. INTRODUCTION

Normal, directional, or progressive freezing is a relatively new and inexpensive technique which has proved to be highly successful in the purification of both organic and inorganic compounds and in the concentration of impurities from relatively pure starting material. The literature on the subject is, as yet, not very voluminous and, until quite recently, treatises on this subject were generally found as a subordinate chapter in the discussion of the complementary technique of zone melting (1,2). Recent developments in scope and technique have, however, made normal freezing sufficiently important to be considered as a separate technique.

The modern investigator, confronted with a problem of separation and purification, has at his disposal a bewildering array of possible separation methods. Many of these techniques have limited applicability or are impractical for large-scale operations. The most versa-

57

tile method has been, and probably still is, distillation. Until quite recently, this technique was employed almost exclusively to purify large batches of material and to concentrate trace impurities. Distillation is still being used in many cases (3) even though the method has some inherent disadvantages. The possible heat sensitivity of the investigated compounds and the danger of contamination by lower boiling materials, which have prewet the take-off path in the still, are factors which rule unfavorably against this technique. Meticulous care is also necessary in the cleaning of a distillation column which has been used to distil a wide variety of materials if contamination is to be avoided. Whenever possible, however, distillation is still used for prepurification and preconcentration purposes. It is almost a necessary preliminary step if the major component is not present in high concentrations. For the preparation of pure compounds, distillation can thus be used to obtain the necessary heart cuts, which can then be subjected to a final purifying step.

The choice of the final purification technique obviously depends on the products processed. If, however, one desires to make use of the solid–liquid phase equilibrium as the basis of this final purification stage, then the choice is more or less limited to either normal freezing or zone melting.

For large-scale operations, the capital outlay for a normal-freezing apparatus appears to be much less than for a zone melter of comparable capacity. Normal freezing also allows better results than zone melting for a one-phase operation (1) although multipass zone melting gives a higher yield of purified material in relation to multiple normal freezing with cropping (4), i.e., discarding the last portion which freezes. This is partially offset by the fact that in normal freezing, stirring of the liquid phase is much easier and will result in a more favorable practical distribution of the impurities between the solid and liquid phases. This may mean that only a small number of runs are necessary to achieve the desired separation purity, and that normal freezing would allow results in a much faster and less elaborate manner. Schildknecht and Schlegelmilch report, for instance, that in the purification of o-cresol, doped with 0.9% phenol and 0.2% m- and p-cresol, two normal freezing runs taking several hours gave an o-cresol of comparable purity to that obtained by zone melting after 20 passes taking 15 days (5).

Modern normal freezing techniques call for progressive crystallization of a solid from the melt under vigorous stirring of the liquid phase. Crystals are formed in layers on top of each other and a well-defined solid–liquid interface moves through the liquid as the run progresses. Although normal freezing can take place even in the absence of stirring, agitation is necessary for improved operation, and we will, therefore, emphasize in this discussion only those forms where stirring is an integral part of the system.

II. LITERATURE REVIEW

Purification by crystallization from solution has been known for several centuries. Purification by crystallization from the melt is of much later date but still several decades old. The first controlled experiments using this technique are those related to the preparation of large single crystals (6). Purification was then observed as an incidental concurrent phenomenon. The first studies in which purification was the deliberate goal were described by Schwab and Wichers of the National Bureau of Standards (7). In this experiment a tube with benzoic acid was slowly lowered through a heating coil. Resolidification of the purified material took place as the tube emerged from this coil with the impurities preferentially segregated in the liquid phase. These experiments demonstrated that for benzoic acid two recrystallizations from the melt were comparable to eight recrystallizations from benzene.

Normal freezing is obviously closely associated with the general technique of purification by crystallization from solutions and in a broader sense with the general field of crystallization itself. There is a voluminous and growing literature on the different aspects of crystallization characterized by increasingly lengthier reviews (8–14). It is beyond the scope of this paper to cover these subjects, however summarily, even though crystallization is the main phenomenon related to normal freezing.

It is interesting to note that in normal freezing, development of the practical aspects have progressed at a much faster pace than the evolution of theoretical considerations. Although this discussion will reflect the general trend to a large degree, subjects like nucleation from the melt, crystal growth (5), mass and heat transfer in fractional solidification, boundary phenomena, and theoretical studies on zone melting (1) are still of indubitable value for an improved understanding of the normal freezing process. Many of these aspects have been reviewed in a recent monograph (8).

The two main variables influencing the quality of the separation by normal freezing are the crystal growth rate and the diffusion of the impurities from the crystal surface to the bulk of the supernatant liquid. In the classical work by Schwab and Wichers (7,15), there was little emphasis on the control of the crystal growth rate. This was more perceptible in the work by Dickinson and Eaborn (16) who purified benzene, p-bromotoluene, and pyridine by lowering tubes with these compounds a number of times into coolant solutions at controlled rates varying from 1 to 4 cm/hr. Approximately 10% of the product present as supernatant liquid is discarded after each run. Results were exceptionally good for that time although it is obvious by now that still better separations would have been obtained if stirring had been employed.

The use of agitation to improve diffusion of the impurities from the crystalline interface, which was first described by the National Bureau of Standards workers (7,15), was soon copied by other authors (17,18); and the basic normal freezing setup, as we know in current practice, was first described by Matthews and Coggeshall (17) in 1959. Vigorous stirring was employed in this unit, and the tube containing the material to be purified was lowered into a coolant solution at a controlled rate.

The technique of normal freezing showed rapid progress from this point on, especially by the work of Schildknecht and co-workers (5,19–24). Studies have been carried out to determine the effect of initial impurity concentration (23), stirring speed (21), and rate of crystal growth (20). A large number of applications have since been noted, such as the purification of benzene (17,25); H_2O_2, acetamide, and o-cresol (5); cyclohexane, nitrobenzene, o-dichlorobenzene, and a large number of other organic compounds (26); the dehydration of formic acid (5); the concentration of volatile substances from aqueous solutions (27); the concentration of biologically active compounds (28); and the acquisition of fundamental data on solid and liquid phases in equilibrium with each other (9). We have ourselves used this technique to purify large batches of n-hydrocarbons. Using high-efficiency distillation as the prepurification step and normal freezing in the final purification stages, all homologs from n-dodecane to n-docosane, inclusive, could easily be obtained in better than 99.9% purity.

Although normal freezing is at present still very much a laboratory scale, analytical separations tool, large-scale modifications have already been reported in the literature (29). Dewey described the production of 99.98% pure aluminum by a continuous normal freezing process (30). Solid ingots of the purified metal with diameters of 8 or 20 in. and lengths up to several yards are drawn off continuously from the bottom of the "freezing" tube, a mold which is maintained at 660°C, a temperature slightly higher than the melting point of pure aluminum. Fresh liquid feed is continuously introduced just above the freezing interface into the stirred melt, and an overflow at the top of the tube draws off the concentrated impurities.

A multiple-stage continuous freezing unit has been described by Doede (26). Each stage consists of a feed and product chamber connected by a small cooled tube with frozen material. Freezing takes place at the interface between this frozen plug and the liquid in the feed chamber. The plug is continuously extruded into the product chamber where it melts and serves as the purified feed for the next stage. Liquid feed is continuously introduced into the feed chamber, and a side stream takeoff maintains the impurity level in each chamber at a constant level. Purity levels of up to 99.96% are reported for a large variety of organic compounds.

III. THEORY

A. General

Separation by normal freezing is based on the solid–liquid phase equilibrium. The mixtures which will be found in practice will generally consist of more than two components. Except in the special case when a solvent is added (31), it is, for general purposes, less complicated to regard the mixture as a pseudobinary. Usually, neither the nature nor the properties of the minor components are known, and the pseudobinary mixture constitutes the most practical approximation.

The simplest phase diagram of a binary mixture is shown in Figure 1. If a mixture with the composition C is cooled, then pure crystals of A will separate out at C'. Further cooling will allow more pure solid A to crystallize out, while the liquid composition will follow the curve $C'E$. At E the crystallizing solid will have the same composition as the supernatant liquid, which is the composition of the eutectic. In the presence of a solvent, the equivalent phase diagram of the pseudoternary system is given in Figure 2. In this system, where S is the solvent with the much lower melting point, A the major, and B the minor component(s), all pairs form eutectics. Cooling of this mixture will first yield pure crystals of A when the normal line of the initial composition C intersects the surface $T_A e_1 E e_2$. Progressive cooling will continue to yield pure A, while the corresponding liquid phase will stay on the intersecting line between plane $T_A AC$ and the surface $T_A e_1 E e_2$. At the intersection of this line with $e_1 E$, the composition of the crystallizing solid suddenly changes into a eutectic mixture of A and B. This mixture will continue to crystallize

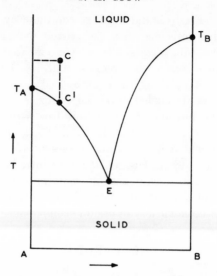

Fig. 1. Phase diagram of binary mixture with eutectic.

out during further cooling while the liquid composition follows line e_1E. As soon as E is reached, the crystallizing solid will consist of the three-component eutectic mixture.

By suspending operations at a certain level and removing the liquid from the solid phase, one achieves separation between A and B. Pure A is obtained as a solid, and B is concentrated in the liquid phase.

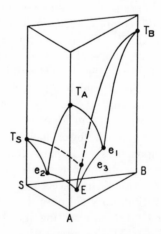

Fig. 2. Phase diagram of ternary mixture with eutectic formation for each pair of components.

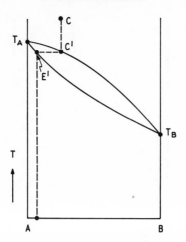

Fig. 3. Phase diagram of binary mixture with mixed crystal formation.

In many cases, however, the impurities present are similar to the major components; and the phase diagram of the binary mixture is better represented by a system in which mixed-crystal formation takes place (Fig. 3).

A liquid with the composition C will, when cooled to C', yield crystals of the composition E'. With progressive freezing the composition of the liquid phase will change according to $E'T_B$. From a mixture corresponding to this type of phase diagram, no pure product can be obtained in one single crystallization operation. A multi-stage operation is necessary to obtain a high-purity product.

In practice one may find a diagram which is a combination of Figure 1, Figure 3, a compound formation, and a peritectic point (Fig. 4). AB is the compound formed, and P is the peritectic point. A liquid with a composition between AB and P will, when cooled, intersect the line $AB–P$. With further cooling, the liquid composition stays on the line $AB–P$ until P is reached. The deposited solid follows the composition of the corresponding solidus curve $AB–F$. With further cooling beyond the peritectic point, the liquid composition follows the liquidus curve $P–T_B$; the deposited solid follows the line $D–T_B$. In the right-hand section of the phase diagram, the impurities are, therefore, preferentially segregated in the solid phase.

If we regard the more "regular" left-hand section of the diagram, it will be obvious that purification is only possible if the initial concentration of A in the mixture is higher than in E. This is the reason

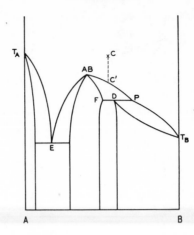

Fig. 4. Phase diagram of binary mixture with eutectics, mixed crystal formation, compound formation, and peritectic point.

why, although in exceptional cases initial purity levels may be as low as 55% (32), it is always advisable to start with as high a purity as possible. It is also clear from the same figure that, although one crystallization run is not sufficient to yield a pure crystalline compound, a few consecutive runs can yield a product of very high purity.

In actual practice, phase diagram data are seldom available for the systems under study. Knowledge of the basic phase diagrams discussed above is generally sufficient to attack the problem in hand successfully. Other possible phase diagrams are described in the literature (2,33).

B. The Solid–Liquid Boundary

The theoretically attainable separation per run, as derived from the phase diagrams, cannot be achieved in practice due to the existence of boundary phenomena which effectively lower the theoretical values to more modest proportions.

a. As the solid contains more of the major component than the liquid from which it has crystallized out, it must eject some of the minor component(s) into the adjacent liquid layer in the process of crystallization. The liquid boundary layer is, therefore, richer in minor component(s) than the average concentration in the bulk liquid. In practice, the crystal growth rate is generally faster than the diffusion rate of the impurities from this boundary layer into the bulk supernatant liquid.

b. Crystallization from a melt does not proceed at a constant rate. There are several factors, such as preferential crystallization sites, temperature fluctuations, and viscosity, which will materially influence the crystal growth. At fast growth rates, especially in a quiescent or slowly stirred liquid, there is a preferential formation of cellular-dendrite type of crystals on the solid–liquid interface. This "rough" crystal surface will result in an increase in δ, the film thickness, and even bodily entrap liquid containing high concentrations of impurities. At slightly lower growth rates, the crystal surface may exhibit a hexagonal network of boundaries that are rich in solute. As these liquid inclusions contain relatively high percentages of the minor components, the separation is unfavorably influenced. This phenomenon is the result of "constitutional" supercooling (33,34)—supercooling because of composition changes in the liquid during freezing—and/or fluctuations in temperature as a result of turbulent or oscillatory free convection (35).

c. If at the relatively low concentration of minor components, which is the general case in normal freezing, we approximate the liquid and solid composition of that part of the relative phase diagram by straight lines (Fig. 5), then the relation between the concentration of impurities in the solid, C_S, and that in the liquid, C_L, can be described by a constant theoretical distribution coefficient, k_0, which is equal to $C_{S(e)}/C_L$. The (e) denotes equilibrium conditions at a negligible growth rate. The impurity concentration in the solid and liquid in equilibrium with each other is given in Figure 6. The left

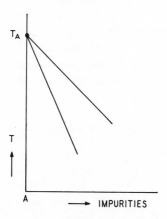

Fig. 5. Enlarged portion of left portion of Figure 4. Both the solidus and the liquidus lines are approximated by straight lines.

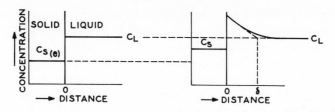

Fig. 6. Impurity concentrations in solid and liquid. At finite growth rates, the impurity concentration in the boundary layer, δ, becomes larger than C_L in the bulk liquid.

of Figure 6 shows the concentration profiles of the impurities at zero growth rate. With a finite growth rate, C_L at the boundary will be higher than C_L in the bulk liquid due to the phenomena described in the previous paragraphs. As k_0 is a constant, C_S will be higher than $C_{S(e)}$. In this case we have an effective distribution coefficient, k, which is equal to C_S/C_L. Burton, Prim, and Schlichter (36) have correlated these two values according to:

$$k = \frac{k_0}{k_0 + (1 + k_0)e^{-f\delta/D}} \tag{1}$$

The dimensionless quantity, $f\delta/D$, may be regarded as a normalized growth velocity. D is the diffusivity, f is the growth rate, and δ is the thickness of the boundary layer. For many liquids, D lies between 10^{-4} and 10^{-5} cm²/sec, and δ ranges from 10^{-3} cm for vigorous stirring to 10^{-1} cm for quiescent conditions.

The influence of the normalized growth velocity on the value of k is graphically depicted in Figure 7 for different values of k_0. For increasingly larger values of $f\delta/D$, the value of k rapidly approaches unity, with concomitant loss of separation efficiency. For good separations it is necessary to maintain $f\delta/D$ at as low a value as practically feasible.

C. Impurity Distribution

Normal freezing is usually carried out in a large-size test tube where a liquid is progressively frozen from the bottom along the axis of the tube. At the end of the run there will be a concentration gradient along the whole length of the rod of the frozen material which may be described by:

$$C_S = C_0 k(1 - g)^{k-1} \tag{2}$$

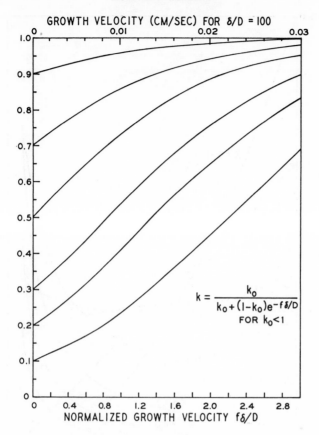

Fig. 7. Influence of normalized growth velocity, $f\delta/D$, on k for different values of k_0.

C_S is the impurity concentration at a certain point along the axis of the tube, C_0 is the impurity concentration in the original liquid, k is the effective distribution coefficient, and g is the distance as a fraction of the whole length of the rod measured from the bottom.

For $k > 1$, the impurities are preferentially concentrated in the crystalline phase. This is the exception rather than the rule. In practice one usually finds the case of $k < 1$, where the minor components are preferentially concentrated in the supernatant liquid phase. To obviate unnecessary complications we have and will tacitly assume $k < 1$ in our discussions unless specifically noted otherwise. For those exceptional cases where $k > 1$, the purification concepts will be reversed.

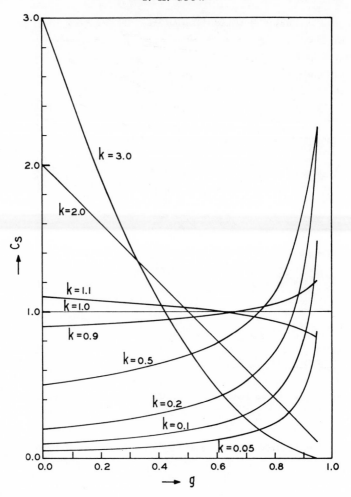

Fig. 8. Impurity distribution in the solidified ingot for different values of k, k being constant over the whole length of the run.

Theoretical impurity distribution curves for different values of k, assuming k to be constant over the whole run, are given in Figure 8. For $k = 1$, there is no change in impurity concentration over the whole rod. For very large and for very small values of k, exceptional separations can be obtained as observed from these graphs.

D. Efficiency Control

For efficient separations it is necessary to decrease both f and δ, as both D and k_0 are already defined by the system to be separated.

It is also evident that once the minimum values have been established by practical aspects, best results would be obtained by maintaining both f and δ constant at the same level. Theoretically, one should use the lowest possible crystal growth rate. There is, however, the necessity to consider a reasonable running time. For a laboratory setup, this would be between 5 and 50 hr. In addition, random fluctuations in the operating parameters would negate any improvement which may be obtained by lowering the tube in the coolant liquid below a certain speed (35).

The thickness of the boundary layer is, in practice, only dependent on the stirrer configuration and the stirring speed. High speed stirring is advisable, and to obviate fluctuations, a synchronous motor or a motor with speed control is used. The value of f is also dependent on the heat flow from the sample to the cooling bath. To minimize fluctuations in the growth rate, bath temperatures have to be thermostatically controlled. Unless this is the case, impurity "bands" perpendicular to the axis will often be observed in the solidified ingots. These visible signs of stratification are formed during periods of high heat transfer and corresponding high crystal growth rates. Loss of efficiency and the formation of impurity bands can also be the result of sharp fluctuations in the "ambient" temperature. The heat flow from the surrounding atmosphere to the tube may be quite considerable, especially for low-melting compounds. The necessity of controlling both bath and "ambient" temperatures can also be derived from the work of Horton and Glasgow (37), where the vessel containing the product to be purified is slowly drawn through the interface of two immiscible liquids separately maintained at constant temperatures. One layer is kept at 25°C above, and the other layer is held at 25°C below the melting point of the crystallizing solid.

The normal-freezing unit should, therefore, preferably be mounted in a constant temperature room. If this is not feasible, adequate precautions must be taken to reduce abrupt fluctuations to a minimum. A relatively simple method is to focus a heat lamp from some distance on the tube. This heat source is connected to a proportional temperature controller which is driven by a temperature probe mounted beside the normal freezing tube.

E. Temperature-Programmed Normal Freezing

The use of a constant temperature bath is sufficient in most cases to attain good separations. It should be realized, however, that in this case there is an increase of heat transfer with time. More surface of the tube is exposed to the cold bath as the run progresses, and there is a decrease in the area of the tube in contact with the "warm" surrounding atmosphere. This accelerated growth is unfortunately noticeable in that section of the run where a sharp increase of the impurity level in the deposited solid is observed. It is really more efficient to decrease the crystal growth rate in this region. This can be accomplished by programming the temperature of the coolant bath over the length of the run. As this entails a system where the temperature of the bath will have to rise in the order of a few degrees per hour, the generally available temperature programmers are inadequate for this purpose. A description of a suitable system is given in Section V.

Results of temperature-programmed runs in comparison to isothermal results can be seen in Figure 9. The data were obtained on an n-hexadecane/n-tetradecane test mixture.

The horizontal axis in Figure 9 shows the fraction of the product crystallized; on the vertical axis is placed the impurity concentration (n-tetradecane) in the most recently deposited crystal layer. The

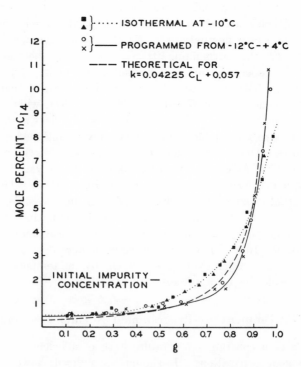

Fig. 9. Impurity distributions at different bath temperatures.

curves, therefore, indicate the purity level along the whole frozen section of the sample.

The theoretical curve for $k = k_0$, i.e., for a 100% efficient normal freezing run, has also been included. The derivation of this curve is given in Section IV.

Up to 50–60% crystallization, bath temperatures apparently do not have any effect, but as more solid crystallizes out, the effect of the programmed bath temperature becomes apparent. Up to 90% crystallization, the solids obtained by using the programmed temperature bath show appreciably less impurities; the obvious result is that the impurity level in the very last section is very much higher in a programmed temperature run in comparison to that obtained in an isothermal bath.

Where sufficient starting material is available and where purification is the main purpose, bath temperatures can be either programmed or run isothermally because one can afford to stop at 60% crystallization. In the majority of the runs, it is generally in the 70% crystallization region that the impurity level of the deposited solid starts exceeding that of the starting material.

In those multipass operations where large yields are desired and in those runs where the objective is the concentration of the impurities, programmed temperatures are preferable because of the higher efficiency in the middle to final regions.

IV. EFFICIENCY MEASUREMENTS

For a comparative study of efficiencies, it would be necessary to have an independent measure to gauge the efficiency of the systems in use. It is obvious that the final degree of purity of a product is no criterion of the effectiveness of the system.

An n-hexadecane/n-tetradecane system can be used as a test mixture to define the efficiency of a normal freezing system. In the 93–100% purity region of hexadecane, both the solidus and liquidus curves are found in the 16–18°C range. This is very convenient as these products are liquid at room temperature. At the same time ice water suffices to yield enough temperature differential to freeze the mixture without difficulties. Sampling of the products during or after a test run can be conveniently carried out without special precautions. Analysis of these mixtures can easily be carried out by conventional gas chromatography.

The phase diagram of this binary system is given in Figure 10. The value of k_0 is obtained by noting the intersections of a

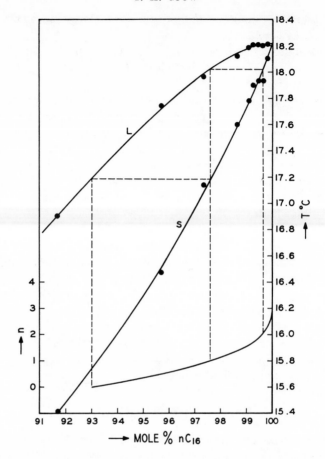

Fig. 10. Phase diagram of n-hexadecane/n-tetradecane.

line of constant temperature with the liquidus and the solidus curves. The ratio of the corresponding tetradecane levels at these points is k_0, the theoretical partition coefficient. Figure 11 shows the dependence of k_0 on the impurity concentration in the liquid. The partition coefficient is not constant; in the 0.5–7% tetradecane region, k_0 is linearly dependent on the C_{14} concentration in the liquid phase.

A regression analysis of these data points yields:

$$k_0 = 0.04225 C_L + 0.057 \qquad (3)$$

where C_L is the concentration of n-tetradecane in mole per cent.

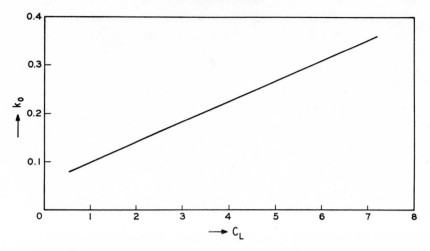

Fig. 11. Dependence of k_0 on the impurity (n-tetradecane) concentration.

Equation (2), which describes the impurity distribution in the ingot, can also be written as

$$\log (C_S/C_0) = \log k + (k - 1) \log (1 - g) \qquad (4)$$

By plotting $\log (C_S/C_0)$ against $\log (1 - g)$, the value of k can be computed either from the slope or from the intercept. Sloan (38) determined the effective segregation coefficient of tetracene, anthraquinone, and other impurities in anthracene at different crystallization rates. He noted that different values of k are obtained, depending on whether he took the slope or the intercept of his graphs. This is probably due to the value of k_0 being dependent on the concentration of the impurities (34).

Equation (2) has been derived assuming k to be a constant. If k_0 is dependent on the concentration, and assuming the efficiency, $\varphi = k_0/k$, to be more or less a constant, k will, therefore, also be dependent on the impurity concentration. Equation (2) will have to be rewritten to take this change into account.

For the general form

$$k = aC_L + b \qquad (5)$$

where a and b are constants, we can derive:

$$C_s = kC_L = (aC_L + b)C_L = aC_L^2 + bC_L \qquad (6)$$

Consider a normal freezing tube of unit cross section when a fraction g of the material is frozen; then a total material balance shows:

$$(1 - g)C_L = C_0 - \int_0^g C_s \, dg \tag{7}$$

The relation between C_s and C_L is given by eq. (6). Substituting this equation in eq. (7) and differentiating to g yields:

$$\frac{dC_L}{dg} = \frac{-aC_L{}^2 - bC_L + C_L}{1 - g} \tag{8}$$

The solution for this differential equation is:

$$\ln C_L - \ln \left| \frac{1 - b}{a} - C_L \right| = (b - 1) \ln (1 - g) + \ln A \tag{9}$$

The constant $\ln A$ can be computed from the following boundary condition:

$$g = 0, \rightarrow C_L = C_0$$

and this leads to

$$C_L = \frac{1 - b}{a + \left(\dfrac{1 - b}{C_0} - a \right) (1 - g)^{(1-b)}} \tag{10}$$

The impurity distribution in the ingots can now be obtained by combining eqs. (3), (6) and (10). Graphs describing this distribution for the ideal case, i.e., for $k = k_0$, are given in Figure 12 for different values of C_0.

Results of a test run can now be superimposed on this chart, and the efficiency at any point can be evaluated from the observed impurity concentration and the impurity concentration on the theoretical curve. A numerical value of the efficiency, φ, can be obtained from the composition of the material at $g = 0$. Practical considerations, however, necessitate the formation of the first crystals under less ideal conditions than during the subsequent part of the run. The value of C_s for $g = 0$ can be obtained by extrapolation from that part of the observed curve which is reliable. The efficiency, φ, is the ratio of the theoretical and the extrapolated concentrations.

V. APPARATUS AND METHODS

A. General

Normal freezing is generally carried out by lowering a tube with the material to be purified into a cooling bath to create the progressive

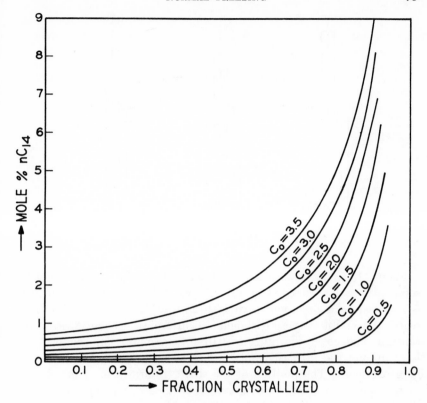

Fig. 12. Ideal impurity distributions after one normal freezing run for different values of C_0.

solidification of the product. Figure 13 shows three phases of a normal freezing run. As the run progresses, the horizontal crystal interface ascends in the tube, and the impurities are concentrated in the supernatant liquid phase. At the end of the run the supernatant liquid is removed either by decanting or suction through a separate liquid removal tube. The run can also be continued until all material is frozen; the impure top section can then be sawed off.

Normal freezing can also be carried out in the annulus of a cooling tube where the crystal layers build up from the cooled wall concentrically (9). An intermediate type is commercially available (39). The product is slowly cooled from the sides and bottom of a stationary receptacle. Hinged stirrer blades are used which rotate at right angles at the beginning of the run and which are forced upwards as the run progresses (Fig. 14).

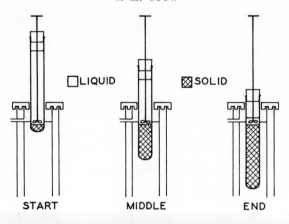

Fig. 13.   Three phases of a normal freezing run.

Figure 15 shows a schematic of our laboratory normal freezing setup for regular work.   The test tube with the sample to be purified is clamped to a precision drive with adjustable lowering rates (Research Specialties, Inc., Richmond, California).   These speeds are constant and cover the range of 0.01–0.4 cm/min.   The cooling bath level around the tube is maintained at a constant height by a large bore overflow.   The position of the stirrer is firmly fixed in relation to the cooling bath with the stirrer head mounted at approximately the same level as the cooling bath overflow.   The stirrer itself has a specially constructed impeller-type head.   The stirrer is driven by either a synchronous motor or by a dc motor connected to an electronic

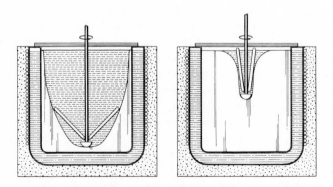

Fig. 14.   The hinged stirrer blades rotate at right angles at the beginning of the run but are forced upwards and up as the run progresses (courtesy The VirTis Company).

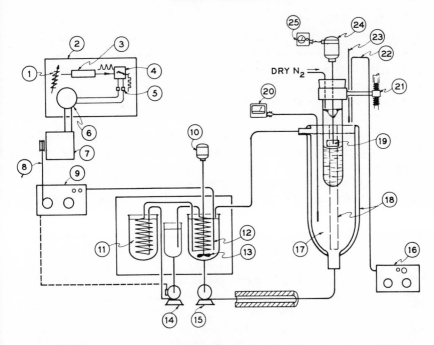

Fig. 15. Temperature programmed normal freezing apparatus: (1) rate potentiometers, (2) motion control unit, (3) unijunction transistor pulse generator, (4) flip-flop circuit, (5) power amplifiers, (6) bidirectional Digimotor, (7) gear box, (8) ratchet, (9) temperature controller, (10) stirrer motor, (11) CO₂-acetone, (14) auxiliary coolant pump, (15) centrifugal circulation pump, (16) environmental temperature controller, (17) acetone–toluene, (18) silvered Dewar with viewing slit, (19) impeller stirrer, (20) temperature readout, (21) adjustable gear drive, (22) environmental temperature probe, (23) thermometer, (24) dc motor, (25) speed controller and tachometer.

controller which maintains the motor on a constant rotational speed irrespective of the necessary torque. A tachometer readout facilitates adjustment of the controller at the desired setting (Servo-Dyne, Cole-Parmer Instrument Company, Chicago, Illinois).

Because of the easy availability, a 1:1 toluene–acetone mixture is employed as the circulating liquid although other liquids or liquid mixtures can and have been used. It should be borne in mind that during prolonged operation at low temperatures, an appreciable amount of atmospheric moisture will dissolve in the cold fluid. Solids may then be formed which would impair the smooth operation of the circulating pump. With an easily available, inexpensive, circulating fluid, frequent change is the easiest solution to this problem.

Good insulation of the coolant lines is necessary to prevent heat uptake and ice deposits from atmospheric moisture. A 1-in. thick foamed-rubber insulation (Armaflex, Armstrong Cork Company, Lancaster, Pennsylvania) has proved to be very adequate down to liquid temperatures of —60°C.

A temperature controller keeps the temperature of the cooling bath at the desired value. This controller switches the coolant pump of the auxiliary cooling circuit on and off as soon as the temperature in the main coolant reservoir exceeds the preset minimum and maximum values.

For work involving temperature programming of the bath, we have found the following system to be useful. A bidirectional Digimotor drives a gear train and gear rack. By adjusting the rate potentiometers in the motion controller, the output linear speed can be varied to any value within a very large range. The rack drives a gear wheel which is mounted on the temperature selector potentiometer of a conventional on–off temperature controller. The programmer rate is obviously dependent on the linear speed of the rack and on the type controller. By connecting the rack to the coarse control of a Resistotrol (Hallikainen Instruments, Berkeley, California) temperature controller, we are able to adjust the rate of temperature change to any value between 0.5×10^{-3} and $6°C/min$. The range most frequently used is 0.02–$0.10°C/min$.

A separate temperature controller maintains the ambient temperature around the tube at a preset constant level.

B. Low-Temperature Operation

With the system described in the previous paragraph, only those compounds melting in the rather narrow range of —5°C to +20°C can be processed without additional precautions or modifications. For compounds melting down to —50°C, the basic setup can still be used, but the interference from atmospheric moisture becomes increasingly troublesome at the lower temperatures. To prevent ice deposits on the tube above the level of the freezing interface, the environmental temperature around the top part of the tube is increased. This is, however, not sufficient for very low temperature work, and the best solution which we have found for this problem is to encase the whole top section in a Lucite cage with dry nitrogen blowing into the case to maintain a dry atmosphere around the tube during the run.

At very low temperatures the problem of solids formation in the circulating liquid because of dissolved moisture also becomes more acute. In addition, special pumps have to be employed which do not "freeze" up at these low temperatures.

For compounds melting below —50°C, liquid nitrogen cooling has been used to cool the tube directly (40). It is obvious that experiments under these extreme conditions pose additional challenges to maintain the operating parameters at the optimum conditions.

An unfortunate property of many low-melting compounds is their poor crystallinity at their freezing points. No definite crystalline form is observed, but an amorphous glass transition stage is formed. This structure is probably comparable to that of petroleum microcrystalline wax. Since crystallinity is an essential requirement for separation by fractional solidification (41), very poor results are obtained. Considering these many adverse factors, construction of an ultralow temperature normal freezing apparatus may very well be only of academic interest in the majority of cases. This is especially the case because these low-melting substances are generally sess relatively large differences in physical properties with respect lower molecular weight compounds where the impurities present posto the major compound. High efficiency distillation or preparative gas chromatography would probably suffice in almost all cases to achieve the desired separation.

C. High-Temperature Operation

For compounds melting in the 10–40°C region, ice water can be used as the coolant medium for isothermal operation. With increasing melting points, hot water or even air will have to be used as the coolant.

The environmental temperature controller around the liquid portion of the product should be adjusted to at least 5–10°C higher than the melting point of the crystallizing product. For higher melting products, extra heaters are, therefore, necessary to keep the liquid from premature solidification. A simple heater for compounds melting up to 100°C is depicted in Figure 16. It can be mounted on most existing normal freezing apparatuses.

Furnaces are necessary for still higher melting products. Considerable refinements in the design of these ovens have been noted because of the interest in the technique of large crystal growing, which is already several decades old. Judging from the literature, the Stockbarger-type furnace is probably the most suitable for high-temperature normal freezing work (42), although other heater types have been advanced (6). For the highest temperatures encountered so far,

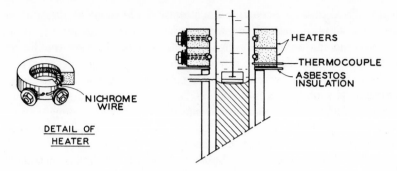

Fig. 16.   Heaters for high-temperature normal freezing.

direct heating of the container to above the melting point of the product is used (30).

There is, therefore, little problem in maintaining a high temperature around the liquid portion of the product; and the higher limit of the applicability of normal freezing techniques seems to be only the heat sensitivity of the processed compounds.

VI. RESUME

Normal freezing is a new and relatively inexpensive purification technique which has been used for the concentration of impurities and for the purification of a very large variety of both organic and inorganic compounds. The number of compounds which can be successfully handled by this technique is certainly a magnitude larger than the products listed so far in the literature. A wide range of melting points has been encompassed so far, from organic compounds melting below —50°C to aluminum which melts at 659.7°C. This temperature range will be expanded in the future, probably not so much at the lower end but more at the higher temperatures. The only criteria are the stability of the compounds at these elevated temperatures and the manifestation of crystallinity in the processed materials.

The necessity of having to have relatively pure starting material is a drawback to the general applicability of the method, and normal freezing will almost always be used in conjunction with another pre-purification method. It is especially applicable as the final purification step in the preparation of a highly purified compound.

Although much still seems to be lacking in the understanding of the phenomena at the liquid–solid boundary, sufficient information is already known to maximize the practical efficiency of the process. It seems that all possible parameters have now been accounted for and that no major breakthroughs in the method as such can be expected in the near future.

The technique is currently still very much a batch analytical laboratory-scale tool. Because of the successful purification of aluminum by continuous normal freezing on an industrial scale, large-scale applications to other compounds can be expected in the near future. As an analytical tool, however, it will always maintain a place in the laboratory. The further developments of both industrial continuous normal freezing and laboratory-scale batch normal freezing will continue along almost parallel lines; they will sometimes cross and enrich, but they will not supersede each other.

References

1. W. G. Pfann, *Zone Melting,* Wiley, New York, 1959.
2. H. Schildknecht, *Zoneschmelzen,* Verlag Chemie, Weinheim, 1964.
3. R. T. Leslie, *Ann. N.Y. Acad. Sci.,* **137,** 19 (1966).
4. W. R. Wilcox, *Separation Sci.,* **1,** 147 (1966).
5. H. Schildknecht and F. Schlegelmilch, *Chemie Ing. Tech.,* **35,** 637 (1963).
6. J. C. Brice, *The Growth of Crystals from the Melt,* Interscience, New York, 1965.
7. F. W. Schwab and E. Wichers, *J. Res. Natl. Bur. Std.,* **25,** 747 (1940).
8. M. Zief and W. R. Wilcox, *Fractional Solidification,* Vol. 1, Dekker, New York, 1967.
9. J. E. Powers, *Hydrocarbon Process. Petrol. Refiner,* **45,** No. 12, 97 (1966).
10. J. A. Palermo and K. H. Lin, *Ind. Eng. Chem.,* **58,** 67 (1966).
11. D. B. Wilson, *Chem. Eng.,* **72,** 119 (1965).
12. J. A. Palermo and G. F. Bennett, *Ind. Eng. Chem.,* **57,** 68 (1965).
13. J. W. Mullins, *Brit. Chem. Eng.,* **9,** 438 (1964).
14. G. Nitschmann, *Ver. Deut. Ing. Z.,* **106,** 1458 (1964).
15. F. W. Schwab and E. Wichers, *J. Res. Natl. Bur. Std.,* **32,** 253 (1944).
16. J. D. Dickinson and C. Eaborn, *Chem. Ind. (London),* **1956,** 959.
17. J. S. Matthews and N. D. Coggeshall, *Anal. Chem.,* **31,** 1125 (1959).
18. J. Shapiro, *Science,* **133,** 2063 (1961).
19. H. Schildknecht, *Z. Anal. Chem.,* **181,** 256 (1961).
20. H. Schildknecht, *Journées Intern. d'Etude des Méthods de Sépn. Immédiate et de Chrom., Paris,* **1961,** 37.
21. H. Schildknecht, G. Rauch, and F. Schlegelmilch, *Chemiker Ztg.,* **83,** 549 (1959).
22. H. Schildknecht and U. Hopf, *Z. Anal. Chem.,* **193,** 401 (1963).
23. H. Schildknecht and U. Hopf, *Chem. Ing. Tech.,* **33,** 352 (1961).
24. H. Schildknecht, K. Maas, and W. Kraus, *Chem. Ing. Tech.,* **34,** 697 (1962).
25. C. W. Beck and J. Buck, Abstracts, Meeting of the American Chemical Society, St. Louis, 1961, A-135.
26. C. M. Doede, *Ind. Chemist,* **38,** 462 (1962).

27. J. Shapiro, *Anal. Chem.*, **39**, 280 (1967).
28. H. Schildknecht and G. Rauch, *Z. Naturforsch.*, **166**, 422 (1961).
29. Anon., *Chem. Eng. News*, **43**, 42 (March 1, 1965).
30. J. L. Dewey, *J. Metals*, **17**, 940 (1965).
31. I. A. Eldib, *Ind. Eng. Chem., Process Design Develop.*, **1**, 2 (1962).
32. D. L. McKay and H. W. Goard, *Ind. Eng. Chem., Process Design Develop.*, **6**, 16 (1967).
33. B. Chalmers, *Principles of Solidification*, Wiley, New York, 1964.
34. G. J. Sloan, *Mol. Crystals*, **1**, 161 (1966).
35. K. Morizone, A. F. Witt, and H. C. Gatos, *J. Electrochem. Soc.*, **113**, 51, 808 (1966).
36. J. A. Burton, R. C. Prim, and W. P. Schlichter, *J. Chem. Phys.*, **21**, 1987 (1953).
37. A. T. Horton and A. R. Glasgow, Jr., *J. Res. Natl. Bur. Std. C*, **69**, 195 (1965).
38. G. J. Sloan, *Symp. über Zonenschmelzen und Kolonnenkristallisieren*, Karlsruhe, 1963, p. 270.
39. The VirTis Company, Inc., Gardiner, New York.
40. A. R. Glasgow, Jr., and G. Ross, *J. Res. Natl. Bur. Std.*, **57**, 137 (1956).
41. J. S. Ball, R. V. Helm, and C. R. Ferrin, *Refining Eng.*, **30**, C36 (1958).
42. J. N. Sherwood and S. J. Thomson, *J. Sci. Instr.*, **37**, 342 (1960).

Preparative Gas Chromatography

MAURICE VERZELE

Laboratory of Organic Chemistry
State University of Ghent, Belgium

I. INTRODUCTION

In analytical gas chromatography, in which the separated substances are not collected, the sample size varies from a fraction of a microgram (for flame ionization, electron capture, or β-ray detectors, etc.) to milligram amounts (for hot-wire detectors, etc.).

When the separated components are collected for the purpose of further experimentation, the operation is called "preparative" gas chromatography. If the subsequent experimentation is the recording of a spectrum (UV, IR, or mass spectra), or perhaps a color test, the sample to be collected can be in the milligram range. On the other hand, if the sample is to be used for other purposes such as synthesis, the amount needed may be very much greater. Preparative gas chromatography may eventually be used to prepare quantities ranging from grams to kilograms. As the scaleup successfully approaches this upper limit, the method will become a useful industrial technique for the separation and purification of certain mixtures.

Because of the broad range of sample sizes involved in these various applications of gas chromatography, it is evident that there is no single or even simple solution to the attendant technical problems. When the variable of separation is imposed on this situation the degree of complexity grows and one is faced with compromise. Because of the versatility in instrumentation, it turns out that the situation can be channeled into four distinct categories.

1. Separations of mixtures which are not too complex and are in milligram quantities. Here normal analytical gas chromatography can be used, and the major problem is that of collection of the separated components.

2. Separations of mixtures containing a large number of components having a rather wide boiling point range, where quantities from 10 μl to a few milliliters are desired.

3. Separations of mixtures containing only a few components but where quantities from 100 μl to 20 ml are desired. When larger quantities of a multicomponent mixture have to be separated, as in *2* above, it is preferable to fractionate the complex mixture and refractionate according to the technique *3*.

4. The separation of mixtures as discussed in *3* above but for even larger quantities.

These points will be presented and discussed in this paper which is based mainly on studies from the author's laboratory. Preparative gas chromatography is an extension of analytical gas chromatography and it will be assumed that the reader is familiar with the theoretical and technical fundamentals of this instrumental method.

II. THEORETICAL CONSIDERATIONS RELEVANT TO PRACTICAL PREPARATIVE GAS CHROMATOGRAPHY

A. Maximum Allowable Sample Size

When the sample size of a single substance is gradually increased, it is well known that the peak band width also increases. A two-component mixture that can be separated when using small samples will, with larger and larger samples, eventually show peak overlap and only partial separation. There is therefore a well-defined "maximum allowable sample size," or MASS for the mixture on a particular column which can just be separated. This MASS depends on the proximity of the two peaks to each other, or in other words, on their relative retention or r value, since $r = V_A'/V_B'$, where V_A' and V_B' are the adjusted retention volumes for components A and B (or time or distances on the chromatogram). "Adjusted" here means measured from the air peak or its equivalent. The r value reflects the difficulty of the separation of A from B. For a multicomponent mixture, the pair of peaks closest together or with the lowest r value will determine the MASS of the whole sample. In this case only the lowest r value of the mixture has to be considered.

This MASS is dependent on three factors: (*1*) the difficulty of the separation, (*2*) the efficiency of the column, and (*3*) the time required for the separation. Each of these factors is discussed in turn.

1. *The difficulty of a separation* is indicated by the r value. This is a measure of the relative volatility of the components and also of the selectivity of the stationary phase for that particular mixture as shown by eq. (1)

$$r = V_A'/V_B' = \gamma_A p°_A/\gamma_B p°_B = k_A/k_B = K_A/K_B \qquad (1)$$

where V_A' and V_B' are adjusted retention volumes for components A and B (or time or distances as measured on the chromatogram), γ_A and γ_B are activity coefficients, $p_A°$ and $p_B°$ are vapor pressures, k_A and k_B are partition ratios, and K_A and K_B are partition coefficients.

On another stationary phase, a new γ value can give a higher r value and therefore a larger MASS. Even a small increase in the r value has a large influence and it may therefore be necessary to try several stationary phases for optimization of this factor. The r values are easily determined on analytical size columns. Generally, components having r values greater than 1.5–2 on nonpolar liquid phases can be separated by distillation. The lowest r values which will give reasonable separations by preparative gas chromatography are about 1.09–1.1. Lower r values should be avoided if possible by a choice of another stationary phase. Indeed, rarely is it necessary to try more than 2 or 3 stationary phases of sufficient polarity difference before finding one which will give at least an $r = 1.10$. This exploitation of different activity coefficients (γ) on different stationary phases is possible only with mixtures containing a few components (e.g., *cis–trans* isomeric mixtures). With more complex mixtures there are so many r values to consider that at least one pair of the components will always happen to give a low r value, regardless of the nature of the stationary phase. In choosing the stationary phase for multi-component mixtures, general polarity is therefore the only factor to consider (polar substances on polar stationary phases and vice versa).

2. *The efficiency* of a preparative column is best expressed in plate numbers denoted by the letter n. Although this has many drawbacks, because of the variations of n with temperature, k value, and stationary phase etc., it is still a useful basis of comparison. In analytical gas chromatography plate numbers are usually calculated according to eq. (2)

$$n = 16(V_R/W_b)^2 \qquad (2)$$

where V_R is the retention volume of the peak (or time or distance on the chromatogram measured from start to peak maximum and W_b is the band width at the peak slope tangent intercepts at the base line.

The equation is a simplification of the more correct one which is

$$n = (V_R/W_b)^2(1/k + 1) \tag{3}$$

Since the k value is usually large in analytical work, the simplification is warranted. In preparative gas chromatography on larger columns, the time element often forces us to use smaller k values, the factor $1/k + 1$ is then no longer negligible and it is better to use eq. (3) to calculate plate numbers.

In another form, eq. (3) may be expressed as

$$n = 16(V_R V_R'/W_b^2) \tag{4}$$

which is the more practical form. With the increased sample sizes used in preparative gas chromatography, band width W_b is also increased and n values calculated with these peaks reflect only the efficiency of the column for that particular sample size. The plate number calculated for the peaks of preparative scale samples is therefore called "preparative scale plate number" and is symbolized by n'. These n' values decrease with increasing sample size, which means that n' and sample size must always be given together. The resulting n' values for a 6 m $\times$ 9 mm coiled column charged with Chromosorb W, 30–60 mesh, containing SE 30 silicone gum at 25% w/w, hydrogen gas rate of 200 ml/min and at a temperature of 145°C are shown in Table I.

3. The time of separation is determined mainly by temperature and is measured by the partition ration k. This is the ratio of the retention volumes (time or distance) of the peak considered and the air peak. Low k values mean that the substances are swept rapidly through the column and it is obvious that in this case the MASS is smaller than could be found for the same mixture at higher k values found at lower temperatures. The k value should not be larger than about 20, otherwise the separations last too long. It should also not be smaller than about 2–3.

B. The Use of r, k, and n'

In analytical gas chromatography eq. (5) relates r, k, and n as follows:

$$n = 16(r/r - 1)^2(1/k_2 + 1)^2 \tag{5}$$

This equation is used to find the necessary column length to give a 4σ separation of a mixture with given r and k. When k is between 3 and 20, this equation can be simplified to:

$$n = 16(r/r - 1)^2 \qquad (6)$$

In preparative gas chromatography, eq. (6) will not be used to find column length, but to find n' and the maximum allowable sample size (MASS) of a specific sample. From the relative retention of the two compounds and the capacity ratio of the second (k_2), the number of theoretical plates, n, which are required to give a 4σ separation is calculated. Then from Table I the maximum allowable sample

TABLE I

Preparative Scale Plate Numbers, n', for Samples of Cumene

Sample size, μl	n'
10	2340
50	1380
100	710
200	400

size giving a value of n' equal to this required n is found and used. An example will serve to illustrate this computation. Consider the $6\,m \times 9\,mm$ column of Table I and a separation time of about 45 min. This is equivalent to a partition ratio in the range of 20. With such a partition ratio the column will have its maximum separation power, and the second term in eq. (5) can be neglected as already indicated. For an r value of 1.25 (*trans–cis* decalin mixture at about 160°C), eq. (6) leads to 400 required plates. From Table I it can be seen that the column gives this preparative scale plate number for sample sizes of 200 μl. For a 1:1 decalin mixture this means then that the MASS which will give complete separation is 400 μl. This is found to be the case experimentally, and each peak gives about 400 plates.

Most mixtures contain components in unequal amounts or a larger number of components than the example given above. When this is the case, the two components in the mixture giving the lowest r values are used to calculate n'. Of these two components in the mixture the percentage x of the one present in largest amount is calculated. We then have for this general case:

$$\text{MASS} = 100S_{n'}/x \tag{7}$$

where $S_{n'}$ is the sample size of a single substance giving the calculated n value on the preparative scale column to be used. Consider, for example, a five-component mixture with lowest r value equal to 1.09 between the second and third peaks, the second peak being the largest and representing 20% of the total mixture. The chromatographic column No. 1 of Table V could separate a MASS of $100 \times 500 \, \mu\text{l}/20 = 2.5$ ml.

From the foregoing it should be clear that the MASS is variable and is determined by the value of r and the composition of the mixture. Values of n' necessary then for all possible r values are easily calculated. Some calculated values neglecting the second term of eq. (5) are listed in Table II.

TABLE II

Plate Numbers Necessary to Separate Mixtures with the r Values Indicated

r value	n' values	r value	n' values
1.03	17496	1.12	1384
1.04	10816	1.13	1221
1.05	7056	1.15	940
1.06	4960	1.20	576
1.07	3745	1.25	400
1.08	2915	1.30	300
1.09	2342	1.40	195
1.10	1936	1.50	144
1.11	1632	2.00	64

Table II shows clearly why r values below 1.1 should be avoided. The separation power of the column has to be very high and from Table I we see that only low MASS will be permitted. For separations involving an $r = 1.10$, about 2000 plates are needed (i.e., a MASS $= 25 \, \mu\text{l}$). Considering the unavoidable drop in the number of plates needed with larger samples, the plate number for preparative gas chromatography must be substantially higher than 2000 plates. This can only be achieved with rather long columns.

III. COLUMNS

A. Column Design and Variations of n' Values

It is obvious that the MASS should be as large as possible and that this is a very important point in preparative gas chromatography. The simplest and the most common method to obtain large MASS values is to increase the column diameter. However, this has an adverse effect on band width and, because of this, separations which are, for example, easily achieved on a $2 m \times 4 mm$ column with a $1 \mu l$ sample, cannot be obtained on a $2 m \times 4 cm$ column with $100 \mu l$ samples (equivalent increase in column cross section and in sample size).

The repetitive injections of small samples on a relatively smaller column is therefore a useful approach. This technique is the basis of commercial instruments, one of which is the Varian Aerograph Autoprep 700. When really large samples are needed this technique, although automated, takes too much time.

Another solution is the simultaneous use of a number of parallel columns, and this idea has also been exploited in commercial instruments as, for example, the Beckmann Megachrom Unit. A set of exactly identical columns, necessary for the success of a multiple parallel column instrument, is difficult to obtain, and this technique has largely been abandoned.

A recent solution in preparative gas chromatography is the continuous flow technique of Dinelli, Polezzo, and Taramasso (2) where a cylindrical stack of columns revolves before the injection port. This continuous operation resembles somewhat the liquid–liquid continuous separation process developed in this laboratory (3). Technically it must be much more difficult than a more traditional approach, and the future will have to show what this technique can do.

To return to columns with increased diameter, the band-widening factor has been the subject of many investigations. While earlier efforts were mainly directed toward the use of more uniform packing, recent work has been devoted to the control of the flow of the gas phase through the column by devices which are now known as "flow homogenizers" (4,5). The successful utilization of these devices indicates that wide bore columns of normal length (2 m) provided with such flow homogenizers are probably the immediate answer to industrial-scale gas chromatography (6).

It is, however, more precise to say that larger MASS on these increased diameter columns is the result of increased column volume than of increased column diameter. Now, column volume cannot only be increased by using wider bore columns, but also by using longer columns of a more conventional gas chromatographic diameter. Both, short, wide and long, narrow columns present advantages and disadvantages. From a purely technical standpoint it can be said that:

1. Short, wide columns are faster, more expensive, consume more gas (helium), and can be used at slightly lower temperatures. They must be operated at high gas rates.

2. Long, narrow columns are slower, less expensive, consume less gas, and have to be used at slightly higher temperatures. They can be operated at low gas rates.

These points are however of secondary importance to the MASS and this as far as the column is concerned will depend on the n' values.

Values of n decrease with increasing sample size but the rate of decrease is determined by a number of factors, namely, the influence of the nature of the carrier gas, the particle size of the support, and the amount of stationary phase used. These will be discussed further. Another factor is the column shape. We can indeed put the same amount of support material to work in a short, wide bore column or in a long, narrow bore column as already explained. Values of n' for such columns are given in Table III. The short, wide bore

TABLE III
Values of n' for Short, Wide and Long, Narrow Columns

	Values of n'	
Sample size	Short, wide 2 m × 7.5 cm column	Long, narrow 19 m × 3 cm column
100 µl	177	2900
200 µl	149	2390
500 µl	132	1632
1 ml	183	760
2 ml	161	480
5 ml	125	225
10 ml	125	96

column was charged with Chromosorb A, 30–40 mesh, coated with Carbowax 20M at 5 g/100 ml. Cumene samples were chromatographed at 140°C using 2000 ml/min of hydrogen. The total volume of the column was 12 liters. The long bore column consisted of 19 sections each 100 cm long, filled with Chromosorb A of 10–20 mesh and coated with SE 30 silicone gum at 5 g/100 ml. Benzene samples were chromatographed at 80°C using 600 ml/min of hydrogen. The total volume of the column was 8.8 liters. The long, narrow column shows large variations of n'; it can handle problems having sample sizes which vary enormously but which are large enough for a great number of investigations. The short, wide column shows practically no variation of n'; it cannot handle difficult problems, but permits the use of very large samples for easy separations. From Table III it can be deduced that long, narrow bore columns are to be preferred in all cases. This should not, however, necessarily be true because the introduction of "flow homogenizers" (5) should improve the n' values of short, wide columns markedly. Such a column with dimensions of 2 m × 10 cm equipped with flow homogenizers showed, according to the published chromatograms, about 600 plates for milliliter samples (7). The short, wide bore column of Table III also contained flow homogenizers, of our own design, but obviously, although we tried a large number of variations and combinations, none were successful.

It remains true, therefore, that long, narrow columns can handle a broader range of problems, namely, those involving r values of 1.08 and greater and sample sizes from 0.1 to 20 ml. Short, wide columns, on the other hand, are best used for volumes of 1 ml and upwards and for mixtures having r values of at least 1.20. From the foregoing it is evident that for every preparative gas chromatographic column, n' values should be determined as was done for Table III. These values permit the calculation of the MASS and in general the comparison of different columns.

B. Choice of Column Dimensions

Column dimensions should be chosen for optimum or good results for a large variety of problems rather than for perfect results for limited applications. This is especially advisable for the research laboratory. Therefore, as stated before, long, narrow columns are to be preferred. Further reasons for this choice are based on considerations for collection and for flow rates. In laboratories where samples are used for identification and for exploratory chemical purposes, the needed sample size can be narrowed to the range of 10 μl–10 ml. To make the separated substances easily available they should be collected in a vessel of about 1–25 ml volume. Larger volumes of the collectors will lead to losses for obvious reasons. In a collecting device of 1–25 ml volume, the flow rate should be rather low, since otherwise good recovery would be impossible. While some substances can be collected adequately with flow rates up to 1 liter/min in the collectors under consideration, other substances need flow rates as low as 25 ml/min, as discussed in later sections. Although the optimum flow rate in preparative work is not so pronounced as in analytical gas chromatography, gas rates of 25–1000 ml/min can only be used in rather narrow columns, with equally good results. Since, on the other hand, high sample capacity can only be obtained with larger volumes of column-filling material, rather long and narrow columns will be needed.

IV. PRACTICAL CONSIDERATIONS

A preparative gas chromatograph contains the same elements as a normal chromatograph and in addition has a collection device. The sample is *injected* and travels into a *column* filled with a *support* coated with a *stationary phase*. The sample is forced through the column with a *carrier gas* into a *detector* and is collected in *fractions*. At the same time the chromatogram is *recorded*. Each of the component parts of the chromatograph underlined above will now be discussed in detail.

A. Injection

In analytical gas chromatography the sample is small enough so that vaporization is not a problem. This is not true for preparative gas chromatography. For samples in the milliliter range it is a major difficulty to completely vaporize them before they are allowed to enter the column. The simplest solution is to avoid this vaporization and to use "on-column" injection. This technique gives excellent results as is shown in Figure 1.

The results of Figure 1 were obtained by varying the injection time in the injection block from a fraction of a second to much longer periods. The results shown are the best obtained. The result that on-column injection gives superior results is not surprising; it should indeed best approach the ideal case, i.e., that of putting the sample onto the column as a plug.

A drawback of on-column injection of large samples is that the stationary phase could be stripped rather quickly from the first part of the column. This stripping possibility was investigated by on-column injection of 100 successive 5-ml aliquots of a mixture containing benzene, toluene, ethylbenzene, and isopropylbenzene in a 20 m × 9 mm coiled column. The support was coated with SE 30

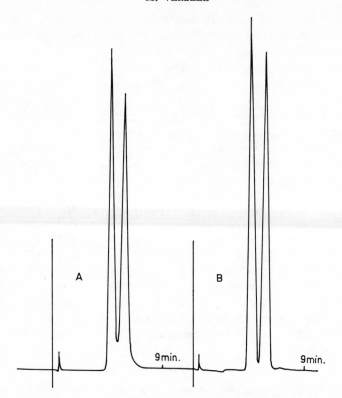

Fig. 1. Injection of 15 μl decalin on Aerograph 90P, 3 m length × 4 mm diameter column (A) in injector, (B) on-column.

silicone gum. After the 100 injections the column packing was removed in small portions and the amount of stationary phase on each portion was determined by extraction and weighing. The results showed that stripping is negligible and that the stationary phase distribution had not changed. This result could have been expected since chromatographic results before and after the injections were practically the same. The separating power of the column for small samples dropped slightly after a few of the injections but thereafter remained remarkably constant, as can be seen in Table IV. For the large sample injections the "preparative scale plate number" changed very little, if at all, considering the poor reproducibility of plate number determinations.

It must of course be remembered that this experiment was performed using SE 30 and that other more soluble stationary phases

TABLE IV
Analytical and Preparative Plate Numbers for Benzene, Toluene,
Ethylbenzene, and Cumene

		Number of 5-ml injections at 165°C			
Sample size	Material	First injection	After 30 injections	After 60 injections	After 100 injections
		Analytical plate numbers n			
50 μl	Benzene	952	792	796	680
	Toluene	1180	905	1032	841
	Ethylbenzene	1471	1098	1220	1098
	Cumene	1530	1154	1368	1126
		Preparative plate numbers n'			
1 ml	Benzene	64	63	62	70
	Toluene	164	149	142	172
	Ethylbenzene	346	279	309	317
	Cumene	576	443	468	491

may have given different results. Then, too, stripping of the stationary phase from a few inches of a column which is 10 m or more in length must have little or no effect on its overall efficiency. This is not true for a wide bore column where stripping of a few inches or channelling in the upper part of the column must certainly have a detrimental effect on the results. Vaporization of the sample in the inlet is necessary in this case.

Injection of large, volatile samples into a heated column can cause a rather large increase in pressure. Backflash of part of the sample into the cold carrier gas lines must of course be avoided. Normally this is achieved by installing a "back-flash valve." However, in our experience even the best back-flash valves tend to become dirty and fail after some time. Therefore back-flash valves are better replaced by needle valves. This valve is closed just before injection and re-opened about 1 min after the injection. The effect of this operation is barely discernable on the chromatograms obtained with the large volume columns under consideration.

An injector heater is not needed for on-column injection. On our instruments the injection ports are provided only with a small analytical-type heater for use with very small samples. All separations discussed in this paper were obtained with on-column injection.

B. Columns

1. COLUMN MATERIAL

Columns are usually made of metal. Copper has to be avoided because it is easily coated with an oxide layer which catalyzes the decomposition of organic substances. Stainless steel is preferred but is rather expensive. Aluminum columns are a good compromise, although some halogenated substances requiring high working temperatures cannot be chromatographed on aluminum columns. Decomposition is minimal in glass columns, but glass breaks rather easily. An attractive advantage of glass columns is their transparency permitting easy observation of the condition of the packing and stationary phases.

2. SHAPE OF COLUMNS

All types of column shapes, such as straight, sectioned, and coiled, can give good results. Values of n' for a number of different columns are shown in Table V. The columns have to fit in the oven of the chromatograph. Coiling is the best way to achieve compactness and to fit large-volume, long columns in as small an oven as possible. With column diameters above a few millimeters, coiling produces an ovalization of the cross section of the column tube. This ovalization is a beneficial factor, since it turns out experimentally that coiled columns show about 10% more plates than equivalent nonovalized columns. Ovalized columns have been advocated before (8). The reason for this better performance must be better temperature equilibration and minimization of the "race track effect." This term refers to the fact that gas molecules at the outer wall of a coiled column have to travel a longer distance than those at the inner wall. This behavior causes band broadening, and the effect becomes more pronounced as the coil diameter is made smaller.

Finally, long metal columns and certainly glass columns must be supported in a frame which is easiest when the columns are coiled. Therefore, coiled, ovalized columns are recommended.

3. FILLING THE COLUMNS

Relatively short, metal columns can be filled upright and coiled afterwards. Long columns and glass columns have to be coiled and then filled. Even with extremely long columns like No. 4 of Table V,

TABLE V
Preparative Scale Plate Numbers for Various Columns

n'	1[a]	2[b]	3[c]	4[d]	5[e]	6[f]
17496					10 μl	
10816						
7056					50 μl	
4960						
3745	10 μl	10 μl	10 μl			
2915	100 μl	100 μl	50 μl		100 μl	
2342	500 μl		100 μl	100 μl		10 μl
1936			200 μl	200 μl		
1632	1 ml	500 μl	500 μl	500 μl	200 μl	
1384						50 μl
1211						
940	2 ml		1 ml	1 ml		
		1 ml			500 μl	100 μl
576			2 ml			
		2 ml		2 ml		
400	5 ml					200 μl
300		5 ml	5 ml		1 ml	
195				5 ml		
144						
64						
Column volume, liters	7.7	8.4	5.1	4.8	1.3	0.39

[a] Oval section, 1.2 cm × 2.5 cm; coiled aluminum; 21 m long; Chromosorb A 20–30 mesh; Carbowax 20M, 5 g/100 ml support or 10% w/w; cumene samples at 145°C, 600 ml H_2/min. Coil diameter = 30 cm.

[b] Round section; straight stainless steel column 19 m × 3 cm in 19 parallel sections of 100 cm each with tapered ends; Chromosorb A 10–20 mesh; SE 30 5 g/100 ml support or 10% w/w; benzene samples at 80°C, 600 ml H_2/min.

[c] Oval section, 1.2 × 2.5 cm; coiled aluminum column 15 m long; Chromosorb A 20–30 mesh; SE 30, 5 g/100 ml support or 10% w/w; cumene samples at 145°C, 500 ml H_2/min. Coil diameter = 30 cm.

[d] Round section; coiled glass column 75 m × 9 mm; Chromosorb G 10–20 mesh; SE 30, 5 g/100 ml support or 10% w/w; cumene samples at 145°C, 400 ml H_2/min. Coil diameter = 30 cm.

[e] Round section; coiled glass column 20 m × 9 mm; Chromosorb W 30–60 mesh; SE 30, 5 g/100 ml support or 25% w/w; cumene samples at 145°C, 200 ml H_2/min. Coil diameter = 12 cm.

[f] The same as column 5 but only 6 m in length. Coil diameter = 12 cm.

this is not so difficult and only requires time to accomplish. The coated support is simply poured through a funnel into the upright column. The filling operation is aided by a water pump vacuum and a vibrator applied to the column frame. For some long columns having large volumes and using certain support materials the filling operation may require several hours. This is also the case for glass columns where great care has to be taken to avoid breakage. The columns can be emptied by a reverse procedure. We have filled and emptied the same 75 m × 0.9 cm coiled glass column often.

The most important consideration in column filling is that this operation should be carried out in such a way as to achieve the best possible permeability of the carrier gas. To accomplish this, the support material should be free of fine particles which may be generated during the coating procedure or by rough handling of the support. Sieving is not recommended, since this can also produce fine particles. Instead, a blowing process is preferred which is carried out by causing the support material to pass through an inclined glass tube. The angle of inclination determines the rate at which the support falls through the tube. The support is made to drop a distance of about 10–15 cm on to a clean sheet of paper while air is blown through it. The finer particles will be blown away from the coarser particles.

The column should be filled as lightly as possible. In small bore columns the effect of packing density seems relatively unimportant. For a 6 m × 9 mm glass column (Autoprep 700) filled with Chromosorb W 60–80 mesh and coated with SE 30, we could vary the packing weight for the full column between 68 and 81 g. For each of these two packing densities, for example, the separations of cis–trans decalin in terms of either preparative plate numbers or analytical plate numbers were remarkably the same when carried out at 160°C and using 200 ml/min of hydrogen gas. There was, however, a big difference in retention times because of the difference in inlet to outlet pressure ratio (compressibility factor). The loosely packed column is nearly twice as "fast" as the most densely packed column. It seems therefore that for the smaller bore preparative gas chromatographic columns, maximum packing density should not be sought, but rather lightly packed columns should be used because of the advantage of shorter analysis time.

For a precoiled column it is important which end is chosen for charging the support. Columns can indeed be filled from either the injection end or from the detector end. The highest packing density will result at the end opposite the one chosen for filling. As the highest

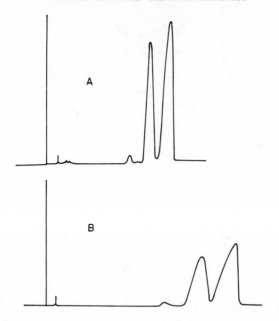

Fig. 2. Separation of 4 ml of 1:1 *cis–trans* isomers of 3,4-dimethylcyclo-
hexanone on column 1 of Table V. 700 ml H₂/min at 150°. Bridge current
150 mA. Recovery was better with *A* than with *B*.

density should be at the inlet port where the high inlet pressure occurs,
it is advisable that columns be filled from the detector end.

An example will demonstrate the effect of charging the column
from either end. Two columns identical to column No. 1 of Table
V, which is ovalized and 21 m in length, were filled by each of the
two possible ways. Each column was then used to separate a 1:1
mixture of *cis–trans* 3,4-dimethylcyclohexanone. The packing was
Carbowax 20M on Chromosorb A 10% w/w. The r value for this
mixture is 1.15 and from Tables II and V it can be deduced that
sample charges of 4 ml each can be handled by these columns. In
operation under an inlet gas pressure of 700 mm Hg the hydrogen
flow rates were 630 ml/min for the detector-end filled column, shown
in Figure 2*A*, and 430 ml/min for the injector-end filled column in Figure 2*B*. The elution time for *A* was 80 min and for *B* was 120 min.

Each time a column is filled, it has to be conditioned overnight
regardless of whether the filling is freshly prepared or has already
been used in another column. It is recommended that conditioning
be done at 250°C for SE 30 and 190°C for Carbowax 20M.

C. Support

1. MATERIAL

A list of some of the more useful support materials with comments on properties and uses is given in Table VI.

For analytical gas chromatography the support material should be of the highest possible quality. In preparative gas chromatography much larger volumes of support material are needed and there may be a tendency to be less critical in its quality. This is a false premise and its practice should be avoided. In preparative gas chromatography the temperatures employed are usually higher, and the substance resides at this temperature longer than in analytical gas chromatography. Thermal decomposition of the chromatographed substancs and of stationary phases is therefore a major difficulty of preparative gas chromatography. This thermal decomposition is catalyzed by the support material, and high-quality supports are indeed least active in this respect. In fact the following "activity sequence" can be advanced for the materials which we have had the

TABLE VI
Support Materials

Glass beads (density = 2) give small capacity and low resolution. Their use will be discussed more extensively below.

Chromosorb W and Gas Chrom P (Johns Manville, Applied Science, State College, Pennsylvania) (density = 0.22) cannot be obtained in grain size coarser than 30–60 mesh. They can only be used in relatively short columns otherwise the pressure drop becomes excessive. They are best suited for columns like nos. 5 and 6 of Table V. Chromosorb W has slightly larger capacity than Gas Chrom P while the latter gives slightly better results with analytical separations (10).

Chromosorb G (Johns Manville) (density = 0.55) can be obtained in any mesh size. It gives a lower MASS, but it is so hard that powder formation is no problem.

Chromosorb A (Johns Manville) (density = 0.45) can also be obtained in any mesh size. It gives better separations and therefore larger MASS than Chromosorb G. It is more destructive to chromatographed compounds and is somewhat brittle.

Chromosorb P (Johns Manville) (density = 0.45) is very similar to Chromosorb A but is even more active for substance destruction.

Sterchamol (Sterchamol Werke, Dortmund, Germany) is a low-cost support but it is very active.

---

occasion to study (9): glass beads < Gas Chrom P ~ Chromosorb W < Chromosorb G < Chromosorb A < Chromosorb P < Sterchamol.

The supports can also be ranked as to capacity defined as the size of MASS for equal volume of support with the same amount of coating and in the same mesh size. According to our experience this sequence follows the order: Chromosorb P > Chromosorb A ~ Sterchamol > Chromosorb W > Gas Chrom P > Chromosorb G > glass beads. Capacity differences, however, are not so large that this point should determine the choice of the support. For Chromosorb A capacity is only about 2.5 times as large as for glass beads. Catalytic activity is therefore the most important factor, and the materials to be recommended are glass beads, Gas Chrom P, Chromosorb W and G, or any similar material of a different brand.

2. SUPPORT MESH SIZE

In analytical gas chromatography, band widths are determined by such factors as flow rate, gas properties, support particle size, percentage of stationary phase, viscosity, etc. In preparative gas chromatography the large size of the sample used is the dominating factor which determines band width, and the factors just mentioned are not critical. Particle size can thus be increased without greatly reducing the effectiveness of the column. This is demonstrated by the experimental results shown in Figure 3.

The advantage of coarse support materials is that they allow the use of very long columns without appreciably increasing the pressure drop. Although small, there is an influence of mesh size on column performance with coarse supports. This is shown in Table VII where the results of comparative experiments with Chromosorb A in mesh sizes 10–20, 20–30, and 30–40 are given for a 75 m × 9 mm coiled glass column. The stationary phase used was SE 30 at 5 g/100 ml of support (10% w/w). The time required to fill the column with each of the three supports was about the same, namely, around 4 hr. To empty these columns required from 1.5 to 2 hr. At the highest gas pressure permitted in this glass column, 3 kg/cm^2, the hydrogen flow rates through the 20–30 and 30–40 mesh supports were 250 and 150 ml/min, respectively. The retention time for the air peak was about 20–30 min and, therefore the analyses were of long duration. With the 10–20 mesh support the results were much better. Hydrogen flow rates as high as 200 ml/min were obtained with only 1 kg/cm^2 inlet pressure. The retention time for the air peak was only 15 min. Comparative results of the plate numbers for these columns are given in Table VII.

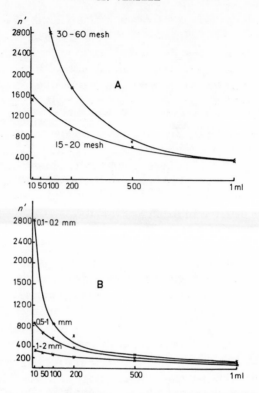

Fig. 3. Shows that n' values for large samples are relatively independent of the particle size of the support material. *A* is for Chromosorb W, 30–60 mesh and an experimental white support 15–20 mesh with *trans*-decalin as sample at 180°C for a 20 m × 9 mm coiled glass column. *B* shows this for Sterchamol in the given particle dimensions with isooctane as sample at 100°C and in a 6 m × 9 mm coiled glass column. Gas rate in all cases 200 ml H_2/min.

Absolutely identical conditions cannot be realized. While the flow rate was about the same for the first three columns of Table VII, the pressure drop was much larger for the first two columns (about 3 kg/cm² inlet pressure) than for the 10–20 mesh support of the third row (1.2 kg/cm² inlet pressure). Deviations from the expected values in Table VII must be explained by this effect. For identical inlet pressures the flow rate was much larger for the 10–20 mesh support (fourth column). Table VII shows again that the ratio of plate numbers for fine particles versus coarse particle supports is higher for analytical samples than for preparative samples sizes.

A number of separations were carried out with the columns of Table

TABLE VII

Effect of Mesh Size on Plate Numbers[a]

| | Comparative plate numbers | | | |
| | 200 ml H_2/min | | | 500 ml H_2/min |
Isopropylbenzene	30–40 mesh	20–30 mesh	10–20 mesh	10–20 mesh
25 μl	12416	7120	5024	4736
250 μl	2176	1944	2304	2048
1.25 ml	496	416	560	496
2.5 ml	240	256	229	208
5 ml	172	170	157	115

[a] Temperature 140°C, partition ratio ~5; packing: 5 g/100 ml support or 10% w/w SE 30 on Chromosorb A; column: coiled glass 75 m × 9 mm.

VII. With decalin at about 190°C and with a hydrogen flow rate of 200 ml/min, the largest samples giving practically complete separation of the *cis* and *trans* isomers are 7.5 ml on the 30–40 mesh column and 6 ml on the 20–30 mesh column. On the 10–20 mesh column it is only 4 ml. Nevertheless the 10–20 mesh support is the one to be recommended in these exceptionally long columns. This column has nearly the same separating power as the other supports but has the advantage of being much faster. With 3 kg/cm² inlet pressure (hydrogen flow rate of 400–500 ml/min) and at 190°C, it separates 4 ml of the decalin mixture in 45 min, while the other columns require more than 2 hr.

Although these very long columns are interesting, better and faster results can be obtained with the same volume of support material in slightly enlarged columns of only 10–20 m length (see the columns of Table V and the separation of Fig. 2). In such columns the pressure drop is not so severe and the mesh size may be 20–30 or even 30–40. These mesh sizes are therefore the recommended ones for long and large volume columns.

D. The Stationary Phase

1. NATURE OF THE STATIONARY PHASE

The number of useful stationary phases in analytical gas chromatography is still increasing at a surprising rate. The job of choosing a suitable stationary phase is quite complicated. However, in preparative gas chromatography the choice is much simpler because it is

TABLE VIII
Useful Stationary Phases

SE 30 is a nonpolar stationary phase and can be used up to temperatures of 300°C.

Apiezon L is an alternative nonpolar liquid phase. It can be used up to 250°C but is rather expensive.

Carbowax 20M [poly(ethylene glycol)] is a polar stationary phase and can be used up to 210°C.

PTMO [poly(tetramethylene oxide)] was made in our laboratory from tetrahydrofuran. It is also a polar phase, but separates substances differently than Carbowax 20M. It can be used up to 230°C.

Versamid is a polyamide and also a polar stationary phase. It can be used up to 250°C. Versamid appears to vary in quality in that some samples are more stable than others.

---

not economically practical to have a large number of large-volume columns. It is therefore obvious that "universal stationary phases" with a high maximum allowable operating temperature (MAOT) will be most useful. We have used SE 30 silicone gum in columns having volumes of 2–10 liters. Carbowax 20M, Apiezon L, Versamid, and polytetramethylene oxide (PTMO) have also been used in columns of this size. In smaller volume columns, 6 m × 9 mm, a greater choice is possible and the large number of separation examples published by K. Dimick (11) provide a useful guide for selecting the correct phase material. Some of the more useful phase materials with MAOT's are presented in Table VIII. The maximum allowable operating temperatures (MAOT) mentioned are only a guide. Variations are possible on different support materials. Also, the values are for a katharometer (hot wire) detector. With a flame ionization detector, for example, the MAOT may be lower.

2. COATING PERCENTAGE

The amount of stationary phase placed on the support is usually expressed on a weight/weight (w/w) basis. The different supports, however, have variable densities as indicated above and therefore expressing the coating on a w/w basis is not the best way. On glass beads, for example, this w/w basis is normally below 1% and this might be thought to be much too low to permit preparative separations. The density of 2.0 for glass beads, however, is high compared, for example, to Chromosorb W whose density is 0.22. This means that comparable column lengths filled with glass beads at 1% w/w stationary phase and Chromosorb with 10% w/w stationary phase contain about the same amount of stationary phase. For this reason it is preferable to express the coating amount on a weight/volume basis. For example, 2-g stationary phase/100 ml Chromosorb W is about 10% w/w while 2-g stationary phase/100 ml glass beads is only about 1% w/w.

In an experiment to show the effect of the quantity of stationary phase used, a 6 m × 9 mm glass column filled with Gas-Chrom P was tested successively with 1,2,3,4,5,6, and 7-g of SE 30 silicone gum/100 ml of support in an Autoprep 700 instrument. These values are equivalent to 5,10,15,20,25,30 and 35% w/w, respectively. Test substances were isooctane and cyclohexane. While there is a maximum in the plate number around the 4–5 g/100 ml value for analytical size samples, as is well known, it turned out that the preparative plate number was still increasing at 7 g/100 ml for the larger samples.

It might be concluded from these results that liquid loadings should be further increased for preparative scale work, but this is not true. Greater liquid loadings increase the separation time and especially increase the required temperature. This is often detrimental to the capacity of a column, since r values generally diminish with higher temperatures. In fact, since the difference is not so very great between 2 and 7 g/100 ml coating, we can use the lower coating amounts with advantage. Similar experiments with high-boiling mixtures confirm the above results.

There is however another and even more important reason why, in general, a low weight/volume packing should be adopted. With homologous series, separation results mainly from differences in boiling points since activity coefficients are similar, and increasing the amount of liquid increases the MASS. Because of large r values, separations of homologous series are easy and even with a low weight/volume packing the MASS will be relatively large. With most isomeric mixtures, on the other hand, boiling points are more similar than for members of homologous series, and separation results mainly from differences in polarity. The polarity of the support material plays a definite role in this instance and with thick coatings this influence of the support is not as evident as it is with thinner layers. These assertions, proved by the results presented in Table IX, have been confirmed repeatedly. Mixtures listed in Table IX were separated in the sizes given on 4-liter coiled glass columns filled with Chromo-

TABLE IX
Effect of Liquid Load on MASS

| MASS on SE 30 | | | |
5 g/100 ml or 10% w/w	12.5 g/100 ml or 25% w/w	Mixture	Column temperature, °C
10	25	Benzene, toluene, ethylbenzene, cumene	100
4	8	Methylpropylketone, methyl-isobutyl-ketone, cyclopentanone, methyl-iso-pentylketone	122
15	30	Ethylacetate, ethylpropionate, ethyliso-butyrate	100
1	1	The four isomeric butanols	100
21	35	Dichloromethane, chloroform, tetra-chloromethane	65
5	10	Methanol propanol	90
1.5	0.3	cis- and trans-β-Decalone	230
4	1	cis- and trans-Decalin	190

sorb A coated with silicone gum. The carrier was hydrogen gas used at 500 ml/min.

Instead of compromising it is better to choose a low weight/volume packing. This will give highest MASS for difficult isomeric separations, while the MASS for easier homolog separations will be sufficiently high. Except for special applications a coating of 5 g/100 ml support can thus be recommended.

The influence of the support material on the separation is also shown by the fact that r values of isomeric pairs change with different support materials if all other conditions remain equal. This is thought to be due to residual absorption and it is normal that this is more conspicuous with lower coating amounts (see Table IX). Some r-value changes due to the support material are presented in Table X. The r values of Table X were calculated from results using small samples and for approximately identical temperatures. A drop from $r = 1.14$ to $r = 1.08$ as in the last group of Table X reduces the MASS considerably.

E. Carrier Gas

1. CARRIER GAS AND DETECTION

In analytical gas chromatography with katharometer detection, hydrogen or helium is the best carrier gas. The use of nitrogen results in a considerable loss of sensitivity, and the nonlinearity of the detection together with possible peak reversals are all additional drawbacks found with this gas. In preparative scale gas chromatography, however, gas consumption is an important factor. Helium is ideal, but its cost is high. Hydrogen is dangerous, although it may be pointed out here that an air–hydrogen mixture only becomes inflammable at hydrogen concentrations above 4%. This means that in a fairly large room ($5 \times 6 \times 4$ m), an entire cylinder of hydrogen containing 5 m^3 may be emptied without danger of explosion. Undoubtedly, however, in preparative gas chromatography the use of nitrogen should be considered.

With nitrogen and katharometer detection the troublesome effect of peak inversion can occur. Bohemen and Purnell(13) have shown that the signal of a katharometer detector is influenced by several factors. The thermal conductivity drop heats the elements when a substance passes and this gives a "positive" deflection. The heat capacity of the substances, however, also plays a part and has a cooling effect. If the heat-capacity factor is the more important, peak inversion occurs. High gas flow rates, spiral filaments, and high filament and detector temperatures favor peak inversion.

With hydrogen and helium, thermal conductivity far outweighs the other factors, but with nitrogen this is not the case and peak inversion can easily occur. Commercial katharometer blocks are usually designed for analytical purposes and have as small a volume as possible.

TABLE X

r Values for Various Support Materials

cis- and trans-β-Hydrindanone have:
$r = 1.25$ on Chromosorb A 5 g/100 ml SE 30
$r = 1.30$ on Chromosorb W 12.5 g/100 ml SE 30

cis- and trans-β-Decalone have:
$r = 1.23$ on Chromosorb G 5 g/100 ml SE 30
$r = 1.29$ on Chromosorb A 5 g/100 ml SE 30

Three isomers of 2,4,5-trimethyl-2-phenyldioxolan I, II, and III:
For I/II, r is 1.16; for II/III r is 1.08 on Chromosorb G 5 g/100 ml SE 30
For I/II, r is 1.16; for II/III r is 1.10 on Chromosorb W 12.5 g/100 ml SE 30

cis- and trans-2,4-Dimethyldioxolan:
$r = 1.14$ on Chromosorb W–Carbowax 20M, 5 g/100 ml
$r = 1.08$ on Chromosorb G–Carbowax M, 5 g/100 ml

The use of such a detector block for preparative purposes with nitrogen at high flow rates produces just the conditions where peak inversion is most easily observed. For example, this is the case for the Varian Aerograph Autoprep 700 preparative gas chromatographic unit. The chromatograph has also analytical uses and in fact the detector block is the same as in analytical chromatographs produced by this firm. Therefore inversion occurs easily. This seemingly annoying phenomenon can, however, be utilized in such a way that inversion is the normal state of affairs. This is achieved by setting the detector oven at a high temperature (250°C) and by using high flow rates and high detector currents. Most substances then show completely reversed peaks with a strong signal over the whole possible concentration range. All that is required is that the polarity of the recorder be reversed.

This is the procedure to be followed using the instrument mentioned above for all concentrations of acetone, ethyl acetate, and benzene, substances for which badly shaped peaks and inversion trouble have been reported frequently in the literature. The highest reversed signal for a peak can be obtained with a nitrogen flow of 250 ml and a detector current of 200 mA. The efficiency of the column is, however, impaired by such a high nitrogen flow rate, and the high katharometer currents considerably shorten the detector filament life time. The same detector current will heat a katharometer wire to a much higher temperature in a nitrogen stream than in a hydrogen or helium stream. With more normal conditions, e.g., flow rate of 200 ml/min and a detector current of 150 mA, the response is only slightly less and the separation is better.

An important point with all detectors is the linearity of the response with sample load. With helium and hydrogen the response of a katharometer detector is linear for very small sample loads and shows decreasing signals for the large samples used in preparative chromatography. The reverse is true for the inverted peaks obtained with nitrogen. For small samples there is no linearity and the response first falls off with increasing sample, size, but remains linear for the larger sample loads. With a flow rate of 300 ml and a 150-mA bridge current the sensitivity of the detection in hydrogen and nitrogen for large samples is nearly the same.

This behavior is shown in Figure 4. To obtain an easily readable diagram, all measured peak areas were recalculated as peaks for 2-ml samples and the sample size ordinate was compressed for larger sample loads. The absolute value of the areas cannot be compared in Figure 4 because measurements were obtained on different recorders and with

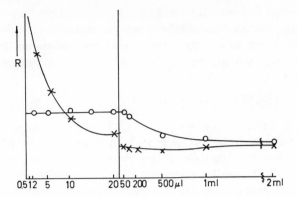

Fig. 4. Response factor of katharometer detection in hydrogen: [(○) normal peaks] and nitrogen [(×) inverted peaks] as explained in text.

different flow rates and detector currents. The scales were adapted to show that for really large samples the areas of the inverted peaks in nitrogen are about the same as for the normal peaks in hydrogen. For small samples the inverted peak area with nitrogen is larger, but the peaks are lower and broader than for the positive peaks obtained with hydrogen. This is a normal result with regard to the higher nonlinear response in nitrogen (response is lower at peak maximum). Although it is possible to work with nitrogen on a katharometer detection instrument as shown, this cannot be recommended since quick breakdown of the detector is unavoidable. With a hot wire detector, hydrogen or helium is far the better.

With a flame ionization detector, nitrogen can be used without great difficulty if a splitter is employed so that only a fraction of the separated substance is burned. We have had little experience with this technique, but several commercial instruments use this approach (Varian Aerograph 705 and 712, F and M model 776 for example) and the predictable difficulties with the splitter seem to have been solved satisfactorily.

2. CARRIER GAS AND EFFICIENCY

With a katharometer, detection is not linear and the peaks drawn on the chromatogram are not the true measure of the real peak coming out of the chromatograph. This distortion is different for hydrogen which is not linear for large concentrations and nitrogen which is not linear for small concentrations. Because of this, comparison of n' values obtained with the two gases is deceiving. Experiments with large samples indicate that differences in n' values are negligible. Here again, sample size is the most important factor determining band width, and it is only normal that the nature of the carrier gas is relatively unimportant as far as n' values go. For more details about this point see reference 14.

3. CARRIER GAS AND SPEED OF SEPARATION

Hydrogen travels more easily through a column than does nitrogen and therefore the inlet pressure with hydrogen can be much lower than with nitrogen. Because of this, columns operated with hydrogen are faster. With smaller preparative scale gas chromatographic columns as, for example, a 6 m × 9 mm column, the time difference is not so important, but with larger columns this difference becomes large enough to be taken into account.

4. CARRIER GAS RATE

A normal analytical column has a volume of about 5–25 ml. Flow rates used in such columns vary between 20 and 100 ml/min. A small preparative scale gas chromatographic column of 6 m × 9 mm for example contains about 500 ml. Optimal gas flow rates, however, are not increased by a factor of 100 but are only around 200–300 ml/min. This can be deduced from Figure 5 which shows that the optimum

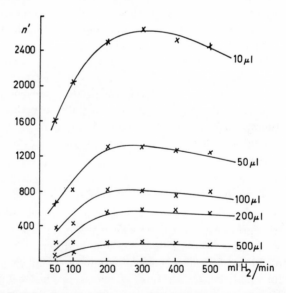

Fig. 5. Values of n' versus flow rate with hydrogen on a 6 m × 9 mm coiled glass column filled with Chromosorb W, 30–60 mesh, 5 g SE 30/100 ml support.

flow rates for large samples is about the same as for small samples. As might be expected the flow rate is again not so critical in preparative as it is in analytical gas chromatography. With 5–10 liter, long, narrow columns (columns 1–5 of Table V), normal carrier flow rates vary between 200 and 1000 ml/min. With short, wide columns the gas rate is more critical and has to be increased approximately according to the increase in cross-sectional area. A 2 m × 100 mm column has a cross-sectional area which is about 250 times as large as that for analytical columns. Normal gas rates are then also about 5–10 liters/min.

In concluding this section on carrier gas it is to be noted that hydrogen and helium are excellent with all instruments. Nitrogen can be used on instruments with a flame ionization detector but the separation time will be longer.

F. Detector

Detectors have already been partially discussed in the preceding section on carrier gas because these factors are interrelated.

In preparative gas chromatography the sensitivity of the detector may be thought to be rather unimportant. With really large samples which are readily separated, this is indeed true, and strong attenuation of the signal will assure very good base line stability. With samples which are harder to separate, the situation is different. To begin with, the larger flow rates of preparative gas chromatography reduce detector sensitivity and stability from that obtained with the smaller gas rates of analytical gas chromatography. Because of this decreased stability a reduced signal has to be used.

In addition, the limit of detection with a katharometer in analytical work is close to the 0.01 μl range, and with compounds having high k values, about 10 times this amount can present difficulties. Since a concentration is measured rather than the absolute amount and since the 5–10 liter volume of our preparative gas chromatography columns is 200–1000 times as large as that in analytical columns, the limit that can be detected is only 10–100 μl. Trial separations will have to be carried out with such sample sizes. This may be reason enough to prefer flame ionization detection (FID) for preparative gas chromatography. With FID a splitter has to be used and only a few per cent are normally burned, although in some cases up to 10–20% of the sample has to be burned because of sensitivity considerations.

With katharometer detection, really large samples, especially of high-boiling substances can cause kicking of the pen. This can be so bad that the chromatogram becomes useless. An example is shown in reference 12. A special large-volume detector may be necessary or, alternatively, a bypass can be drilled in the detector block (12). By using a drilled bypass, pen kicking is avoided and the sensitivity of the detector is practically unchanged.

G. Substance Recovery

Recovery of the separated substances in preparative gas chromatography is a problem for which there is no single solution. It can be very good but it can also be very bad. Low-boiling substances give only partial condensation because their vapor pressures, even at low temperatures, are still high or simply because sufficient rapid cooling of the vapor-loaded carrier gases is impossible to obtain. High-boiling substances can produce aerosols which are also easily swept away by the carrier gas. Various methods have been advanced to solve this problem, such as electrostatic precipitation (15,16), centrifugation of the collector bottle during chromatography (17), the use of gradient-cooled collectors to prevent aerosol formation (18,19), and filling the collecting bottles with a glass wool trap (20). None of these methods seems to have found general acceptance. It should also be noted that only a simple collector bottle can be practical, since it turns out in a laboratory doing preparative gas chromatography that a very large number of these collector bottles are needed. The shape of the collector bottles for work as shown in this paper does not seem to be very important. We have tried several models, but were unable to improve upon the commercially available Autoprep 700 collectors (21).

The efficiency of these collector traps is often improved markedly by filling their upper part as tightly as possible with glass wool. The first drops of recovered substance form a liquid layer on the glass wool and this acts as a stripping column. The glass wool scrubber, however, does not always improve recovery. Its resistance to gas flow can cause leaks at the connection between the column exit and the collection bottle. This is indeed a delicate point to which adequate attention must be given. Another unavoidable problem which is encountered in the collection of vapors with glass-wool plugs is the clogging of the glass wool by solid condensates, as with chloroform collected at very low temperatures. For these reasons collectors are more often used without glass wool scrubbers than with them. The efficiency of recovery is limited by a number of factors as discussed in the following paragraphs.

1. VOLATILITY OF THE SUBSTANCES

While the effects of volatility on the collection of a sample from a gas stream are perhaps obvious, a practical experiment to emphasize this property is probably worthwhile. Samples of 200 μl of ether, cyclohexane, isooctane and *trans*-decalin were chromatographed using a flow rate of 200 ml/mm of hydrogen at their respective boiling points. The collector was cooled to −20°C. The recoveries were 0, 39, 45, and 88%, respectively, and the improvement is in direct order of the decreasing vapor pressures of these substances. The effect of further cooling of the collector on the recovery of ether was studied further. Under conditions similar to those used above, 200-μl samples of ether were chromatographed isothermally at 70°C with the collection bottles held at 0, −8, −14, −25, −65, and −80°C. Recoveries were 0, 2, 17, 24, 83, and 89%, respectively. Adequate cooling of the collector is therefore very important for efficient recovery in preparative chromatography.

2. CONCENTRATION

The concentration of the substances in the gas phase plays an important role in the recovery; the efficiency of recovery increases with increased concentration. This concentration is determined by column parameters such as the volume of mobile phase in the column and the separation efficiency of the column; more plates per meter gives narrower bands and higher concentrations. For a given column, however, these parameters are only variable to a limited extent. Other factors directly related to the concentration are the amount of material put on the column, and, more particularly, the temperature of elution. Large samples chromatographed at a high temperature will show the highest concentration in the eluate. Unfortunately, the most interesting components are nearly always in the rarest and most complicated chemical mixtures, which are therefore difficult to separate, so that only small samples and long retention times (low temperatures) can be used. The recovery percentage will be low although it is just with such mixtures that high recovery is most important. It may be obvious that the sample size has a direct influence on the concentration of the eluted substances, but the influence of the temperature of elution is not so obvious and may be illustrated by the results given in Table XI where the percentage recoveries of *trans*-decalin is presented as a function of elution temperature under isothermal conditions. The values in Table XI are the averages of five

TABLE XI

Recovery of *trans*-decalin as a Function of Elution Temperature

Quantity chromatographed, μl	Recovery, %	Column temperature, °C
200	93.1	200
	87.8	180
	72.7	160
	71.3	140
	66.7	120
	47.0	100
100	69.0	160

determinations. The collecting bottles were tared and the collected sample weights were compared with the weight of the injected samples as delivered by a syringe. Other conditions of the experiment involved on-column injection into an Autoprep 700 unit, hydrogen flow rate of 200 ml/mm, and collection bottles cooled in ice water. Good separation, long retention times, and high temperature at elution can be obtained by temperature programming, although this is not always possible and not as easy as isothermal operation.

An alternative is to run the chromatograph at a lower temperature for the desired time, stop the flow of carrier gas, increase the temperature of the column oven, and then restore the gas flow. In this way the substances are chromatographed long enough to be separated and are eluted at a high temperature. The improvement in recovery that is possible by this technique is shown by the values obtained for the 200-μl samples of ether, cyclohexane, isooctane, and *trans*-decalin which were run as described above, but with an additional temperature increase of 70°C with the gas flow stopped just before elution. The recovery was, respectively, 14.4, 42.4, 61.5, and 95.4%, which is decidedly higher than without the temperature increase.

H. Gas Flow Rate

High speeds of flow through the collecting trap could be detrimental to the recovery. This can be ascertained by diminishing the gas flow rate just before elution of a peak. In another recovery experiment on 200-μl ether, cyclohexane, isooctane, and *trans*-decalin, as described above, the temperature was increased by programming, and the gas flow was reduced to 25–30 ml/min just before elution of the peaks. This was very easily done without changing any control setting of the instrument by inserting a suitable capillary tube through the injection gasket at the appropriate moment to bleed off the incoming elution gas.

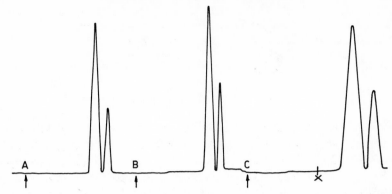

Fig. 6. Chromatograms of decalin on Autoprep 700: sample load 200 μl, bridge current 150 mA; attenuator 32 for *A* and *B*, 64 for *C*. Other conditions: (*A*) isothermal operation at 170°C—200 ml/min H_2; (*B*) programmed from 150 to 190°C at peak elution; 200 ml/min H_2; (*C*) programmed from 150 to 200°C at peak elution. At point × the gas rate was changed from 200 to 25 ml/min H_2.

The recovery results were now, respectively, 31.8, 73.4, 84.6, and 97.0%. While such a drastic flow-rate reduction would have a marked influence on baseline stability in normal analyses with high sensitivity, this is not the case in preparative work, since the sensitivity of the instruments in greatly reduced with high sample load. Comparative gas chromatograms on decalin are shown in Figure 6.

As explained earlier in this paper, with preparative scale samples, such reduction of the gas rate is possible in narrow columns without harmful effects to the separation as shown in Figure 6. This is once more shown by the n' values in Table XII.

So far, the experiments in this section about recovery were carried out with rather small columns (6 m × 9 mm) and with small samples for which recovery is essentially difficult. For larger columns with larger samples, recovery is much better.

With aerosol-forming substances, increasing elution temperature or reducing the flow rate does not always help. For substances of low

TABLE XII
Values of n' for 500 ml Benzene at 80°C ($k = 5$)

Flow rate, ml/min	Column 2 of Table V	Column 6 of Table V
250	1345	1447
500	1455	1497
750	1304	—
1000	1416	1460
1250	1555	1427
1750	1579	—
2400	—	1505

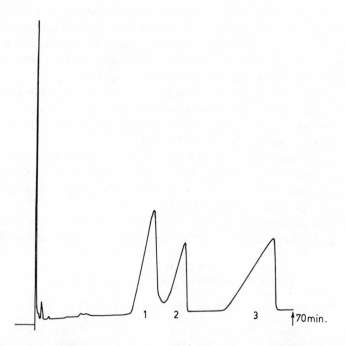

volatility (and this is mostly the case with these aerosol formers) adsorption on silica gel can be the solution. This is shown in the separation of Figure 7 of the substances of structures **1–3.** Collection of five injections of 150 μl was obtained by passing the effluent gas through traps of 2 g of silica gel (10–20 mesh) in glass tubes of 12 × 40 mm attached to the chromatograph outlet. Each of the three tubes were eluted with ether and gave 133 mg of **1,** 78 mg of **2,** and 173 mg of **3.** Substance **2** is isomeric with **3** but its structure is not known. The same silica gel tubes can be used repeatedly. Although this point was not checked specifically, we think collection was quantitative.

Fig. 7. 150 μl sample solution on coiled glass column of 3 m × 9 mm, filled with Chromosorb W, 30–60 mesh, coated with Versamid 5 g/100 ml support. Conditions: 150 ml H₂/min at 230°C; Autoprep 700 bridge current 150 mA; attenuation × 4; collection on silica gel.

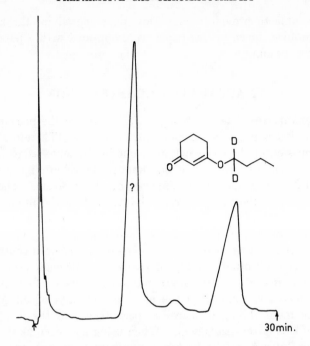

Fig. 8. Separation of 30 μl sample on 2 m × 4 mm column, Chromosorb 60–80 mesh, coated with 5 g/100 ml PTMO at 200°C with 25 ml H₂/min: Conditions: Varian Aerograph A 90P; bridge current 150 mA. Collection in simple capillary gave a sizable drop of the desired substance.

Collection of milligram samples for subsequent mass spectra or infrared spectra analyses, for example, can often be achieved very simply. For solids and high-boiling liquids insertion of a capillary in the chromatograph outlet is often sufficient. An example of this technique is shown in Figure 8. A large number of such deuterated enol ethers and analogs were thus purified for spectral investigation. When working with more volatile substances, merely passing the chromatograph effluent through a U tube cooled in liquid air is usually sufficient. An example of a separation where collection was achieved in this way is shown in Figure 10.

I. Recording of Preparative Separations

For preparative chromatography the recorder must be as sensitive as for analytical scale gas chromatography, i.e., full scale deflection for 1 mV. As the preparative scale separations are of long duration, it is advisable to provide a very slow paper speed for the recorder. Most examples shown in this paper were obtained with a paper speed of 1 in. for 10 min.

V. APPARATUS CONSTRUCTION

Excellent instruments, applying, in fact, most of the principles discussed in this paper, are commercially available. Therefore a complete discussion of the construction of our instruments seems unnecessary. A schematic drawing of our unit is shown in Figure 9. The oven is 1.2 m long and has a diameter of about 40 cm. The heater is at one end of the tubular oven. By regulating the input of fresh air a temperature gradient can be created if desired. Up to 150–200°C, the temperature can be held remarkably constant over the entire length of the oven. Because fresh air is blown continuously into the oven, eventual buildup of carrier gas through leaks is prevented and unsafe conditions are minimized when using hydrogen.

The detector is coupled to the electrical circuits of an Autoprep 700. The mere closing of a switch permits use of the 6 m × 9 mm column of this latter instrument. When using hydrogen as the carrier gas and a 150 mA as the katharometer heater current, the detector hot wires have a lifetime of 6 months to a year. With nitrogen and at higher current this lifetime may be much shorter.

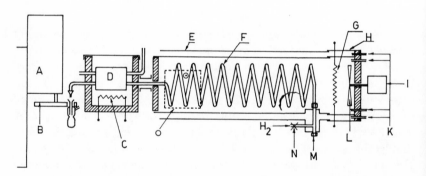

Fig. 9. Apparatus for preparative gas chromatography on long columns. Not to scale. (A) Autoprep 700, (B) analyzer collector table, (C) detector oven heater, (D) katharometer detector oven, (E) double mantle column oven, (F) coiled column (actually the coils are as close as possible to one another), (G) column oven electric element, (H) detachable heater fan motor block, (I) fan motor, (K) inlets for air under pressure, (L) fan blade, (M) injector, (N) valve, (O) trap door.

The coiled ovalized columns can be made relatively inexpensively from aluminum tubing of which columns 1 and 3 of Table V are examples. The following procedure can be used to make these columns: Tubes, 6 m long × 24 mm diameter, are filled with sand; flattened until the smallest diameter is 15 mm; wound in spiral form to a coil diameter of 30 cm; the sand removed; sections joined by welding to give the desired length; and the assembled coil checked for leaks, then washed with water, acetone, dried, and finally filled with the coated support by vibration and placed in the oven described above.

Coiled and ovalized glass columns of the same dimensions can also be made. To do this a 1 m × 30 cm cylinder of 2-mm sheet steel is mounted horizontally so that it can be rotated very slowly. Pyrex glass tubes of 1.5-m length are coiled on the cylinder by softening the glass with a gas torch and welding the pieces together until the desired length is obtained. With tubes of 1.5-m lengths and 8–9 mm bores this procedure can produce columns up to 75 m in length (23). With tubes of 2-cm bore the coiled column will have an ovalized section, but this is a favorable factor as already discussed above. These glass columns are conditioned against stress by baking overnight in an oven at 500°C before they are removed from the steel cylinder. The coil is finally mounted on a frame which will immobilize the coils to prevent breakage. To facilitate the connection of glass columns to the chromatograph, a 30-cm piece of stainless steel tube having a 1.5-cm bore is fixed to the column ends. Alternatively, leakproof connections can also be obtained with fittings of injection septum silicone rubber. These are better than Teflon fittings which are too hard and can only be used once.

VI. EXAMPLES OF SEPARATIONS

In the introduction of this paper it was stated that about four different preparative gas chromatographic techniques can be discerned. These depend on the nature of the mixture and on the amount of substance to be chromatographed. A discussion of each of these categories follows.

A. Simple Mixtures—Spectrometric Analysis

A normal analytical gas chromatograph with katharometer detection and a column of 1–3 m in length by 4–6 mm in diameter will do this job easily. Examples are presented in Figures 1 and 8 and another is given in Figure 10. The sample size of Figure 10 is rela-

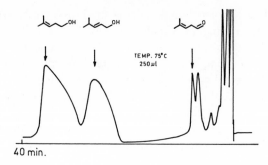

Fig. 10. 250 μl of the reaction mixture produced by LiAlH₄ reduction of N-isohexenoylaziridine. 3 m × 6 mm column—Chromosorb W, 60–80 mesh coated with PTMO at 5 g/100 ml support, in Aerograph 90P instrument. Working temperature 75°C. Structure of samples shown in the figure. Collection of aldehyde and alcohols in cooled U tubes.

tively large although the column is small. Collection of the substances of this separation was mentioned in the section on recovery.

B. Complex Mixtures—Spectrometric Analysis

For mixtures containing a large number of components, say 10 or more, r values below 1.1 are unavoidable and the column will need exceptionally high separating power. In this case the boiling range will also be rather large and programming is indicated. A characteristic of preparative scale separations is that they are of longer duration than analytical separations. Therefore the programming rate must be slower and will normally be in the range of 0.5–2°C/min. Suitable columns for this kind of problem have the following characteristics: 20 m × 9 mm filled with Chromosorb W 30–60 mesh and coated with 5 g/100 ml. One such column with Apiezon L as stationary phase produced 20,000 plates with 10 μl samples, (see column V of Table V). An example of such a separation is shown in Figure 11. Poly(tetramethylene oxide) is used because it is a "universal" phase, giving equally good results for substances having different polarities. Poly(tetramethylene oxide) separates the components differently than does Carbowax. It is therefore useful as an alternative polar phase and it presents the advantage of being slightly more stable towards temperature effects than Carbowax 20M. The permeability of the column charged with 30–60 mesh support used for this experiment is not high and the duration of the separation shown in Figure 11 lasted for 4 hr.

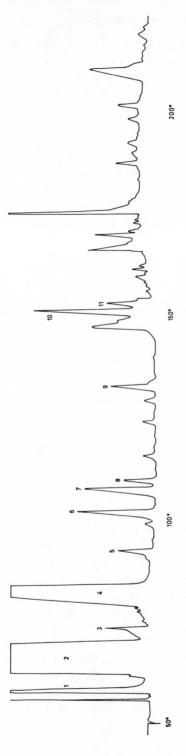

Fig. 11. 4 ml of volatile hops aroma components on 20 m × 9 mm coiled glass column using Chromosorb W, 30-60 mesh, coated with 5 g/100 ml of PTMO and using 250 ml H₂/min. Programmed at 1°C/min from 50 to 200°C and isothermal operation afterwards. Modified Autoprep 700. Peak 2 is ether solvent, peak 10 is myrcene. Most other peaks were collected in sufficient amount to be identified by normal NMR and other spectrometric techniques.

Such columns are also indicated when the search for the best stationary phase failed to produce r values of at least 1.1. This is the case for example, for the isomeric *cis–trans* mixtures of 4-methyl-2-alkyldioxolans with alkyl equal to isopropyl and n-butyl. The r values on Carbowax 20M and on poly(tetramethylene oxide) (PTMO) were similar and between 1.07 and 1.09. A large volume 20 m $\times$ (1.5 $\times$ 2.5 cm) ovalized column, filled with Chromosorb G, 20–30 mesh coated with Carbowax 20M failed to separate even 100 μl of the mixture. The maximum n' value for this column was only 1750 and for an r value of 1.09 at least 2300 plates are necessary. A 20 m $\times$ 9 mm coiled glass column (similar to column 5 of Table V) filled with Chromosorb W 30–60 mesh and coated with PTMO separates a 100-μl sample completely, giving n' values for the peaks between 3000 and 5000. In this instance then, the MASS for this column lies between 150 and 200 μl.

C. Simple Mixtures—Larger Amounts

Columns 10–20 m in length and having a bore of about 2 cm are used for this application. With coiled columns, ovalization of the sections is preferred. The support can have a mesh size of 20–30 or 30–40. An example of such a separation has already been demonstrated in Figure 2. The r value for these *cis* and *trans* 3,4-dimethylcyclohexanones is 1.15. With eq. (6) or from Table II it can be deduced that 940 theoretical plates are needed. According to Table V this column still produces 940 plates with 2-ml samples. Since the mixture contains about equal amounts of both isomers, the MASS is 100 $\times$ 2/50 = 4 ml as calculated by eq. (7).

Other examples, showing the MASS values as a function of r and conditions in general are found in Table XIII.

A similar series of *cis–trans*-dioxolan mixtures was also separated. It is interesting to note that the dioxolans could not be chromatographed on Chromosorb A but readily on Chromosorb G.

This sort of preparative gas chromatography allows the separation of sufficient material for subsequent chemical work. The general procedure to do this is illustrated in the example of Figure 12 concerning the separation of a mixture of dimers obtained from a 2-methyl-2-pentene and 2-methyl-1-pentene mixture on column 1 of Table V. The r values of the sample are first investigated on available analytical columns. A preparative column is chosen in accordance with these results and a test chromatogram is run at a temperature of about 20–50°C higher than the temperature used for the analytical

TABLE XIII

MASS Values as a Function of r Values

No.	Mixture	r, min	MASS, ml	Total separated amount, ml	Columns	Temp., °C	Support mesh 20–30	Stat. phase, g/ml	H_2 gas rate, ml/min
	meta dioxans *cis/trans*				Aluminum 21 m × (1.5–2.5)	100	Chrom. A	Carbowax 20M 5/100	700
1	*meta* Dioxan benzene		10	150					
2	4-Methyl-6-isopropyl	1.36	8	25		140			600
3	4,6-di-*n* propyl	1.25	6	30		140			600
4	4-Methyl-6-*t*-butyl	1.30	6	20		140			600
5	4,6-Dimethyl	1.35	8	8	Aluminum 15 m × (1.5–4.0)	100	Chrom. G	SE 30 5/100	1000
6	4,5-Tetramethylene +1,3-diolacetate +2 other peaks	2 / 1.32 / 1.22	2	15		100			600
7	4,6-Di-isobutyl	1.1	0.2	0.2	Aluminum 21 m × (1.5–2.5)	170	Chrom. G	Carbowax 20M 5/100	700
8	4,5-Pentamethylene	1.11	0.5	8		170			600
9	4,6-Di-isopropyl	1.25	4	28		150			700
10	4-Methyl-6-*n*-butyl	1.3	5.5	12		150			700
11	4,6-Diethyl	1.3	2.5	12		170			600
12	4-Methyl-6-isobutyl	1.31	6	15		170			600
13	4-Methyl-6-ethyl	1.38	10.5	10.5		130			600
14	4-Methyl-6-*n*-Pr	1.36	7.5	21		135			700
15	4-Methyl-6-*sec*-butyl	1.36	5	42		165			700
16	Di-*n*-butyl-4,6	1.14	2	18.5		180			700
17	Di-*sec*-butyl-4,6	1.12	0.5	19		180			700
18	5-Ethyl-5,5-diethyl	2.3	3.8	3.8		150			700

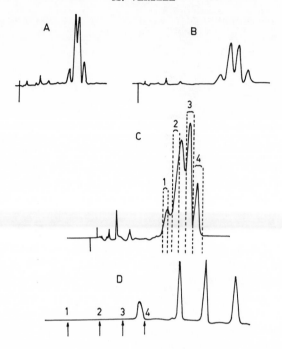

Fig. 12. Separation of methylpentene dimers on a 21 m × (1.5 cm × 2.5 cm) coiled aluminum column using Carbowax 20M Chromosorb A, 20–30 mesh. 700 ml H₂/min.

separation (trace *A* in Fig. 12 obtained at 130°C). Carbowax 20M is often most useful. With this result a final temperature is chosen and a trial separation is run (trace *B* in Fig. 12 at 110°C). Analytical results *A* and *B* permit the deduction of the approximate MASS which is here only 1 ml since the smallest *r* value is 1.08–1.09. A trace of one of the 5 separations carried out with 1-ml samples is shown in Figure 12*C*. Also indicated in Figure 12*C* are the fractions of the composite peaks collected as *1, 2, 3,* and *4* (total recovery for *1, 2, 3,* and *4* was 60%). Finally an analysis was run on each of the collected fractions on the same column. The injections were made every 10 min as shown in Figure 12*D*. Peak *1* is too broad and is still a mixture. Peaks *2,3,* and *4* are sufficiently pure for further work although this seems not so predictable from trace Figure 12*C*. This is however something which we have experienced repeatedly, namely, that the result is better than could be hoped for from the appearance of the chromatograms. Subsequent IR and NMR analyses showed that the substances are *cis-* and *trans*-4,6,6-trimethyl-3-nonene and 3-ethyl-2,4,4-trimethyl-1-heptene.

The type of column under discussion in this section is also the one to be chosen for enrichment of trace amounts to a more useful concentration. Very large volumes of complex mixtures are injected (considering the earlier experiments about stripping, an SE 30 column seems best indicated), and although the chromatogram shows minimum separation, collection is carried out at the retention time of the desired trace component. The enriched mixture can then be treated in a normal way.

A further application for such a column is the recovery of large amounts of pure components. One example is the recovery of deuteroform from the reaction mixture obtained by exchanging the proton of chloroform by deuterium. The deuteroform was contaminated with 0.25% of 1.1-dichloroethane. This impurity made the deuteroform unusable as a solvent for NMR spectroscopy. The boiling point difference is only 4°C, and fractional distillation was unsuccessful. Preparative scale gas chromatography in column 1 of Table V was possible with samples of 40 ml. At a hydrogen flow rate of 500 ml/min, recovery was poor. Clogging of the glass wool scrubber was partly at fault. By reducing the gas rate at the moment of peak elution to 100 ml H_2/min and in an Autoprep 700 collection bottle of 25 ml capacity without any scrubber, the recovery goes up to 80–90%. Even the impurity could be completely collected with an elution flow rate of 25 ml/min. In this way the 100 mg present in one 40-ml charge of the mixture was condensed and its nature was ascertained by NMR spectroscopy. Elution of the deuteroform peak in this way requires several hours.

Another example of recovery can be demonstrated with cyclohexane which was recovered to the extent of about 80% when employing an ovalized column 20 m long × (1.5 × 2.5 cm) containing Chromosorb G of 20–30 mesh coated with Carbowax 20M. In one experiment using a 2-ml sample at a hydrogen gas rate of 800 ml/min at 80°C, a recovery of 81% was obtained in a 5-ml Autoprep 700 collector. In another experiment using a 5-ml sample and the same gas rate and temperature, but collected in a 25-ml Autoprep 700 collector, the recovery was 77%. When the collectors were cooled to −40°C and with lower flow rates the recoveries increased to over 90%.

D. Very Large Samples

Scaling up the technique described in this paper by a factor 2 or 3 seems quite possible, although unnecessary for research laboratory purposes. To scale up by a factor of 10–100, or even more, will require another approach. This, if possible, would bring gas chromatography into the field of industrial scale separations. The introduction of "flow homogenizers" seems to be a possible answer. In section C on column design, this point has already been discussed. We have little experience with this technique and therefore refer the reader to the literature (5–7).

VII. FURTHER POINTS OF INTEREST

A. Use of Glass Beads in Glass Columns

The possibility of catalyzed thermal destruction has been referred to repeatedly in this paper, and it has also been stated that glass is one of the best materials for avoiding this effect. The coating of the stationary phases on glass beads has also been mentioned and practical amounts turn out to be 1–2 g per 100 ml of glass support material. The coating of glass beads with such quantities of an oily stationary phase gives a rather sticky material which makes them very difficult to charge into a column, although certainly not impossible. Stationary phases which are solid at room temperature present no great difficulties and Carbowax 20M on glass beads gives a satisfactory column filling.

Glass beads of different sizes were tested. The finer the beads, the greater the efficiency of the column, but also the greater is the pressure drop. A satisfactory compromise turned out to be glass beads with a diameter of 0.8 mm (Ballotini shot lead glass beads no. 11 obtained from Jencons, Hemel Hempstead, England).

Similar experiments were carried out with glass wool. This was cut in a mixer until the longest strands were about 4–5 mm. The resulting material was purified by frequent decantation and by washing with strong acid, soap water, and then chloroform. Coating with stationary phase was carried out as usual, but the charging of columns with such coated glass wool is much more difficult than with coated glass beads. The efficiency of glass wool columns is better than that for glass-bead columns, but the pressure drop is excessive. Therefore further work with glass support was carried out exclusively with glass beads having a 0.8-mm diameter.

About 6 liters of such glass beads, coated with 2 g/100 ml Carbowax 20M, were charged into a 20 m $\times$ 2 cm coiled glass column. The glass column was made from 1.5-m glass tubes with a 2.0-cm bore as described earlier.

To facilitate connection of this column to the gas inlet and outlet, a 30-cm piece of stainless steel capillary with 1.5-mm bore was fixed to its ends. The permeability of the column was good, allowing a hydrogen flow rate of 500 ml/min with a 1250-mm Hg inlet pressure at temperatures up to 200°C. The separation power of this column for small samples is much below the results obtained with conventional supports. For large samples it compares well with the earlier columns developed in this laboratory for large-scale preparative gas chromatography. A comparison of the performance of glass beads with Chromosorb A is given in Table XIV. All results were obtained with cumene at 145°C which has a K value of about 5 and uses a hydrogen flow rate of 500–700 ml/min. Column 1 contained 0.8-mm glass beads coated with 2-g Carbowax 20M per 100 ml of support in a 20 m $\times$ (1.5 $\times$ 2.5 cm) coiled glass column. Column 2 was 15 m $\times$ (1.5–2.5 cm) filled with Chromosorb A, 20–30 mesh, and coated with 5 g SE 30 per 100 ml of support. Column 3 was 21 m $\times$ (1.5–2.5 cm) filled with Chromosorb A, 20–30 mesh, coated with 5-g Carbowax 20M per 100 ml of support. An example of a separation on the glass-bead columns is shown in the chromatogram of Figure 13. The mixture was obtained by alkaline treatment of dichloromethyl-isobutylketone. Only peaks *6*, *7*, and *8* were identified, being $(CH_3)_2CH—CH=CHCOOMe$, $(CH_3)_2CHCH_2COCHCl_2$, and $(CH_3)_2—CHCH(CH_2Cl)COOMe$, respectively.

Similar mixtures obtained from other dichloromethyl alkyl-ketones were also separated on this column. Purification of *n*-butyliodide and *n*-butylbromide by preparative gas chromatography was also only possible on glass beads. A series of deuterated *n*-butylhalides was thus isolated from the reaction mixture, using a 1.5 m $\times$ 9 mm column

TABLE XIV
Comparison of Glass Beads with Chromosorb

	Values of n'		
	Glass beads	Chromosorb A	
Sample size	Column 1, volume 6.2 liters	Column 2, volume 5.1 liters	Column 3, volume 7.7 liters
100 μl	508	2626	3075
200 μl	542	1913	3301
500 μl	541	1803	2584
1 ml	456	1093	1731
2 ml	314	673	976
5 ml	174	270	373
10 ml	99	110	154

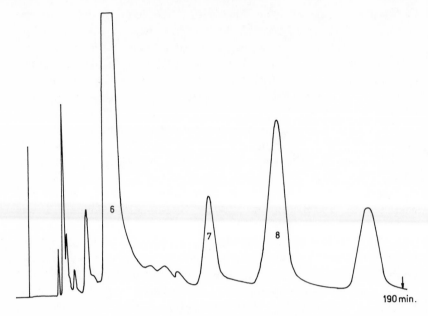

Fig. 13. 2-ml reaction mixture—isothermal operation at 150°C, 600 ml H₂/min. Coiled glass column 20 m × (1.2–2.4 cm) filled with 0.8 mm glass beads coated with 2 g Carbowax 20M/100 ml support.

and with SE 30 as stationary phase. The trend in our laboratories is toward the increasing use of glass beads in glass columns.

B. Retention Times

Reproducible retention times can only be obtained with small samples. With the larger samples of preparative scale gas chromatography, retention times are not reproducible and the peaks can show typical deviations. Early peaks (low k) tail while late peaks (high k) show leading edges (see Figs. 7 and 8). Intermediate peaks can be perfectly symmetrical. Tailing and leading of peaks is often discussed as a function of nonlinear partition isotherms of the substances. As inferred above, however, with larger samples all early peaks tail and all late peaks lead, regardless of the nature of the substance. Typical changes also occur in peak front retention. This effect is shown in Table XV for benzene, toluene, ethylbenzene, and cumene where the peak front retention is given in mm. A 20 m × 9 mm coiled glass column was used, charged with 15–20 mesh support, covered with SE 30 silicone gum and a hydrogen flow rate

TABLE XV
Peak Front Retentions

Sample size in μl	Benzene, mm	Toluene, mm	Ethylbenzene, mm	Cumene, mm
50	62	89	127	165
100	62	89	129	168
200	61	91	135	176
500	57	89	135	179
1000	55	89	136	181
2000	53	88	140	190

of 200 ml/mm was employed at 165°C. Reproducibility here is also only achieved with small samples.

For low partition ratios peak front retention decreases with increasing sample size and for high partition ratios it increases. In-between values showing no change are possible (toluene). Peak front retention with increasing sample size is influenced by several counteracting factors.

1. It *decreases* because of exponential sample introduction. This is most strongly observed for substances with small partition ratio (higher temperatures) and low flow rates on short columns with small pressure drop (coarse support).

2. It *increases* because of increased resistance to flow rate with large samples. This is most strongly observed with high flow rates and in longer columns with high pressure drop (small grained support).

Another reason for peak front retention increase is probably found in temperature effects. Evaporation at the rear end of a first band cools the column locally and this retards a second substance. High partition factors should enhance these temperature effects and this is indeed true. This effect is general and occurs also on analytical size columns with larger samples. As far as practical results go, these effects are not important (**22**).

C. Miscellaneous

1. TEMPERATURE

Mixtures with only a few compounds are best separated isothermally. For complex mixtures temperature programming may be needed. We have also studied "chromathermography" or the technique where a temperature gradient is applied along the column, but this has only a minor influence on preparative gas chromatography (9).

2. PRESSURE

Using about 5–7 kg/cm² inlet pressure and 4–6 kg/cm² outlet pressure does not increase the capacity of preparative gas chromatography to any appreciable extent. In general it only increases the separation time.

In contrast, when reducing the pressure at the outlet so that $p_i = 600$ mm Hg and $p_o = 200$ mm Hg, a reduction in separation time was noted. To reestablish the same retention time to normal conditions, the oven temperature can be reduced by about 20–50°C. This approach may be of some importance in connection with stationary phase stability, but the technical difficulties are large and we do not foresee much interest in this approach.

3. COLUMN OVALIZATION

We have also tried ovalization of the column so that the axes ratio in one case was 5:1 (very small volume) and in another, 4:2.5 (larger volume), but found that the results were not significantly better than for the ratio of 2.5:1.5 used for the columns described in Table V.

4. CATALYTIC EFFECTS OUTSIDE THE COLUMN

The separation of *cis*- and *trans*-2,4-dimethyldioxolan gives a normal chromatogram with $r = 1.08$ on Chromosorb G covered with Carbowax 20M. A "multiple process" (11) separation of these two isomers, however, gave products which showed the same composition on analytical gas chromatography as the starting mixture. The reason of this isomerization of the pure compounds after the separation was an acid-fouled detector. Washing the detector with tetralin and bicarbonate water removed the acidic contamination and subsequent separations proceeded in a normal way.

VIII. CONCLUSIONS

Problems in preparative gas chromatography are related to speed, resolution, recovery, and decomposition of stationary phases and chromatographed substances. Any one of these factors can only be optimized at the expense of the others. Compromising is therefore necessary. Ways to do this under certain situations have been advanced in this paper. The column is the heart of preparative gas chromatography. Best results can be expected with column length-to-diameter ratios between 300 and 2000. For a research laboratory, column volume of 25 ml, 2 liters, and 5–7.5 liters are indicated. With smaller volume columns, best performance is obtained with supports having a mesh size range of 30–60, but for the larger volume columns a mesh range between 20 and 40 is recommended. The coating percentage is best around 5 g per 100 ml of support.

For each column, values like those given in Tables I, III, V, VII, and XIV should be determined. Separation problems are then first investigated on analytical columns to determine r values and sample composition. The r values lead to the necessary plate number required for the separation using eq. (6).

$$n = 16 \ (r/r - 1)^2 \qquad (6)$$

From these tables can be found the component size $S_{n'}$ which gives this plate number. The maximum allowable sample size (MASS) of the mixture is $100 \ S_{n'}x$ where x is the percentage of the compound in the mixture.

In the multitude of problems to be solved by preparative gas chromatography in the research laboratory, we can distinguish:

1. 10–200 μl of mixtures containing only a few compounds and with all r values above 1.1. For this a normal analytical column and isothermal operation is sufficient.

2. 100 μl–2 ml of multicomponent mixture with large boiling range. Temperature programming with a column 20 m $\times$ 1 cm filled with a 30–60 mesh support is indicated.

3. 100 μl–20 ml of mixtures containing only a few compounds and with all r values above 1.1. For such separations a column of the type 20 m $\times$ (1.5 $\times$ 2.5 cm) filled with a coarse support (20–30 or 30–40 mesh) is best. Good cooling of the collector bottles is essential for recovery. For very volatile substances the gas rate must be reduced during collection (this is possible during the separation, with the narrow bore column advocated above). For aerosol-forming substances, adsorption on coarse silica gel is very often a good solution.

References

1. S. D. Nogare and R. S. Juvet, *Gas Liquid Chromatography,* Interscience, New York, 1962.
2. D. Dinelli, S. Polezzo and M. Taramasso, *J. Chromatog.,* **7,** 477 (1962); *J. Gas Chromatog.,* **2,** 150 (1964).
3. (a) F. Alderweireldt, *Anal. Chem.,* **33,** 1920 (1961); (b) F. Alderweireldt and M. Verzele, *Bull. Soc. Chim. Belges,* **70,** 703 (1961).

4. G. Frisone, *J. Chromatog.*, **6**, 243 (1961).
5. F. Debbrecht, F. and M. Company tech. paper No. 28, 1966.
6. *Chem. Eng. News*, **43**, 46 (1965) ; *Chem. Week*, March 20, 1965.
7. *Facts and Methods*, Vol. 6, No. 5, F. and M. Company Publication, 1965.
8. E. Taft, Conference on Analytical Chemistry, Pittsburgh, March 1964, Varian Aerograph Documentation, W-116.
9. M. Verzele, K. Van Cauwenberghe, and J. Bouche, *J. Gas Chromatog.*, **1967**, 114.
10. M. Verzele, J. Bouche, A. DeBruyn, and M. Verstappe, *J. Chromatog.*, **18**, 253 (1965).
11. K. P. Dimick, *Gas Chromatographic Preparative Separations*, Varian Aerograph laboratory manual, 1966.
12. M. Verzele, *J. Gas Chromatog.*, **1966**, 180.
13. I. Bohemen and J. H. Purnell, *J. Appl. Chem.* (*London*), **8**, 433 (1958).
14. M. Verzele, *J. Chromatog.*, **15**, 482 (1964).
15. P. Kratz, M. Jacobs, and B. Mitzner, *Analyst*, **84**, 671 (1959).
16. A. Thompson, *J. Chromatog.*, **6**, 454 (1961).
17. A. Wehrli and E. Kovats, *J. Chromatog.*, **3**, 313 (1960).
18. R. Teranishi, J. Corse, J. Day, and W. Jennings, *J. Chromatog.*, **9**, 244 (1962).
19. R. Stevens and J. Mold, *J. Chromatog.*, **10**, 398 (1963).
20. K. Dimick and E. Taft, *J. Gas Chromatog.*, **1**, 7 (1963).
21. M. Verzele, *J. Chromatog.*, **13**, 377 (1964).
22. J. Morizur, B. Furth, and J. Kossanyi, *Bull. Soc. Chim. France*, **1967**, 1422.
23. M. Verzele, *J. Chromatog.*, **19**, 504 (1965).

Zone Refining of Chemical Compounds: Problems and Opportunities

W. G. Pfann

Bell Telephone Laboratories, Inc.
Murray Hill, New Jersey

It is about a decade and a half since the techniques of zone *melting* were introduced (1). Perhaps the most notable of these has been zone *refining,* the technique of repeated passage of liquid zones in one direction through a solid. A main object of zone refining is to concentrate impurities at one or both ends of the solid charge so as to purify the remainder.

A photograph of one of the first zone refiners used to purify germanium for transistors is shown in Figure 1. A 1-lb ingot of germanium in a graphite boat is pulled slowly through six induction-heating coils. Record purity (for any substance) was achieved in the

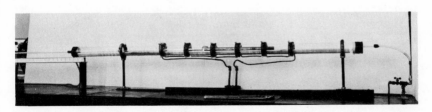

Fig. 1. Inductively heated zone refiner used for ultrapurification of germanium.

first 80% or so of the ingot. Apparatus just like this is still widely used today for certain semiconductors and metals.

Even though germanium was the first material to be extensively zone refined, it is "damning with faint praise" to say, as so many authors have, that "zone refining was introduced to purify semiconductors, but it is applicable also to metals (organic compounds, inorganic) as well, and in this article we . . ." A careful reading of the first paper (1) will fail to reveal the mention of any particular substance, but will show the complete generality of this then new concept. And in the present paper we explore, in particular, aspects of zone refining especially pertinent to organic compounds.

At this writing, I must admit to not knowing exactly how widely zone refining has been used, but a fairly recent review of published information (2) indicates that about one-third of the elements, about 100 or more inorganic compounds, and several hundred organic compounds have been raised to their highest purity by zone refining.

This is very exciting! I believe that many scientists and engineers, even today, continue unaware of the magnitude of the revolution in *purity* that is taking place, not only through zone refining, but through many other techniques.

The main point is that elements and compounds exhibit properties—physical, chemical, electrical, mechanical, optical—that were not even suspected before the materials were obtained in really pure form.

I think that zone refining offers great promise to organic chemists in particular, for there are so *many* organic compounds, and most of them are impure! Yet many of them can be melted without decomposition, and crystallize readily. In this article are described new techniques of zone refining especially suited for organic compounds.

I. PERTINENT PARAMETERS OF THE MATERIAL

Substances amenable to purification by zone refining can be grouped in a few major classes: metals, semimetals and semiconductors, inorganic compounds, and organic compounds. We do not include polymers. However, by special techniques, these too are amenable to zone refining, and can be classed with organic compounds. Each of these classes embraces a wide range of physical properties, and to that extent they overlap insofar as zone refining is concerned. Nevertheless, the optimal design of a zone refiner is, in general, characteristic for each class because each class is characterized by certain physical parameters pertinent to zone refining.

What are some of these parameters? Among the most important are: thermal conductivity, heat of fusion, melting temperature, viscosity of the liquid, ease of nucleation of crystals, stability against decomposition, volatility, chemical reactivity of the liquid, volume change on melting, hardness of the crystalline form.

Of course, the distribution coefficient, k, of the impurity to be removed or concentrated is important to success. k is defined as the ratio of the impurity concentration in the just formed solid to that in the main body of the molten zone. Zone melting will be successful (with certain rare exceptions) only if k differs from unity. However, since the value of k has little bearing on apparatus design, I do not discuss it further here.

A. Thermal Conductivity

Thermal conductivity is one of the most important parameters. It is high, in general, for the metals. This means that a large heat flux is required to establish reasonable temperature gradients at the interfaces of the zone—say, 10–50°C/cm, which in turn means that there cannot be too great a thermal impedance between the heat source and the heat sink. Thermal impedance might be comprised of air gaps, tube walls, and boat walls. Tin and copper, which have unusually high thermal conductivities, require more careful heat control, than, for example, bismuth or antimony, which have relatively lower thermal conductivities. The problem of getting heat into the zone is often circumvented, for metals, by direct induction heating of the metal. However, induction heaters are not cheap, and not always suitable for long-time, continuous operation.

At the other extreme, very low thermal conductivity, both in the solid and the liquid, also presents a problem in zone refining. This is particularly true of inorganic compounds, whose thermal conductivity is typically of the order of one-thousandth that of metals.

The problem is not so much to produce a high temperature gradient to establish the molten zone, but rather to provide heat of fusion, H, at the leading, or melting, interface and to withdraw it at the trailing, or freezing, interface, particularly the latter.

Because of this difficulty, early designs of zone refiners for organic compounds have been characterized by unduly large molten zone length, l, and long interzone spacing, i. Since reducing l strikingly improves the ultimate purification attainable in a charge of given length (3), and since reducing l and i reduces the time needed to pass a given number of zones through the charge, a design that achieves these objectives is most desirable.

A large step in the right direction was taken by Süe et al. (4), and particularly by Schildknecht et al. (5), who used closely spaced arrays of alternate heaters and coolers around a charge in a horizontal tube. Zone lengths as short as about a tube diameter were achieved.

Recently a further major advance was made by Pfann et al. (6) with their rotation–convection refiner, to which the rest of this paper is devoted. The discussion, in large part, is taken from reference 6.

II. ROTATION–CONVECTION REFINER

The rotation–convection refiner employs a new heating technique, which uses natural convection to achieve very short zone lengths, and a new cooling technique, which uses direct contact between a liquid coolant and the container to produce a flat liquid–solid interface.

A. Heating Technique

The heating technique utilizes a horizontal glass tube filled with the substance to be zone refined. The heater is a single wire of O- or U-shape. The tube is slowly rotated about its own axis as the heater advances along the tube.

To explain how the heating technique works we first describe what happens if the tube is not rotated. Imagine a glass tube of 1.2-cm bore filled with a compound such as resorcinol (melting point 110°C), with one turn of 0.5 mm diameter Nichrome wire about it. A current of 4 or 5 A produces a molten zone. Natural convection causes an upward sheet-like flow of hot liquid, more or less in the plane of the wire. When the rising hot liquid reaches the top of the tube (or the surface of the molten zone if the liquid does not fill the cross section of the tube) its flow pattern is broken, and hot liquid accumulates at the top of the zone. The hot liquid melts the adjacent solids, thereby lengthening the top of the zone and giving the zone a shape something like an upside down trapezoid (Fig. 2a).

Because of this convective behavior it is difficult with such an arrangement to achieve a zone length shorter than the tube diameter. Moreover, the shape is undesirable in several respects: mixing is suppressed near the top of the zone; the interface is usually curved in a complex fashion, which makes growth of a single crystal more difficult; and the convective flow pattern tends to produce cross-sectional nonuniformity of solute concentration.

If, however, the tube is rotated slowly (for example, at 1 or 2 rpm for tubes of 1.2–2.5 cm bore), the zone assumes an almost cylindrical

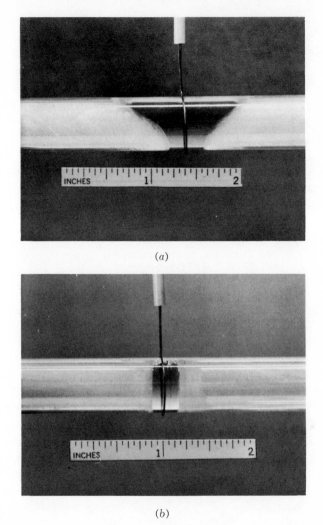

(a)

(b)

Fig. 2. Molten zones in ½-in. bore horizontal tubes containing resorcinol; (a) unrotated, and (b) rotated at 5 rpm.

shape (Fig. 2b). By reducing the heater current, and, if necessary, by providing auxiliary cooling of the adjacent solid, molten zones as short as one-fourth of a tube diameter are readily produced and controlled. Such zones have been produced in tubes up to 7.5 cm bore.

We refer to this heating method as "rotation–convection heating," for natural convection not only heats the zone efficiently, but also, in combination with slow rotation, achieves a short zone length and a desirable zone shape.

B. Cooling Technique

Cooling methods used previously for organic compounds have been less than satisfactory, especially at large tube diameters. A ring of air jets behind the zone has been used, but air is inefficient as a heat transfer fluid, and the cooling effect of an air stream is rather diffuse. An internally cooled metal container placed around the tube has also been used, but the air gap between tube and container presents an unduly high thermal impedance.

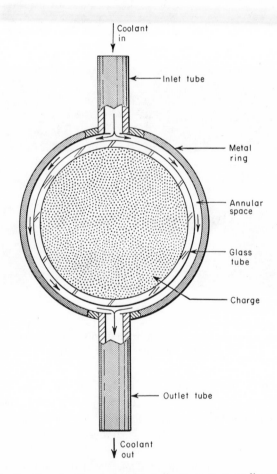

Fig. 3. Cross-sectional sketch of direct contact cooling ring.

Fig. 4. Multielement apparatus using the rotation–convection heating technique and the direct contact–cooling technique. Naphthalene in 1-in. bore Pyrex tube. The molten zones are the dark regions inside the heating wires.

An ideal cooling means for a cylindrical tube is a narrow, ring-form, heat sink in direct contact with the tube. The problem is to move such a heat sink along the tube with the heater. A practical solution to this problem has been developed. It is illustrated in Figure 3, which is an axial view, in cross section, of the charge tube and cooler. For a 2.54 cm o.d. charge tube, a 2.60 cm i.d. ring about 0.6 cm long is cut from a length of brass or copper tubing. Inlet and outlet copper tubes for coolant are fitted to the ring, and the ring is mounted concentric with the charge tube. Water flows down through the space between ring and tube, on both sides of the tube, and leaves through the outlet tube at the bottom of the ring. If the inner surface of the ring is wetted by water, surface tension confines the water to the annular space between the ring and the charge tube. Also, the exit tube exerts a siphoning effect which helps to confine the water. A typical flow rate for the cooler described is about 70 cm^3/min. Part of the heat removed from the charge is carried off by the water, and part is removed by conduction to the metal ring and attached tubes.

C. Apparatus

A multielement zone refiner utilizing the heating and the cooling means described above is shown in Figure 4. There are ten heaters and ten coolers, at 2.5-cm intervals. The charge tube is 2.5 cm o.d. Pyrex tubing. The material is naphthalene. The mean zone length is about 0.6 cm; hence the 25-cm charge is about 40 zone lengths

Fig. 5.   Closer view of apparatus of Figure 4 showing three molten zones.

long.   Cooling water flows from a constant-head overflow tank.   The heaters are connected in series and powered by 60 cps current.   Minor adjustments can be made for each heater by a variable shunt resistor across the heater.   Once the heater powers are set, the apparatus can be run without attention for days at a time, without further control of the heaters.   The usual zone travel rate has been about 0.8 cm/hr or less.   For substances that melt closer to room temperature than naphthalene it was found helpful to shield the apparatus from drafts.   The usual rotation rate has been about 1.3 rpm.   In this particular setup, the charge tube and rotation motor are mounted on a carriage that travels with reciprocating motion, while the heaters and coolers are stationary.

A closer view of a 1-in. bore tube containing naphthalene and showing three molten zones appears in Figure 5.   The purifying action is rather striking when seen in natural color (the impurity is a dye, oil red), but is evident by a darkening of successive zones in the figure.

III. DISCUSSION

In zone melting with vertical tubes, natural convection tends to increase the zone length.   Typical zone lengths are several tube diameters long, and it has been stated that the smallest feasible zone length for organic compounds is about one tube diameter (7).   In contrast, by the rotation–convection heating technique, zone lengths of one-fourth tube diameter or less are readily achieved in tubes of 2.5 cm diameter.

The translation rate of the zone must be tailored to the effectiveness of the heat sink and the natural convection in the zone. If the rate is increased beyond the proper value, the zone lags behind the heater, the melting interface becomes convex toward the liquid, and the freezing interface becomes concave toward the liquid. For 2.5-cm tubes and for a number of organic compounds, about 1 cm/hr has been found to be a suitable maximum rate with respect to control of zone shape and zone size.

In view of the very short zone lengths and inter-zone spacings achieved for charges of sizable diameter by the techniques described here, it is expected that these techniques will find widespread application, particularly in the zone refining of compounds of low thermal conductivity. Furthermore, the techniques are applicable to other operations, such as zone leveling, normal freezing, and the growth of large diameter single crystals, and to a certain extent to materials of high thermal conductivity (8).

Other important details of the rotation–convection technique, such as void formation and matter transport (its benefits and its difficulties) are discussed in reference 6.

A very useful analytical technique, devised by J. D. Hunt and used on the rotation–convection refiner described here comprises judging purity by examining the formation of a cellular substructure while the refiner is in operation. It is particularly applicable to substances that exhibit a low entropy change on freezing and hence form cells and dendrites just as metals do (9). If the freezing interface is examined through a microscope, cells may be evident in the solid. This shows that the solid is not yet pure. At a further stage of refining, the solid may *not* exhibit cells, but still may not yet be pure. This is demonstrated by blowing on the tube for a few seconds, thereby causing a temporary increase in growth rate. If cells are seen then, still further refining is indicated. Finally, if blowing produces cell-free solid, the limits of this very simple but very sensitive (for low-k impurities) (10) test will have been reached. This method revealed, for example, impurities in cyclohexanol which were undetectable by gas chromatography.

References

1. W. G. Pfann, *Trans. AIME,* **194,** 747 (1952).
2. W. G. Pfann, *Zone Melting,* 2nd ed., Wiley, New York, 1966.
3. Reference 2, p. 42.

4. P. Süe, J. Pauly, and A. Nouaille, *Soc. Chim. 5th Series, Memoires,* 593 (1958).

5. H. Schildknecht, *Zone Melting* (transl. from German), Academic Press, New York, 1966, pp. 100, 114.

6. W. G. Pfann, C. E. Miller, and J. D. Hunt, *Rev. Sci. Instr.,* **37,** 649 (1966).

7. W. R. Wilcox, R. Friedenberg, and N. Back, *Chem. Rev.,* **64,** 187 (1964).

8. C. E. Miller, unpublished experimental results.

9. K. A. Jackson and J. D. Hunt, *Acta. Met.,* **13,** 1212 (1965).

10. W. A. Tiller and J. W. Rutter, *Can. J. Phys.,* **34,** 96 (1956).

Membrane Permeation: Theory and Practice

ALAN S. MICHAELS, *President*

and

HARRIS J. BIXLER, *Vice President, Research*

Amicon Corporation
Lexington, Massachusetts

PART 1. FUNDAMENTALS OF MEMBRANE PERMEATION

I. Introduction

If a multicomponent fluid mixture is brought in contact with one
surface of a membranous barrier (e.g., a solid, homogeneous film)

under conditions where the other side of the barrier is in contact with another fluid mixture within which the partial molar free energies (chemical potentials) of one or more components common to both fluids differ across the barrier, there will generally be a transfer of components through the barrier in the direction of declining chemical potential for each. If the "upstream" and "downstream" fluids are gas or vapor mixtures at differing hydrostatic pressures, for example, components pass through the barrier in the direction of declining partial pressure by a process usually termed *gas permeation*. If the "upstream" fluid is a liquid mixture and the "downstream" fluid is a gas in which the partial pressures of the permeating components are lower than the vapor pressures over the components in the upstream liquid, transmembrane permeation occurs by a process often termed *pervaporation*. If liquid phases are in contact with the membrane at equal hydrostatic pressures, components can transfer across the barrier under the action of a concentration difference between the contracting liquids by the process of *dialysis*. If a liquid mixture is confined at high hydrostatic pressure on one side of a barrier under conditions where lower hydrostatic pressure is maintained on the other side, certain components of the upstream liquid will permeate the membrane by the process of *ultrafiltration* or *reverse osmosis*.

All of these membrane transport processes have the common features of (*a*) transport of mass by the action of a free-energy driving force, and (*b*) the capacity to alter mixture composition by virtue of the ability of the membrane barrier to pass one component more rapidly than another *despite equality of driving potential*. It is this latter unique characteristic of membrane separation which differentiates this process from most common separation operations. It is the purpose of this review to examine the properties and characteristics of membranes which give rise to their unique separation capabilities, and to relate these properties to the simple physicochemical properties of matter with which most chemists and engineers are familiar. From this examination, it is hoped, will appear a rather obvious rationale for explaining membrane permeation and for predicting how various membrane materials will behave with respect to their ability to permeate specific components, and to discriminate between such components in mixtures.

Virtually all permselective membranes relevant to this discussion can be shown to be homogeneous solids devoid of "pores" or "holes" in any conventional sense. Most are amorphous or at best partially crystalline solids; by far the majority are polymeric in their molecular structure. Their capacity to be penetrated by small molecules clearly requires that such molecules be capable of *entering* into the solid phase, and of being *mobile* once they have so entered. The concepts of molecular entry into and mobility within homogeneous phases are traditionally accepted in physics and chemistry as the elements of molecular diffusion, and there is little justification for seeking more obscure explanations for membrane permeation unless phenomena not reconcilable by such a simple mechanism are uncovered. As will be shown here, the treatment of the membrane-transport process as a special case of molecular diffusion is remarkably fruitful and quite adequate to explain most of the experimental facts.

II. Basic Diffusional-Transport Relationships

A. FICKIAN DIFFUSION

If a homogeneous film of uniform thickness separates two fluid phases of differing pressure and/or composition (conditions being maintained isothermal), and the boundary conditions on both sides of the membrane are maintained constant, a steady-state flux of one or more components through the membrane will ultimately be established which, at every point within the membrane for component i, will be specified by Fick's first law of diffusion:

$$J_i = -D_i \frac{dc_i}{dx} = \text{const} \tag{1}$$

where J_i is the mass flux (g/cm^2 sec), D_i is the local diffusivity (cm^2/sec), c_i, is the local concentration of component i, and x is the distance through the film measured normal to the film surface.

B. MEMBRANE BOUNDARY CONDITIONS

If conditions at both membrane–fluid boundaries are such that the potential rates of arrival (or removal) of all membrane-permeable components at (or from) the membrane surfaces are large compared with their rates of penetration through the membrane, then the fluid phase compositions will be constant right up to the membrane surfaces, and the membrane substance at these surfaces will be in "saturation equilibrium" with respect to the penetrants. In this case, the boundary conditions with respect to the concentration of each component in the membrane can be specified by

$$c_{i(1)} = f_i(a_{i(1)}) \qquad x = 0 \tag{2}$$

$$c_{i(2)} = f_i(a_{i(2)}) \qquad x = l \tag{3}$$

where the subscripts (1) and (2) refer to the "upstream" and "downstream" surfaces of the membrane, respectively, f_i is the thermodynamic function relating the concentration of component i in the membrane substance to the activity a of that component in the adjacent fluid phases, and l is the film thickness.

C. THE PERMEATION FLUX

Integration of eq. (1) within these boundary conditions yields:

$$J_i l = \int_{c_{i(2)}}^{c_{i(1)}} D_i \, dc_i \qquad (4)$$

The product Jl, frequently referred to as the "normalized flux" Q, is thus seen to be independent of membrane thickness, and dependent solely on the boundary conditions and the specific solution–diffusion properties of the membrane–penetrant pair. Implicit in the above relations is the possibility that both the diffusivity of any component in the membrane, and its concentration, may vary with the composition of the penetrant mixture present in the membrane at any point.

If, in addition, the *sole* source of components appearing in the "downstream" phase adjacent to the membrane is the permeation of components through this membrane (as is the usual case for gas permeation, pervaporation, and ultrafiltration), then a further constraint is provided by a mass balance over the system, viz.,

$$J_i : J_j : J_k \ldots = c_{i(2)} : c_{j(2)} : c_{k(2)} \ldots \qquad (5)$$

That is to say, the ratio of any two components in the permeate is equal to the ratio of the fluxes of those two components through the membrane. Hence, only one intensive property of the downstream phase (e.g., pressure) need be specified to define the component fluxes and permeate composition for a given set of conditions in the upstream fluid.

The preceding relations are more phenomenological than useful since there are no generalized relationships between solubility, diffusivity, and activity which can be utilized to solve these equations explicitly for specific membrane–penetrant systems. However, these relations do show that membrane permeability (i.e., flux per unit overall driving force) will always be *inversely proportional to membrane thickness,* and that permselectivity will always be independent of membrane thickness. To gain insight into how (and how fast) a given penetrant moves through a given membrane, we must consider the molecular processes taking place within the membrane which influence both the sorptive and diffusive events that govern penetration.

III. Noncondensible Gas Permeation

A. "IDEAL" SOLUTION AND DIFFUSION

The above relationships can be greatly simplified for the case of gaseous penetrants which are sparingly soluble in the membrane substance and which do not chemically associate with one another or the membrane material. In such cases, the membrane molecular structure is not perturbed by the dissolved molecules, and thus the diffusivity of a penetrant is essentially constant throughout the membrane. Moreover, the solubility of such species within the membrane is essentially directly proportional to its activity (i.e., pressure) in the equilibrium gas phase; i.e., Henry's law applies:

$$c_i = k_i p_i \tag{6}$$

$$J_i = D_i \left[(c_{i(1)} - c_{i(2)})/l \right] \tag{7}$$

or

$$J_i = k_i D_i \left[(p_{i(1)} - p_{i(2)})/l \right] \tag{8}$$

The product, $k_i D_i$, is usually termed the "permeability," $\bar{p}_i$ of the membrane to i; it is equal to the flux of i through a membrane of unit thickness when exposed to a unit pressure difference of i across it. Most gaseous penetrants and membranes obey eq. (8); the permeability is thus *independent* of the *total pressure* of the penetrating gas.

B. IDEAL PERMSELECTIVITY

For the simultaneous permeation of two gases through a membrane, the relative fluxes of the two permeants is given by:

$$\frac{J_i}{J_j} = \frac{k_i}{k_j} \cdot \frac{D_i}{D_j} \cdot \frac{p_{i(1)} - p_{i(2)}}{p_{j(1)} - p_{j(2)}} \tag{9}$$

We may define a permselectivity of the membrane to i with respect to j by the relationship:

$$\sigma_{ij} = (p_{i(2)}/p_{j(2)}) \cdot (p_{j(1)}/p_{i(1)}) \tag{10}$$

where, for ideal gases, the ratio of partial pressures in a gas mixture is equal to the mole ratio. Hence σ_{ij} is an "enrichment factor" for the membrane.

If the permeate composition is determined by the permeation fluxes of each component, eqs. (9) and (10) can be combined to yield:

$$\sigma_{ij} = \frac{k_i}{k_j} \cdot \frac{D_i}{D_j} \left[\frac{1 - \sigma_{ij}(p_{j(2)}/p_{j(1)})}{1 - (p_{j(2)}/p_{j(1)})} \right] \tag{11}$$

Equation (11) shows that, if the downstream gas pressure is maintained much lower than the upstream pressure (i.e., $p_{j(2)}/p_{j(1)} \to 0$), the permselectivity of the membrane approaches in the limit:

$$\lim_{p_2 \to 0} \sigma_{ij} = (k_i/k_j) \cdot (D_i/D_j) \qquad (12)$$

As the pressure difference across the membrane is reduced, it can be shown that $\sigma_{ij} \to 1.0$, meaning that the permselectivity of the membrane tends to vanish as the pressure difference across the membrane tends to zero.

Equation (12) indicates that the limiting maximum permselectivity of a membrane toward one of two noncondensible gases is determined uniquely by the relative solubilities of the two gases in the membrane (at equal pressures) and their relative diffusivities. Since solubility is a thermodynamic property of a membrane–penetrant system, whereas diffusivity is a molecular–kinetic property, these two properties must be examined independently if membrane permeability and selectivity are to be satisfactorily analyzed and correlated.

C. SOLUBILITY CONSTANTS AND THE JOLLEY-HILDEBRAND RULE

The solubility of a noncondensible gas in an amorphous organic polymer is governed by the magnitude of the van der Waals force field around each gas molecule, and the field intensity associated with individual chain segments of the polymer. The former is reflected in the critical temperature of the gas; the latter, in the so-called "cohesive energy density" (CED) of the polymer. Jolley and Hildebrand (1), by simplified thermodynamic analysis, were able to show that the Henry's law constant (gas solubility at 1 atm pressure) for a variety of gases in simple organic liquids at a reference temperature T_0 could be correlated with the Lennard-Jones force constants for the gases (which are determined from gas PVT data and are roughly proportional to critical temperature) by an equation of the form:

$$\ln k_{T_0} = A(\epsilon/\bar{k}) + \ln B \qquad (13)$$

where ϵ = Lennard-Jones constant, $\bar{k}$ = Boltzmann constant, A = a constant dependent on the temperature, and B = a constant dependent upon the solvent only.

Equation (13) has been shown (2–4) to be equally satisfactory for correlating gas solubilities in polymers, as illustrated in Figure 1. In general, for amorphous polymers, the higher the CED or polarity of the polymer, the lower the solubility of any gas in it. This simple relationship does not apply, however, to very polar gases (e.g., H_2O) in highly polar polymers.

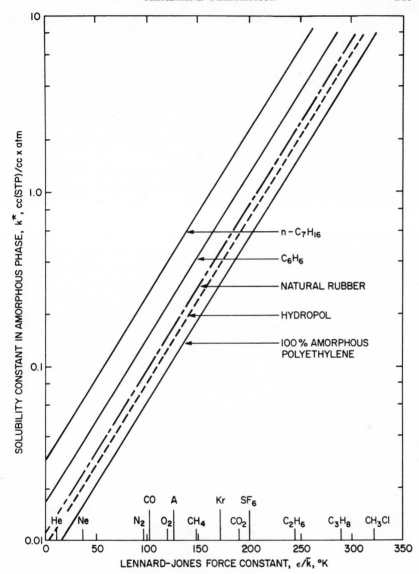

Fig. 1. Jolley–Hildebrand correlation of solubility constants (25°C) of gases in amorphous polymers (2).

Similarly, it can be shown that, for gases obeying the Jolley-Hildebrand rule, the enthalpy of solution of a gas in the polymer, ΔH_s, defined by the van't Hoff equation:

$$\frac{d \ln k}{d(1/T)} = - \frac{\Delta H_s}{R} \qquad (14)$$

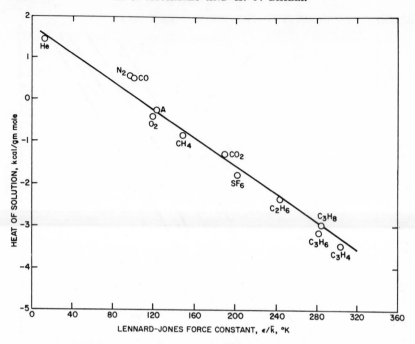

Fig. 2. Jolley–Hildebrand correlation of heats of solution of gases in amorphous polyethylene (2).

is related to the Lennard-Jones constant by the equation:

$$\Delta H_s = B' + A'(\epsilon/\bar{k}) \tag{15}$$

where A' and B' are constants dependent on the solvent. This relationship is also quite adequate for predicting heats of solution of gases in polymers, as shown in Figure 2.

Equations (13) and (15) thus provide a reliable basis for predicting the solubility of any noncondensible gas in any polymer at any temperature, from a single experimental determination.

D. CORRELATION OF DIFFUSION CONSTANTS

The diffusion of a gas molecule within an amorphous polymer matrix is an activated process involving the cooperative movements of the molecule and of the polymer chain segments that surround it. In effect, thermal fluctuations of chain segments must allow sufficient local separation of adjacent chains to permit the passage of a penetrating molecule; it is by this stepwise process that diffusion occurs. It should be obvious that the frequency with which a diffusion step occurs will depend upon (a) the size and shape of the diffusing molecule, (b) the tightness of packing and force of attraction between adjacent polymer chains, and (c) the stiffness of the polymer chains. By rather sophisticated statistical mechanical analysis, Barrer (5), Meares (6), Brandt (7), and others have derived expressions relating the diffusivity and activation energy for diffusion of simple molecules within polymer matrixes to the molecular dimensions of the penetrant and the configuration of the polymer chains. While these relationships have not proved to be particularly accurate for predicting absolute values of these transport functions, they have yielded functional relationships which are surprisingly useful for correlating diffusion data. For example, the diffusivities of a large number of gases in a given amorphous polymer can be satisfactorily correlated by a relation of the form:

$$D_{T_0} = G \exp{(-\omega d^n)} \tag{16}$$

where D_{T_0} is the diffusivity of a given gas in the polymer at a reference temperature T_0, G and ω are constants characteristic of a particular polymer, d is the "effective molecular diameter" of the gas, and n is a constant which is unity for highly flexible chain macromolecules such as natural rubber or polyethylene, and approximately 2.0 for rotationally hindered, stiff-chain macromolecules such as polypropylene, polyisobutylene, poly(ethylene terephthalate), and cellulose. For essentially spherical gas molecules, d is the diameter determined from gas-viscosity measurements; for unsymmetrical molecules, the most satisfactory value of d for correlation purposes appears to be the square root of the ratio of the molecular volume to the maximum linear dimension of the molecule. This implies that, during the activated diffusion event, the most frequent orientation of the molecule is that where the long axis of the molecule lies in the flow direction.

Similarly, the activation energy for diffusion, defined by the Arrhenius relation:

$$\frac{d \ln D}{d(1/T)} = \frac{-E_D}{R} \tag{17}$$

can be satisfactorily correlated with penetrant molecular diameter by the relationship:

$$E_D = E_{D_0} + \omega' d^n \tag{18}$$

where E_{D_0} and ω' are constants characteristic of the particular polymer.

Typical diffusivity and diffusion activation energy correlations for a flexible-chain polymer (natural rubber) and a "stiff-chain" polymer [poly(ethylene terephthalate)] are presented in Figures 3–6. Except for hydrogen and helium, whose diffusivities are anomalously high for their apparent molecular diameters, the correlations defined by eqs.

(16) and (18) are remarkably accurate, thereby allowing prediction of both D_{T_0} and E_D for a given gas in a given polymer (and thus, of D at any temperature) within a probable error of $\pm 20\%$.

E. GLASSY VERSUS RUBBERY POLYMERS

Most polymers exhibit a series of rather abrupt changes in several second-order thermodynamic properties (e.g., thermal expansion coefficient, compressibility, heat capacity) at a characteristic temperature known as their "glass transition temperature," T_g. Such polymers, in membrane form, display corresponding changes in their second-order gas-transport properties (specifically, heats of solution and diffusion activation energies) at or near T_g; below T_g, heats of solution are smaller (more negative) and activation energies smaller (less positive) than above. Nonetheless, the correlations expressed by eqs. (16) and (18) apply below as well as above T_g, but with

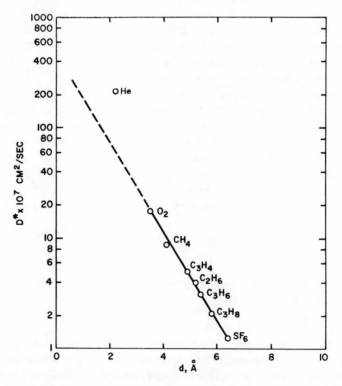

Fig. 3. Correlation of diffusion constants (25°C) of gases in amorphous polyethylene (12).

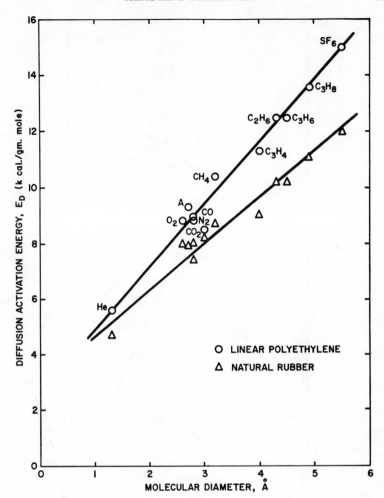

Fig. 4. Correlation of activation energies for diffusion of gases in rubber and polyethylene (12).

different constants. This is illustrated in Figure 5, for poly(ethylene terephthalate). Analysis of sorption and diffusion data below T_g (3,8) suggests that polymers in the glassy state contain "frozen-in" submicroscopic voids which can trap penetrant molecules which contribute little to the diffusive process, and also contain regions of virtually immobilized chains through which activated diffusion is impossible. The anomalous heats of solution and diffusion activation energies observed T_g can be reconciled in these terms.

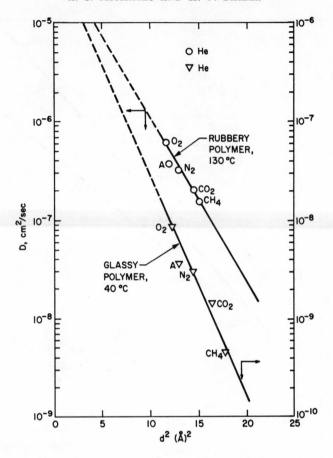

Fig. 5.   Correlation of diffusion constants of gases in poly(ethylene terephthalate) above and below T_q (8).

F. GENERALIZED CORRELATION OF PERMEABILITY AND PERMSELECTIVITY

For noncondensible gases permeating through polymeric membranes, therefore, it can be seen that the *permeability* of any gas through any membrane is given approximately by

$$\bar{P} = kD = BG \exp \left\{ \left[A \left(\frac{\epsilon}{\bar{k}} \right) - bd^n \right] \right.$$
$$\left. + \left(1 - \frac{T_0}{T} \right) \frac{a' + b'd^n + B' - A'(\epsilon/\bar{k})}{RT_0} \right\} \quad (19)$$

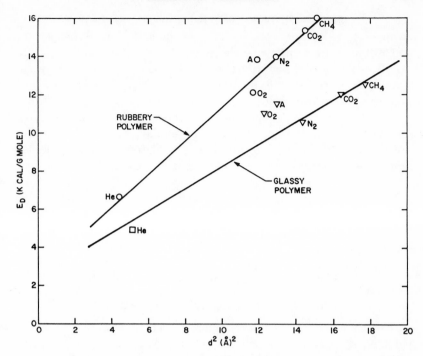

Fig. 6. Correlation of activation energies for diffusion of gases in poly(ethylene terephthalate) above and below T_g (8).

where the quantities $(\epsilon/\bar{k})$ and d are characteristic of the gas only, and the quantities B, G, b, A', B', a', b', and n are characteristic of the polymer only (A is a "universal constant"). For a particular membrane, all of the polymer-determined constants can be established by experimental measurement of the solubilities and permeabilities (or diffusivities) of only two gases at two temperatures (a total of four experiments). It is interesting to note that the *permselectivity* of a given membrane with respect to a specified pair of gases involves many fewer constants:

$$\frac{\bar{P}_i}{\bar{P}_j} = \exp\left[A\,\frac{\epsilon_i - \epsilon_j}{\bar{k}} - b(d_i{}^n - d_j{}^n) \right.$$
$$\left. + \left(1 - \frac{T}{T_0}\right)\frac{b'(d_i{}^n - d_j{}^n) - A'(\epsilon_i - \epsilon_j)/\bar{k}}{RT_0} \right] \quad (20)$$

Examination of eq. (20) reveals important general characteristics of permselective polymer membranes, viz.,

1. A membrane will be selectively permeable to that gas (of two or more in a mixture) which has the highest critical temperature and/or the smallest molecular diameter.

2. Membrane permselectivity invariably decreases with increasing temperature, irrespective of whether selectivity is attributable to differences in solubility or diffusivity.

3. Stiff chain/polymer membranes, while invariably less gas permeable than flexible-chain polymers of similar chemical constitution, are the more permselective toward small relative to large molecules, or toward unsymmetrical relative to symmetrical molecules of equal size.

To summarize, the processes of membrane permeation by noncondensible gases are quite satisfactorily interpreted and correlated by the simple principles of solution and molecular diffusion, the membrane being regarded as a homogeneous solid or quasi-liquid phase.

IV. Permeation by Liquids or Condensible Vapors

A. DISSOLUTION AND SOLVATION

When a liquid or saturated vapor is brought into contact with a polymer, the extent to which the polymer will absorb the compound will be governed primarily by the similarity of chemical constitution of the solute and solvent. If intermolecular attractions within the liquid are much stronger (or weaker) than those in the polymer (e.g., water and polyethylene, or hexane and cellulose), mixing of the two will be energetically unfavorable, and the solubility of the liquid in the polymer correspondingly low. If, on the other hand, the intermolecular attractions are of similar intensity in both phases (e.g., toluene and polyethylene, or water and cellulose), absorption of the liquid or vapor into the polymer will be energetically favorable, and the solubility will be high. Under these circumstances, absorption will be accompanied by "solvation" or swelling of the polymer, which may be unlimited—that is, the polymer may be miscible in all proportions with the liquid. It is, of course, impossible to conduct a membrane-permeation process with such a mutually soluble liquid–polymer system; however, one can impose constraints on the swelling of polymers in such "compatible" liquids by introducing chemical crosslinks or crystallinity into the polymer. In such event, the polymer membrane will swell to only a limited degree in contact with a "compatible" liquid and will display lower swelling or sorptivity for less compatible liquids.

The most convenient measure of "compatibility" of a given liquid

in a given polymer is the difference in the solubility parameters of
the liquid and polymer. The solubility parameter, δ, is the square
root of the cohesive energy density; for simple liquids, this is readily
calculated from the molar enthalpy of vaporization and the liquid
density. For polymers, it can (and has) been estimated from the
atomic constitution of the polymer (9) and can be determined experimentally by measuring the sorption of a series of liquids of known δ
and observing which of the series swells the polymer to the greatest
extent. Qualitatively, the dependence of sorption and swelling of
a given polymer by liquids of differing solubility parameters is represented by Figure 7. For amorphous, noncrosslinked polymers, solubility goes to infinity as $(\delta_p - \delta_L) \to 0$; for crystalline or crosslinked
polymers, solubility reaches a maximum at this point, the limit being
determined by crosslink density or crystalline morphology.

Furthermore, for strongly swelling penetrants, the dependence of
sorption in the polymer upon the *activity* of the penetrant (e.g., its
partial pressure in a gaseous mixture, or its concentration in a liquid

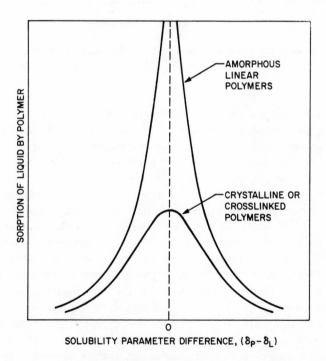

Fig. 7. Sorption and swelling of polymers by liquids as a function of solubility
parameter difference.

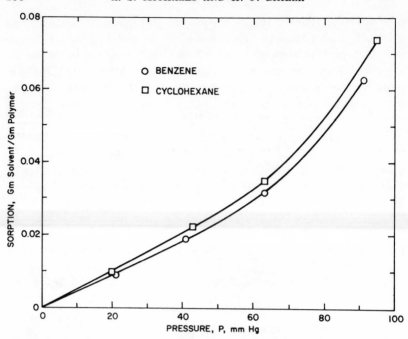

Fig. 8. Sorption isotherms of benzene and cyclohexane in linear polyethylene at 25°C (10).

mixture) is seldom a simple linear function of pressure as with noncondensible gases of low solubility. This is a result of the fact that penetrant–penetrant interactions became important at higher concentrations in the polymer, and that polymer-chain arrangements and interactions are perturbed by the dissolved species. Such polymer–solvent sorption equilibria can be represented by characteristic sorption isotherms as in Figure 8; unfortunately, these must be determined experimentally and cannot usually be predicted from basic thermodynamic parameters. Typically, such sorption isotherms are S-shaped curves or curves convex to the pressure or activity axis.

The sorption characteristics of polymers toward mixtures of swelling penetrants are further complicated by the fact that the sorption of one component by the polymer is influenced by the type and quantity of other components present. As a consequence, the measurement of the sorption isotherm of pure component A by a polymer will not, except under special circumstances, permit accurate prediction of the sorption of A by the polymer from a liquid or vapor mixture of A and B. If, however, A and B are nearly identical in chemical constitution (e.g., position isomers such as o- and p-xylene, or homologs such as heptane and octane), where the sorption isotherms for each component alone are virtually identical, then "ideal mixture" rules apply, and the relative sorptions can be predicted from:

$$c_A = c_A^* x_A \tag{21}$$

and

$$c_B = c_B^* x_B \tag{22}$$

where c_A and c_B are the concentrations of A and B in the polymer in contact with a liquid mixture in which the mole fractions are x_A and x_B, and c_A^* and c_B^* are the concentrations of A and B in the polymer in equilibrium with pure liquid A or B.

In contact with mixtures of dissimilar (but miscible) liquids, such rules are very unreliable; indeed, polymers frequently show greater swelling by liquid mixtures than by the mixture components separately.

B. DIFFUSION IN SOLVATED POLYMERS

When a polymer is significantly solvated by a penetrant ("significant" meaning a penetrant concentration in excess of 1% of the polymer by volume), the diffusion process is catastrophically altered. The reason is simple: as the solute concentration becomes elevated, solute–solute interactions come into play, and polymer chain segments are "loosened" and separated by the solute. As a consequence, the penetrant diffusivity becomes a rapidly increasing function of penetrant concentration, and the activation energy for diffusion rapidly declines with increasing concentration. These trends are approximated by relations of the form:

$$D_c = D_{c=0} \exp (\gamma c) \tag{23}$$

and

$$E_{D(c)} = E_{D(c=0)} - \gamma' c \tag{24}$$

where c is the penetrant concentration in the polymer, and γ and γ' are constants characteristic of the penetrant–polymer pair. The magnitudes of the changes in D and E_D with concentration are impressive; an increase of 1000-fold in D and a decrease of 10 kcal in E_D is not unusual for a change in c from zero to 20 vol. %, as illustrated in Figure 9.

If random thermal fluctuations nonetheless govern diffusion in the swollen polymer then Fick's law can still be applied with allowance

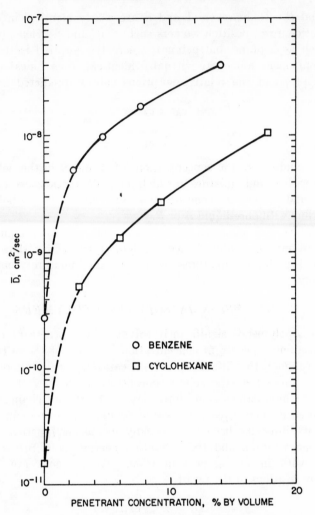

Fig. 9. Diffusivities of benzene and cyclohexane in linear polyethylene as a function of penetrant concentration at 25°C (10).

for concentration dependence of the diffusivity. At steady state:

$$J = -D \frac{dc}{dx} \qquad D = f(c) \tag{25}$$

Integrating over the limits $c = c_1$ at $x = 0$, $c = c_2$ at $x = l$,

$$Jl = Q = -\int_{c_1}^{c_2} D \, dc = \bar{D}(c_1 - c_2) \tag{26}$$

where $\bar{D}$ is the "integral diffusivity" over the concentration interval $c_1 \rightarrow c_2$. For the special case where $c_2 \cong 0$, and eq. (23) applies, eq. (26) can be written:

$$Q = (D_{c=0}/\gamma)(\exp \gamma c_1 - 1) = \bar{D}c_1 \tag{27}$$

or

$$\bar{D}_{0 \rightarrow c_1} = D_{c=0} \frac{(\exp \gamma c_1 - 1)}{\gamma c_1} \tag{28}$$

A little thought will reveal that the concentration profile within a swollen membrane through which steady-state penetration is occurring is far from linear; the concentration gradient increases toward the downstream boundary, giving the impression that the major barrier to transport is localized near the outflow surface. (This is actually not the case, however, since the permeation flux is everywhere constant.) It is further evident that, as the penetrant concentration at the upstream surface of the membrane is increased (by increasing the penetrant activity in the upstream fluid), the permeation flux will increase far more rapidly than linearly with the concentration. This gives rise to the rather surprising observation that a membrane exposed to a swelling vapor at 95% of saturation will frequently show a three- to fivefold *lower* permeability than the same membrane exposed to saturated vapor or liquid at the same temperature.

Because of the pronounced concentration dependence of the diffusivity in these systems, membrane permeabilities to swelling liquid (or vapor) penetrants are two to four orders of magnitude larger than are their permeabilities to noncondensible gases. At ordinary temperatures, permeation fluxes under "pervaporation" conditions are of the order of 0.2–0.2 lb/hr $\times$ ft^2 for 1-mil membranes—well within the range of practicability for industrial separations.

C. PERMSELECTIVITY OF SWOLLEN MEMBRANES

For the same reasons that swelling increases diffusion coefficients and reduces diffusion activation energies, it tends to reduce membrane permselectivity attributable to differences in penetrant molecular size or shape. For example, a polyethylene membrane which, unswollen, shows a permeability to benzene about 20 times that to cyclohexane, when swollen with the liquid hydrocarbon, shows a permeability ratio of only 3.5 (10). As a rule, the gain in permeation flux which accompanies swelling of the membrane is so large in comparison with the sacrifice in selectivity that, as a practical matter, operation of membrane-permeation separation with a swollen membrane is preferred. As a matter of fact, it has been recommended (11) that, when mixtures with limited swelling capability are to be 'separated via membrane permeation, a third strongly swelling component be added to the mixture solely to solvate the membrane and raise the permeation flux.

For liquid mixture permeation through swollen membranes, the actual permselectivity of the membrane will usually not be equal to the predicted permselectivity based on the pure component permeabilities. This is in part due to solubility nonidealities mentioned earlier; a more important factor, however, is that the local diffusion coefficient of a given component in a swollen membrane is dependent upon the total local concentration of *all* the penetrants, as well as on the relative amounts of each. These complexities make the prediction of absolute and relative permeation fluxes of two or more components from single pure-component data nearly impossible. For the special (but important) case of mixtures of isomers (e.g., n- and isoparaffins, the xylene isomers), for which each pure component has nearly the same solvation power for the membrane, essentially "ideal" permeation is observed, viz.:

$$J_i' = \bar{D}_i c_{1i} x_i = J_i x_i \tag{29}$$

$$J_j' = \bar{D}_j c_{1j} x_j = J_j x_j \tag{30}$$

and

$$\sigma_{ij} = \bar{D}_i c_{1i} / \bar{D}_i c_{1j} \cong \bar{D}_i / \bar{D}_j \tag{31}$$

where J_i' and J_j' are the permeation fluxes of i and j through the membrane in contact with a liquid mixture in which the weight (or mole) fractions are x_i and x_j; $\bar{D}_i$ and $\bar{D}_j$ are the integral diffusivities of i and j through the membrane when in contact with *pure* liquid i and j, respectively; and c_{1i} and c_{1j} are the saturation concentrations of i and j in the membrane substance in equilibrium with the *pure* liquid components. Since, for isomers, the saturation concentrations are essentially identical, the membrane permselectivity reduces to the ratio of the integral diffusivities.

Differences in liquid permeation rates of compounds differing but slightly in molecular configuration are surprisingly large. For example, polyolefin membranes permeate p-xylene at about twice the rate of o-xylene, despite a difference of less than 10% in effective molecular diameter between the two isomers. N-Heptane permeates such membranes about 4–5 times as fast as isooctane, and trans-acetylene dichloride, about twice as fast as *cis*-acetylene dichloride. It is this sensitive "size–shape" discrimination of membranes that makes membrane permeation so promising a candidate for separation tasks difficult to accomplish by traditional methods.

V. The Influence of Polymer Crystallinity and Morphology on Membrane Permeation

Inasmuch as molecular transport in polymeric membranes involves separation and movement of polymer chain segments, it is only reasonable to expect that the introduction of constraints upon chain-segment mobility in a polymer will influence its permeability. The influence of such constraints upon permeability is, as will be shown below, profound.

There are two important sources of chain constraint in polymers; one in crystallinity, and the other, strain-induced orientation. A crystalline region in a polymer is a region of high molecular order and (usually) tight molecular packing. A large body of experimental evidence (2,3,8,12,13) indicates that penetrant molecules are insoluble in polymer crystallites and thus unable to diffuse through the crystalline phase. Thus, when gases, vapors, or liquids permeate through crystalline polymers, the penetrating molecules are confined to the noncrystalline (amorphous) regions between crystallites. The crystallites thus reduce the permeability in two obvious ways: they reduce the available volume of the polymer for penetrant solution, and they constrain diffusion to take place through irregular, tortuous pathways between crystallites. The reduction in solubility (for noncondensible gases) is directly proportional to the volume fraction of crystalline phase; the reduction in diffusivity, on the other hand, is a function of the shape and size distribution of crystallites, as well as their volume concentration. For highly crystalline polymers such as linear polyethylene, where the volume fraction of crystallinity exceeds 70% and the crystallites are of lamellar shape, "geometric impedances" to diffusion are of the order of 10, meaning that the diffusivity of the gas in the crystalline polymer is 10 times as low as that in the completely amorphouse polymer. The overall permeability of the crystalline polymer to a given gas would thus be about 1/30 that of the same amorphous polymer.

In addition, crystallites in polymers are usually of such small size that the intercrystalline amorphous regions through which diffusion occurs are themselves of near-molecular dimensions. As a consequence, diffusing molecules, if they are large enough, frequently encounter intercrystalline passages which are too narrow to accommodate them. This means that the fraction of the amorphous phase accessible for diffusion of a given penetrant *decreases* as the molecular size of the penetrant *increases*—that is, the crystalline polymer functions as a much more discriminating "molecular sieve" than the amorphous polymer. Thus, a highly crystalline polyethylene membrane which is 1/30 as permeable to a small molecule such as helium as is an amorphous polyethylene membrane, will be less than 1/1000 as permeable to methane as is the amorphous polymer. Satisfactory correlations have been developed (2,3,8,12) to predict the influence of crystallinity upon solubility and diffusivity of noncondensible gases in various polymers. For example, the relationships between "diffu-

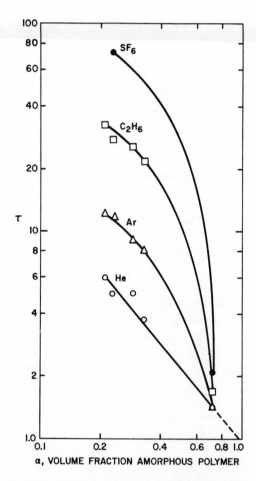

Fig. 10. Diffusion impedances ($\tau = D^*/D$) of gases in polyethylene as a function of amorphous-phase content (α) and gas molecular diameter (12).

sional impedance" degree of crystallinity and gas molecule size for polyethylene are shown in Figure 10.

Because of the influence of crystallite shape upon diffusional impedances in crystalline polymers, and since crystallite morphology is dependent upon crystallization conditions, the thermal history of a crystalline polymer membrane has a marked effect on its gas permeability. It has been found (4), for example, that a linear polyethylene membrane slowly cooled from the melt has much lower gas permeability than the same polymer quenched from the melt and subsequently annealed at a high temperature, even though both membranes have the same level of crystallinity. The differences are attributable to the presence of well-formed, thin, lamellar crystallites in the former polymer, and of thicker, defective lamellae in the latter.

When exposed to strongly solvating penetrants, crystalline polymers show even greater microstructure dependence of permeability than they do to sparingly soluble gases. The explanation lies in the fact that, while the crystalline phase of the polymer is not swollen by the penetrant, the amorphous phase does swell, but swelling is limited by the elastic deformation of chain segments which bridge between crystallites. Thus, low-crystallinity polymers swell far more in good solvents than high-crystallinity polymers, the difference being much greater than the difference of their amorphous phase contents. The greater swelling is naturally accompanied by greatly elevated diffusivities; hence, low-crystallinity polyethylene (amorphous content 50% by volume) is about ten times more permeable to liquid hydrocarbons than high-crystallinity polyethylene (amorphous content 30% by volume) as shown in Figure 11.

Restrictions on swelling imposed by crystallinity also tend to elevate membrane permselectivity arising from molecular size differences; thus, highly crystalline polymer films are more permselective with respect to isomeric compounds than less crystalline films. This can also be seen in Figure 11, where the permeabilities of o-, m-, and p-xylenes through low and high density polyethylene membranes are plotted as a function of temperature.

The extent of intercrystalline "bridging" or "crosslinking" by chain segments in crystalline polymers is very sensitive to thermal history, as is crystalline morphology. High-temperature annealing, for example, causes the polymer to recrystallize in a more relaxed state, thereby reducing the number of "elastically effective" chain tie segments; this is reflected in a significant increase in swelling after annealing and a still much greater increase in permeability (14,15). If the polymer is annealed while swollen with a penetrant, the effects are

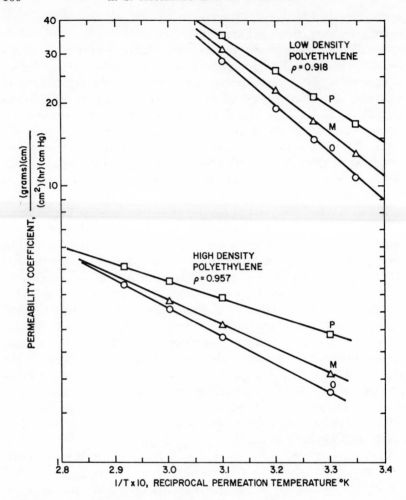

Fig. 11. Permeability coefficients of liquid xylene isomers in high- and low-density polyethylene as a function of temperature (14).

even more dramatic: large increases in sorption and diffusion result, as shown in Figure 12. Five- to tenfold increases in liquid permeation flux are not unusual consequent to such treatments, often with little loss in membrane permselectivity. Indeed, "solvent annealing" of highly crystalline polymer membranes provides a convenient means for obtaining permeation fluxes characteristic of low-crystallinity membranes, but with significantly higher permselectivities.

The effects of *mechanically induced orientation* of crystalline polymers upon their liquid permeation characteristics have only recently been investigated (10,16), and the results are dramatic and provocative. If a polymer film is uniaxially stretched well below its melting point (i.e., "cold-drawn"), the drawn film, despite negligible change in crystallinity consequent to the deformation, displays significantly less (two- to fivefold) swelling by "good" solvents than its undrawn counterpart, and its permeability to such solvents is drastically reduced—by at least two orders of magnitude. Moreover, its permse-

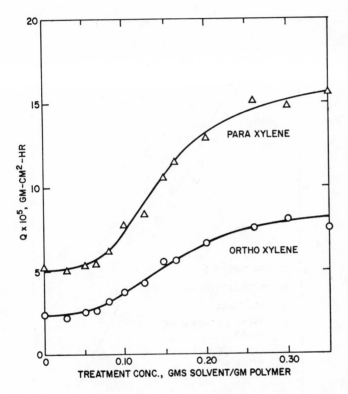

Fig. 12. Normalized permeation fluxes of *p*- and *o*-xylene through linear polyethylene (30°C) as a function of "solvent annealing" of the membrane (annealing solvent *p*-xylene; annealing temperature 97°C) (15).

lectivity for closely related liquid pairs (e.g., *p*- and *o*-xylene, benzene and cyclohexane) is dramatically increased. For benzene–cyclohexane with linear polyethylene membranes, for example, the selectivity increase in from ca. 3 to ca. 35 (10). If the stretched film is subsequently annealed (either dry or in the swollen state, but under conditions where it cannot shrink), the permeability rises to much higher levels, albeit with some loss in selectivity. It is found that, irrespective of the degree of stretch or the conditions of annealing, the membrane permselectivity is uniquely related to its "swellability"—that is, the larger the quantity of liquid hydrocarbon it can imbibe, the lower its selectivity. This is illustrated in Figure 13. Thus, by combining stretching and annealing treatments, it is possible to produce, from the same polymer, a large number of structurally different mem-

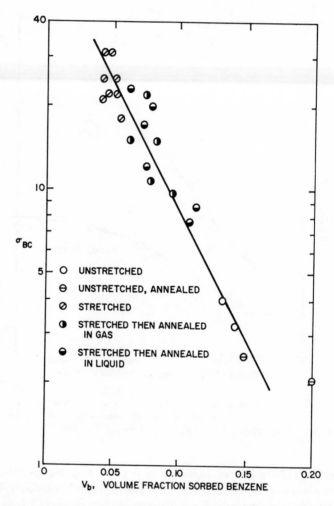

Fig. 13. Dependence of permselectivity of linear polyethylene membranes to benzene versus cyclohexane (25°C) as a function of the benzene sorptivity of the polymer (10).

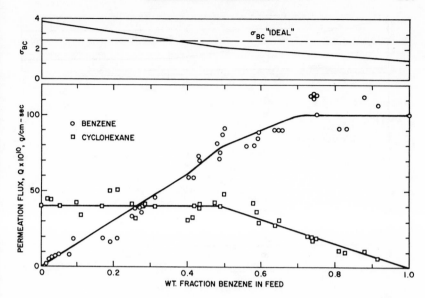

Fig. 14. Binary mixture transport rates and selectives for permeation of liquid benzene–cyclohexane mixtures through a linear polyethylene membrane at 35°C (10).

branes with widely differing permeabilities and selectivities. The implications of these observations to the "tailoring" of membranes for specific separations are self-evident.

The causes for these changes in membrane transport characteristics with orientation are not completely understood. One plausible explanation is that, on cold drawing, the crystalline structure of the polymer is fragmented, and the amorphous polymer chains are pulled taut into parallel alignment. This fragmentation and alignment greatly increase the number of intercrystalline "tie chains" and greatly decrease the mobility of individual chain segments. As a consequence, the capacity of the polymer to sorb liquid is reduced, and the difficulty which a penetrating molecule experiences in passing between polymer chains is greatly increased. Hence, the diffusivity of a penetrant is reduced, and the influence of molecular dimensions on penetrant diffusion enhanced.

Binary liquid mixture separation characteristics of unstretched and stretched-and-annealed polyethylene membranes have recently been measured for benzene–cyclohexane mixtures (10); results are summarized in Figures 14 and 15, where the individual permeation fluxes

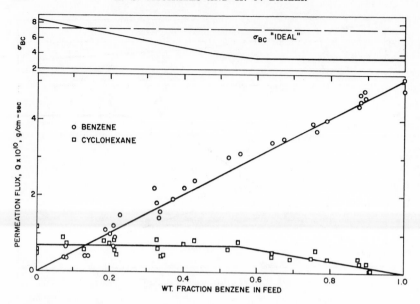

Fig. 15. Binary mixture transport rates and selectivities for permeation of liquid benzene–cyclohexane mixtures through a stretched, annealed, linear polyethylene membrane at 35°C (10).

and the permselectivities are plotted as a function of feed-liquid composition. As expected, the stretched membrane is over twice as selective for benzene relative to cyclohexane, but about 1/50 as permeable as the unstretched film. For this liquid pair, moreover, permeation is far from "ideal"—that is, the permeation flux of each component is nonlinear in the fraction of that component in the upstream liquid, and the permselectivity varies with composition. These anomalies arise from the fact that benzene, being the smaller penetrant molecule, is more effective in "plasticizing" the polymer and facilitating diffusion of both penetrants than is cyclohexane; hence, a membrane swollen with a mixture rich in benzene is less permselective than one swollen with a mixture rich in cyclohexane. By measuring the concentration dependence of the diffusion constants in the membrane for each pure component independently, however, it has been found possible to predict the binary mixture permeation properties quite accurately (10). It will be appreciated that data such as those contained in Figures 14 and 15 are all that is needed for the design of a multistage membrane-permeation process for mixture separation.

VI. Conclusions

The permeation of gases, vapors, and liquids through polymeric membranes has been found to be governed by simple molecular solution and activated diffusion in the membrane substance. Component separation during permeation occurs because of differences in solubilities or diffusivities of penetrants in the membrane.

Permeation rates of sparingly soluble gases through membranes can be predicted from gas *PVT* data and gas molecular size and shape. For gaseous or liquid mixtures which swell the membrane substance, permeation rates are extremely high, and transport kinetics are complicated by nonideal solution behavior and concentration-dependent diffusion. The permselectivity of membranes permeating swelling-penetrant mixtures is appreciable even for components differing but slightly in molecular size or shape.

Polymer crystallinity and chain orientation have profound effects upon membrane permeability and permselectivity, particularly for swelling penetrants. Thermal and solvent treatment, and mechanical deformation of crystalline polymer membranes provide simple means for altering polymer morphology and permeation characteristics. These techniques offer real promise for preparing membranes of "tailored" permeability and selectivity for specific mixture separation processes.

PART 2. ENGINEERING ASPECTS OF SEPARATION BY MEMBRANE PERMEATION

I. Introduction

In recent years, we have witnessed a rapidly growing interest in the potentialities of membrane separation as a new approach to large scale purification of fluid mixtures. In large measure, this interest has been aroused by (*1*) the high level of research and development activity engendered by the Office of Saline Water (U.S. Department of the Interior) in the field of membrane demineralization of water (mainly by reverse osmosis), and (*2*) a growing body of scientific knowledge regarding the nature of mass transport through polymeric membranes and the origins of membrane permselectivity as described in Part 1 of this Chapter. At present, the development of membrane-purification systems for the treatment of water and aqueous process streams (via ultrafiltration, reverse osmosis, and electrodialysis) is being vigorously pursued on a number of fronts, with the prediction of widespread penetration of such processes into the chemical, food, sanitary, and pharmaceutical industries within five to ten years.

Interest in the use of membrane processes for separation and purification in the petroleum and petrochemical industries has not yet gained popularity, despite the curious fact that experimental evidence that membranes useful for hydrocarbon and other petrochemical separations are readily producible has been in the public domain for many years. The primary reason for the time lag in exploitation of this very attractive separations technique appears to be a reluctance on the part of the chemical engineers to face up to and come to grips with the problem of designing devices and systems which can effectively utilize the separative properties of a membrane, and which can practically and economically carry out a separation on a large scale. As will be pointed out below, design considerations for membrane separation systems depart radically from traditional engineering design concepts; the object of this discussion is to point out some of the unique engineering requirements of membrane processes, and to indicate the means by which these requirements can be satisfied.

As has been pointed out earlier, the selection of membranes useful in effecting separation of specific hydrocarbon or petrochemical mixtures can today be made on a rational basis. The reader is also referred to other review articles on membrane permeation (17,18) for a comprehensive treatment of this topic. Suffice to say that, for separation of hydrocarbon mixtures of reasonable volatility, membranes with transmission fluxes of 0.1–0.5 lb/hr $\times$ ft^2 of membrane area are now available, which display useful selectivities under realistic operating conditions. The probability of development of membranes with permeabilities upwards of 1–3 lb/hr $\times$ ft^2 within the next few years is, in our opinion, quite high.

Membrane processes should be considered as a separations technique only for mixtures which are difficult or costly to separate by such conventional methods as distillation, absorption, absorption, extraction, or crystallization. Membrane processes should—and soon will—be competitive with more sophisticated processes currently being explored for separations such as molecular-sieve adsorption and column chromotography. Specific types of mixtures which are attractive candidates for membrane separation include:

1. Mixtures of compounds of similar chemical constitution and nearly identical boiling points (e.g., benzene–cyclohexane).

2. Mixtures of structural or position isomers (e.g., the xylenes; normal and isoparaffins).

3. Azeotropes (e.g., methanol–benzene).

4. Mixtures containing thermally unstable components (e.g., vinyl monomer mixtures).

For the separation of mixtures of organic compounds which are liquids at or near room temperature, membrane-permeation processes such as gas permeation, dialysis, and ultrafiltration are usually unattractive, due to depressingly low permeation rates, and/or poor separation efficiency. Considerably more favorable conditions for rapid permeation and high selectivity are achieved when (*1*) mixtures in the liquid phase are contacted with membranes which are solvated by the permeating species and (*2*) a large activity gradient of the permeating components is established across the membrane. There are two membrane-permeation processes which satisfy these criteria; these are pervaporation (cf. p. 144), and a process for which we have coined the term "perstraction." Perstraction, or membrane-moderated extraction, occurs when a liquid mixture is contacted with one surface of a membrane, and a second liquid phase (the primary component of which is membrane impermeable) capable of dissolving one or more components of the upstream mixture is contacted with the other membrane surface. The ensuing discussion is devoted primarily to the engineering of pervaporation and perstraction into practicable and economic separation systems.

II. Membrane Separation Cascades

Except in very rare cases, the permselectivity of a membrane toward one of a series of mixture components is seldom of sufficient magnitude to permit the production of that component in high purity in a single permeation stage. Consequently, to obtain highly purified components by membrane permeation, it is necessary to subject the permeate to a number of successive permeation stages.

The integration of membrane-separation stages into a continuous operating system designed to yield products of specified purity from a designated feed stock requires the application of cascading principles familiar to most chemical engineers. Of a mixture delivered to a given separation stage (see Fig. 16), a fraction is allowed to permeate; this fraction constitutes the permeate stream, while the unpermeated balance is the retentate stream. The permeate is fed to another "enriching" stage, which yields a permeate still richer in the more permeable component; the retentate is similarly fed to another "stripping" stage, yielding a residue poorer in the more-permeable component. The classical thermodynamic principle applies here as in other cascaded systems: streams issuing from any one

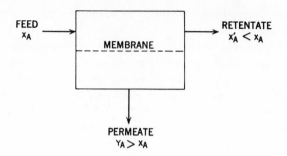

FEED
x_A

MEMBRANE

RETENTATE
$x'_A < x_A$

PERMEATE
$Y_A > x_A$

Fig. 16. Typical membrane-separation stage. If upstream liquid is completely mixed $S = Y_A(1 - Y_A)/X'_A(1 - X'_A) = $ constant.

stage should, for maximum efficiency, be fed into streams of nearly identical composition elsewhere in the system. The resulting membrane cascade resembles in many respects a fractionating column in its flow-line pattern, with one important difference. Since the quantity of fluid to be permeated per stage declines as the mixture becomes richer or poorer in a key component, and since the more permeable component penetrates the membrane faster, stages near the product ends of the cascade can be made smaller than those in the middle. It is thus possible to taper cascades and reduce membrane area and capital equipment requirements (see Fig. 17). Also, one can employ different membranes with differing permeabilities and selectivities in selected sections of the cascade to minimize the number of stages or maximize product purity. These variations indicate the far greater latitude that the engineer has at his disposal for membrane-process optimization, compared with that for distillation.

The mathematical treatment and optimization of membrane-separation cascades is essentially identical to that developed for isotope separation by Benedict and Pigford (19). The reader is referred to this quite thorough presentation for further details. The key parameter required to perform the cascade calculations is the "enrichment factor" or permselectivity of the membrane; analogous to relative volatility, this is the quotient of the mole ratio of the two key components A and B in the permeate to that of the same components in the feed:

$$S = \frac{Y_A/Y_{B\text{(permeate)}}}{X_A/X_{B\text{(feed)}}} \tag{32}$$

For membrane-separation processes involving mixtures of chemically similar species, this permselectivity (under conditions of constant temperature and large driving potential) is essentially constant and independent of feed composition. Thus, a combination of a material balance with eq. (32), yields the basic relations between feed composition, head- or permeate composition, tail- or retentate composition, and the permeate recovery ratio (i.e., fraction of the feed permeated). As in fractional distillation, separation efficiency in a membrane stage is maximized when the permeate recovery ratio is zero (total reflux) and goes to zero (i.e., $Y_A/X_A \rightarrow 1.0$) as the recovery ratio approaches unity.

Cascade calculations make it possible to determine the number of stages required to effect a given separation as a function of the re-

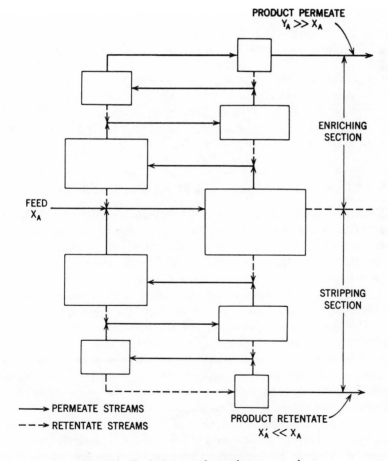

Fig. 17. Typical tapered membrane cascade.

covery ratio. (There is, incidentally, no constraint on the recovery ratio, which can, if desired, be varied from stage to stage.) To optimize the cascade, a second parameter is required—namely, the membrane permeability (flux/unit area). The permeability determines the membrane area requirement per stage for a given production rate. Since the number of stages is relatively insensitive to the recovery ratio when the recovery ratio is small, whereas the area requirement per stage is roughly inversely proportional to recovery ratio, the total area requirement for the cascade drops rapidly as the recovery ratio is increased from some small value, but begins to increase in proportion to the number of stages as the recovery ratio approaches unity. There is, therefore an optimum recovery ratio corresponding to minimum total membrane area. This recovery ratio need not—and usually does not—correspond to the economic optimum cascade design, primarily because the energy requirements for the process do not vary directly with total membrane area. The energy required to drive a unit of material across the membrane is roughly constant for a given operating temperature and pressure; hence the total power requirement for operation of the cascade is proportional to the total permeate traffic in all stages. This is equal to the product of the average permeate rate per stage and the number of stages, and usually is minimized at a recovery ratio different from that corresponding to minimum membrane area.

III. Basic Considerations in Membrane Module Design

Cascade optimization, while an important aspect of the design of a membrane-separation system, is no more than idle mathematical exercise unless certain basic engineering questions relating to events occurring within the *membrane separation unit* can be rationally answered. Such questions include the following:

1. How does one bring liquid in contact with a membrane uniformly, and efficiently remove permeate from the other side?

2. How does one efficiently supply the necessary driving potential for permeation within the unit?

3. How does one support the (necessarily very thin and quite fragile) membrane material within the device so it can withstand mechanical stresses, and minimize or eliminate the undesirable consequences of pinholes or related membrane defects?

4. How can one design and build a membrane-containing module which can either permit simple and inexpensive replacement of membrane, or is itself low enough in cost to be discarded when it has outlived its usefulness?

Answers to these questions must be sought by exploring design concepts which are far from traditional engineering approaches to mass-transfer devices. The problem is put into sharper perspective when one realizes that a square foot of typical permeation membrane one mil in thickness will permeate of the order of only 0.3 gallon of liquid hydrocarbon per day under favorable operating conditions, and that the average useful life-time of such a membrane is of the order of six months. This requires that membrane devices be designed to contain a large membrane area in a small volume (perhaps 200 ft² per cubic foot of total volume), and that costs of membrane and its replacement be kept to very low levels—of the order of one dollar per square foot or less.

A membrane module containing 200 ft² of membrane area per cubic foot of volume, irrespective of its particular geometry, will require a means for spacing membrane with an average separation between membrane surfaces of the order of 0.050 in., and it is through spaces of these dimensions that liquid and/or vapor must flow. Since the membrane will require some kind of physical support on at least one of its surfaces, the actual dimensions of the fluid channels within the unit will be even smaller. Furthermore, the transpiration rate through a typical membrane of ca. 0.1 lb/hr-ft² corresponds to a volumetric throughput of ca. 0.8 cc/min-ft², meaning that the flow rate of liquid over the upstream surface of the membrane will seldom exceed 1–2 cc/min-ft². As a consequence, liquid flow within a membrane separation device will invariably be laminar and confined to a large number of small, parallel channels. To minimize pressure losses, it will also be necessary to keep the liquid flow paths very short.

These requirements can be satisfied by several modular designs, four of which are shown in Fig. 18. The first two designs (one of which requires that the membrane can be fabricated into the form of hollow, thin-walled tubes) allow for controlled supply of feed and removal of retentate, but do not provide for external supply of fluid to the permeate side of the membrane. These designs are thus useful only for permeation processes in which the permeate can be directly collected as a gas or liquid. The latter two designs provide for delivery to and removal of fluid at both sides of the membrane, and are thus useful in processes where a sweep gas or liquid is used to remove permeate from the unit. The membrane, in all cases but the first, is supported by grooved or porous sheet material. The selection of support materials is governed primarily by the hydrostatic stresses to which the membrane is to be subjected. If a large pressure

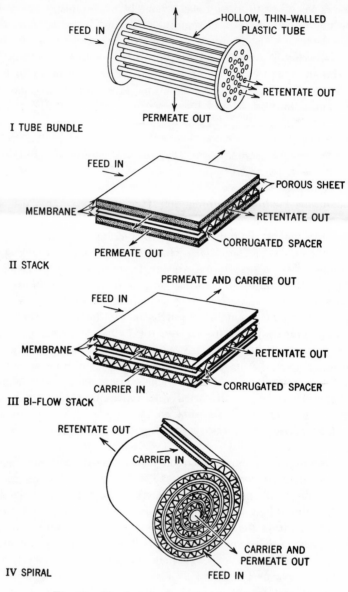

Fig. 18. Membrane-permeation module designs.

difference must be sustained by the membrane (as in gas permeation), the support material must be an extremely small-pore material, such as a felt or woven fabric. Since hydraulic flow resistances of such materials are high, path lengths for fluid flow in the plane of the sheet must be kept short, or these support sheets must in turn be laid down upon coarser-textured materials which provide larger passages for fluid movement. If pressure differences across the membrane are small (as in liquid permeation or dialysis), the membranes may be supported upon grooved or corrugated spacer sheets, wherein the fluid flow channels are large enough to offer relatively little hydraulic resistance.

The dimensions of the channels which carry fluids past the membrane surface must be held as small as possible (consistent with reasonable pressure losses) for three important reasons:

1. The smaller the fluid channels, the larger the membrane area per unit volume of module, and the higher the throughput capacity of the device.

2. It is important, for separation efficiency, that each element of fluid entering the membrane unit have about the same residence time of membrane contact as any other element. This can be accomplished (in laminar flow) by confining the fluids to flow through a large number of small, parallel flow channels of essentially equal hydraulic resistance.

3. During membrane permeation, the components which are retained by the membrane accumulate on its upstream face and must counterdiffuse through the upstream fluid; similarly, the more rapidly permeating components tend to accumulate at the downstream face and must diffuse into the downstream fluid. Thus, during steady-state transport, the upstream and downstream membrane surfaces are exposed to fluids which are, respectively, poorer and richer in the more rapidly permeating components than are the bulk fluid streams passing over the membrane. This "concentration polarization" tends to reduce both the effective permselectivity and permeability of the membrane. Under laminar flow conditions, it can be shown that the loss in efficiency is exponentially dependent upon the thickness of the liquid layers in contact with the membrane, and upon the transmission rate through the membrane. To a first approximation, this loss in efficiency can be minimized if the fluid film thickness is kept below about one-tenth of the product of the membrane thickness and the ratio of the diffusivity of the primary permeate in the fluid phase to that of the permeate in the membrane. Thus, for a one-mil membrane, where (typically) the diffusivity of a permeant is of the order of 10^{-7} cm^2/sec in the membrane and 10^{-5} cm^2/sec in a liquid mixture, the liquid film thickness should be of the order of 10 mils or less. With gas permeation, however, diffusivities in the gas phase are so high relative to those in the membrane that concentration polarization is never of any consequence; hence, channel dimensions for gaseous streams do not influence separation efficiency.

In most other respects, the design and materials-selection criteria for membrane-separation devices are similar to those employed in standard chemical process equipment design. It should be noted, however, that because of the limited useful lifetime of membranes, provision must be made either for (1) rapid and inexpensive disassembly of membrane modules for membrane replacement, or (2) design of integrated membrane/spacer modules which can be easily replaced and are inexpensive enough to discard after use. Because of the high costs of labor, the latter approach seems to be the more economic; in this case, the low-cost, readily formable materials such as paper and plastic must be used for module construction, and the modules must be capable of automated, high-speed machine manufacture.

IV. The Pervaporative Membrane Stage

In pervaporation, a liquid mixture is contacted with the membrane, a fraction of which is allowed to diffuse through and evaporate from the downstream surface of the membrane. Useful separation of the mixture components, and practical permeation fluxes, can be realized only if the partial pressure of the permeating components in the downstream gas phase is small relative to the upstream liquid vapor pressure.

The most obvious pervaporation mode, which involves delivering liquid to the membrane at essentially atmospheric pressure and removing permeate vapor under vacuum, poses very serious operational problems for which there is no obvious solution:

1. Permeate vapor delivered at low absolute pressure will naturally have a low mass density, leading to extremely high volumetric flow rates for quite small mass flow rates. The attendant large pressure drops in the (necessarily small) permeate channels of the membrane module make it difficult to sustain the low pressures near the membrane surface required for efficient permeation. Moreover, the collected low-density permeate vapor must be compressed and condensed to provide feed liquid for the next separation stage. Capital equipment and operating costs of high volumetric capacity vapor compression pumps—one per membrane stage—make this mode of operation virtually impractical.

2. The pervaporative process involves the volatilization of liquid, and thus supply of the necessary enthalpy of vaporization to the membrane–permeate vapor interface where the phase change occurs. This considerable energy requirement must be supplied conductively or convectively to all parts of the membrane module. Since it is nearly impossible to supply the necessary energy as sensible heat in the feed liquid, it can only be supplied by supplemental heating means. This would require that the membrane module be fabricated from high thermal-conductivity materials (i.e., metals such as copper or aluminum), which could be heated electrically or by contact with suitable heat-transfer media. The cost of such a module would, in all likelihood, be prohibitive.

Both of these problems can be effectively circumvented by employing a condensible vapor such as steam as a sweep gas on the permeate side of the membrane. As saturated steam (at a temperature below the boiling point of the upstream liquid mixture) passes over the membrane surface, partial condensation occurs, supplying directly the energy for vaporization of the hydrocarbon, which mixes with the uncondensed steam. As long as the mass ratio of steam fed to permeate delivered is large enough to keep the partial pressure of hydrocarbon in the steam low relative to the liquid-mixture vapor pressure, efficient enrichment and rapid permeation will take place.

The permeate stream delivered from each membrane stage will consist of a hydrocarbon–water vapor mixture with entrained condensate. This stream is then fed to a total condenser and a separator; the liquid hydrocarbon phase is then delivered as feed to the next stage in the cascade, while the aqueous phase is reevaporated and returned to the same stage. A flow sheet for such a stage is shown in Figure 19.

By this means, mechanical compression of permeate vapor is eliminated, as is the necessity to supply evaporative energy directly to the membrane module. By confining all primary heat-transfer operations to equipment external to the membrane module, standard boiler–condenser units can be employed, and membrane-module design greatly simplified.

Despite the plant equipment economies which can be realized by steam-sweep pervaporation, the energy requirements for this type of system are inordinately high. By way of example, the maximum concentration of hydrocarbon allowable in the steam-hydrocarbon vapor permeate will typically be about 2.5% by volume (at atmospheric pressure), or about 10% by weight. Evaporation of 1 lb of hydrocarbon will require of the order of one-half pound of steam.

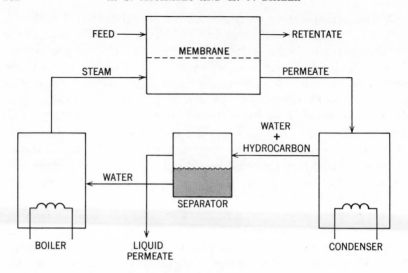

Fig. 19. "Steam sweep" pervaporative stage.

Hence, the boiler and condenser duty *per stage* will be about 10–11 lb of steam (ca. 10,000 Btu) per pound of hydrocarbon permeate produced. At one dollar per million Btu, this corresponds to an energy cost of ca. one cent per pound of hydrocarbon *per stage*. For a pervaporation plant requiring more than a few cascaded stages, it will thus be necessary to gang the boilers and condensers in multiple-effect-evaporator fashion, as shown in Figure 20. This will require that a series of successive membrane stages be operated at gradually decreasing temperatures and pressures, in order to allow an adequate temperature difference between permeate vapor from one stage and recycle water from the next. Triple- or quadruple-effect evaporation would thus bring steam costs down to practicable levels for a 10–15 stage cascade.

V. The "Perstractive" Membrane Stage

Yet more impressive economies can be realized in hydrocarbon separation by membrane permeation if the phase transformation within the membrane module can be eliminated. This can be accomplished by absorbing permeate into a liquid phase from which the dissolved permeate can subsequently be separated by stripping or decantation. The absorbing phase must meet certain stringent requirements if it is to be satisfactory for this purpose: (*1*) it must itself be essentially incapable of permeation through the separation membrane; (2) it should be of substantially lower volatility than the permeating components; and (3) it must be capable of dissolving a substantial quantity of the permeate. Such a liquid can be passed over the permeate side of a membrane, whereupon permeate will dissolve in it. After removal from the membrane unit, it can be heated to volatilize the dissolved permeate and then cooled and recycled to the unit. The vaporized permeate can be condensed and delivered as feed to the subsequent membrane stage.

Several attractive possibilities exist for extractants meeting these requirements:

1. A relatively high-boiling, high molecular-weight hydrocarbon oil. As a rule, the transport rate of such oils through hydrocarbon-permeable membranes is so low that the extent of contamination of the mixture to be separated by counterpermeation will be negligible.

2. A concentrated aqueous solution of an active hydrotrope such as sodium toluene sulfonate, or sodium p-cymene sulfonate. These solutions have the capacity to solubilize significant concentrations

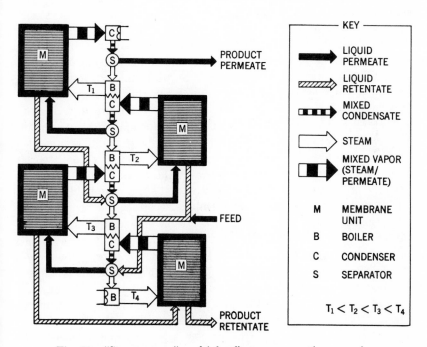

Fig. 20. "Steam sweep" multiple-effect pervaporative cascade.

(ca. 10% by weight) of hydrocarbons, yet have quite low water vapor pressures. In addition, they also exhibit a very high positive temperature coefficient of hydrocarbon solubility, so that when saturated hot and cooled, phase separation occurs. This property makes it possible to recover permeate without vaporization–condensation at any step in the process.

3. A concentrated aqueous solution of soap or micellizing surfactant. Such solutions display high hydrocarbon solubilities and *negative* solubility–temperature coefficients; thus, heating hydrocarbon solutions of this type is usually accompanied by phase separation. These solutions must be carefully formulated to avoid hydrocarbon emulsification, which would make decantation of the separating hydrocarbon phase impossible.

4. Concentrated, stable polymer latices, in which the dispersed polymer phase is readily plasticized by the hydrocarbon permeate. Hydrocarbon can be recovered from such dispersions by stream stripping and condensation; this type of extractant, because of its high hydrocarbon sorptivity (perhaps as high as 50% by weight), is particularly attractive for use with hydrocarbons of high volatility relative to water.

It should be evident that, in perstractive membrane permeation, it is unnecessary to strip the extractant completely free of permeate, since the extractant is returned to the membrane module in a closed loop. This makes it possible merely to "top" the extractant of permeate on each cycle and thereby limit the energy requirement per unit of permeate recovered. While this may require a rather high extractant traffic through the membrane unit (e.g., 20 lb per pound of permeate recovered), since this involves only liquid recirculation through the unit, the operating costs will be low. A flow diagram of a perstractive membrane cascade is shown in Figure 21.

In the opinion of the authors, perstraction appears to hold the greatest promise for adaptation of membrane permeation to large-scale hydrocarbon and petrochemical purifications and separations. Its advantages stem from (*1*) the potentialities for greatly reduced heat and power costs, (*2*) the ability to conduct the permeation process at near-atmospheric pressure without imposing significant stresses on the membrane, and (*3*) the lack of necessity for handling and transporting gases or vapors within the membrane modules. A membrane device capable of operating under these conditions has already been developed (20), and its utilization for large-scale hydrocarbon purification with commercially available membranes is now under evaluation.

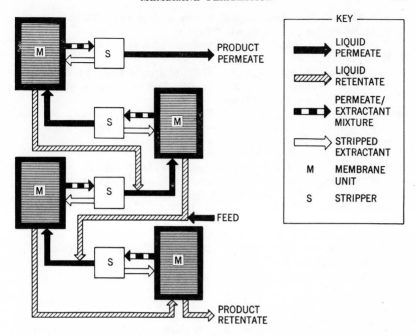

Fig. 21. "Perstractive" membrane separation cascade.

VI. Conclusions

Imaginative application of sound engineering principles to membrane permeation has made it evident that operative and economic large scale processes for hydrocarbon and petrochemical purifications can be developed around this unique separation technique. Membranes with the proper permeability and permselectivity for specific separations can now be selected and manufactured on a rational basis. Efficient membrane-containing devices of inexpensive construction have been developed, as have the principles of cascading of membrane stages for process optimization. Membrane-moderated extraction ("perstraction") promises to be the most economical system for large scale use of membranes. It is expected that the next decade will see membrane permeation emerging as a new and commercially significant unit process in the chemical and petroleum industries.

References

1. J. E. Jolley and J. H. Hildebrand, *J. Am. Chem. Soc.*, **450**, 1050 (1958).
2. A. S. Michaels and H. J. Bixler, *J. Polymer Sci.*, **50**, 393 (1961).
3. A. Michaels, W. R. Vieth, and J. A Barrie, *J. Appl. Phys.*, **34**, 1 (1963).

4. A. S. Michaels, H. J. Bixler, and H. L. Fein, *J. Appl. Phys.*, **35**, 3165 (1964).
5. R. M. Barrer, *J. Phys. Chem.*, **61**, 178 (1951).
6. P. Meares, *Trans. Faraday Soc.*, **53**, 10 (1959); **54**, 40 (1958).
7. W. W. Brandt, *J. Phys. Chem.*, **63**, 1080 (1959).
8. A. S. Michaels, W. R. Vieth, and J. A. Barrie, *J. Appl. Phys.*, **34**, 13 (1963).
9. H. Burrell, "Solubility Parameters for Film Formers," *Offic. Dig., Federation Soc. Paint Technol.*, 726–728 (1955).
10. W. B. Krewinghaus, "Solution and Transport of Organic Liquids and Vapors in Structurally Modified Polyethylene," Sc.D. Thesis, Chem. Eng., Massachusetts Institute of Technology, Cambridge, April 1966.
11. J. M. Stuckey, *U.S. Pat.* No. 3,043,891 (1962).
12. A. S. Michaels and H. J. Bixler, *J. Polymer Sci.*, **50**, 413 (1961).
13. S. W. Lasoski, Jr., and W. H. Cobbs, Jr., *J. Polymer Sci.*, **36**, 21 (1959).
14. A. S. Michaels, R. F. Baddour, H. J. Bixler, and C. Y. Choo, *Ind. Eng. Chem. Process Design Develop.*, **1**, 14 (1962).
15. R. F. Baddour, A. S. Michaels, H. J. Bixler, R. P. deFilippi, and J. A. Barrie, *J. Appl. Polymer Sci.*, **8**, 897 (1964).
16. H. J. Bixler and A. S. Michaels, "Effects of Uniaxial Orientation on the Liquid Permeability and Permselectivities of Polyolefins," National Meeting, American Institute of Chemical Engineers, 53rd, Pittsburgh, 1964, Preprint 32d.
17. N. N. Li, R. B. Long, and E. J. Henley, *Ind. Eng. Chem.*, **57**, 18–29 (1965).
18. R. N. Rickles, *Ind. Eng. Chem.*, **58**, 18–35 (1966).
19. M. Benedict and T. H. Pigford, *Nuclear Chemical Engineering*, McGraw-Hill, New York, 1957.
20. A. S. Michaels, *U.S. Pat.*, 3,173,867 (Sept. 28, 1962); Re. 26097 (Oct. 11, 1966).

Separation and Purification of Plasma Proteins: Analytical and Preparative Separation, Purification, and Concentration Methods

CAREL JAN VAN OSS

Director, Serum and Plasma Departments
Milwaukee Blood Center, Inc., and
Associate Professor of Biology
Marquette University
Milwaukee, Wisconsin

Plasma is the protein solution which remains after all the red and white cells have been removed from whole blood. When the main clotting protein of plasma, fibrinogen, has been removed (generally by transforming it into insoluble fibrin) the remaining clear protein solution is serum.

I. SEPARATION ACCORDING TO CHARGE

One of the main characteristics of the vertebrate serum proteins is their electric charge, which varies from protein to protein. At physiological pH, which is quite close to neutrality (approximately 7.2–7.5 in most cases), all serum proteins are from quite strongly to rather faintly negatively charged, albumin (with the exception of some "prealbumins") having the highest charge, and the various globulins decreasingly lower charges.

A. Electrophoresis

The electrophoretic mobility (u) (per unit potential gradient of the electric field) of all substances is, on the whole, independent of their size, and is only proportional to their electrokinetic potential ζ at the slipping plane, the dielectric constant ϵ of the medium, and the fluidity $1/\eta$ (the reverse of viscosity) of the liquid:

$$u = \zeta\epsilon/k\pi\eta \tag{1}$$

The factor k in the denominator depends on the shape of the migrating molecules and on the thickness of the diffuse electrical double layer as compared to the size of the migrating particles or molecules (1). For molecules surrounded by a double layer with a thickness of less than 1% of their diameter, $k = 4$. But when, as is sometimes the case with protein molecules in solutions of low ionic strength, the thickness of the double layer surrounding them is of the same order of magnitude as their radius, $k = 6$. For very long cylindrical molecules with a thick double layer, $k = 4$ when they move parallel to the electric field, and $k = 8$ when they move perpendicular to it; $k = 6$ is probably the best value when they are randomly oriented, although in that case a wider spectrum of electrophoretic mobilities should be expected than with more spherical molecules. When the thickness of the double layer is of the same order of magnitude as the dimension of the particles or molecules, a further correction factor, due to lagging behind of the double layer, which tends to slow down the molecules and is called the "relaxation effect," has also to be taken into account (2). This relaxation effect does not arise with very thin and firm double layers (at high ionic strengths), nor with very extended but rarified double layers (at very low ionic strengths), nor does it occur at a ζ potential below 25 mV, because in that case the electrophoretic mobility is too slow to deform the double layer to any significant extent.

1. MOVING BOUNDARY ELECTROPHORESIS

Moving boundary electrophoresis is the transport of charged molecules, dissolved in a buffer, in a U-tube, in contact in both branches with overlying buffer connected to electrodes which furnish a constant voltage gradient.

The Tiselius U-tube (3), although quite complicated because of optical as well as thermal and mechanical requirements (it should be made of optically plane glass or quartz, and it should be capable of forming sharp boundaries through sliding together of top and bottom parts), has not materially changed in the last 25 years. The U-tube, or cell, is normally immersed in a water bath, kept at approximately 1°C, so that the temperature inside the cell remains at 4°C, the temperature of maximum density of water, at which small temperature variations create the least convexion. As the proteins in the U-tube owe their continued stability to the density gradient created by the redistribution of the proteins due to the electrophoretic transport (4) (see also Sect. I-A-4-b), there is a minimum protein concentration ($\approx$0.2%) below which no stable electrophoretic patterns can be obtained. The optics most generally used for visualizing the protein concentration gradients caused by the electric field are schlieren optics (5). These show, as a continuous curve, the change in protein concentration versus migrated distance, which, in these overlapping concentration gradients, nevertheless makes the various fractions appear as distinct Gaussian peaks (see also Sect. II-A-1). Most Tiselius electrophoresis apparatus are also equipped with Rayleigh interference optics (5), which show, in a series of parallel vertical interference bands, the relation of total protein concentration to migrated distance. Although these patterns are not so readily interpreted at first sight, they have their use, when combined with the schlieren pattern, in determining the relative protein concentration of each peak.

For analytical purposes, moving boundary electrophoresis of serum has no longer any particular advantage over zone electrophoresis (see Sect. I-A-2), which explains why these Tiselius machines (costing from 6500 dollars to well over 22,000 dollars) are fairly rare. In the laboratories that do have one, the apparatus is perhaps as much used for measuring diffusion coefficients, for which the Tiselius cell, the constant temperature bath, and the optics are ideally suited. Indeed, for this latter application there exists no better type of apparatus.

In 1937 Tiselius (6) applied moving boundary electrophoresis to serum and found that about 50% was rather strongly charged albumin, and the other 50% was composed of three distinct fractions of diminishing electrophoretic mobility, called, in order of decreasing negative charge, α, β, and γ globulins. For subdivision of these groups see Section I-A-3.

Apart from the diffusion applications, for which the apparatus was *not* originally created, moving boundary electrophoresis serves two other purposes:

1. To obtain small amounts of the fastest, as well as of the slowest migrating protein from a mixture (with serum albumin and γ globulin), in a rather pure state, by removing them with syringes, at the end of a run. The needle can be made visible, and superimposed on the schlieren image so that the syringe can be exactly placed in the optimal position. The fractions of intermediate electrophoretic mobilities always remain, in a U-tube, mixed with one another or with a fraction at one of the extremities, and thus can never be obtained in a pure state by this method.

2. For studying the *interaction* between two proteins, e.g., the interaction between antigen and antibody molecules during the electrophoresis run itself (7). Interactions generally are visualized by irregularities in the schlieren pattern at the place of encounter of the reagents. The electrophoretic mobility of the reaction product can also be determined and compared to the mobilities of the reagents.

Apparatus

In the United States the available moving boundary electrophoresis apparatus in the order of increasing cost are: Perkin-Elmer Model **238**, with Polaroid camera back; American Instrument "Portable" Electrophoresis, a compact machine in which the fairly long optical light path is tucked away by the aid of a multitude of mirrors; Beckman–Spinco Model H, a most accurate but expensive machine. In Europe the principal available apparatus are made by Hilger & Watts, London, England (contains many mirrors); Phywe, Göttingen, Germany (contains many mirrors); and Strubin, Basel, Switzerland. The Fokal instruments of Strubin's have full length optical benches with adjustable optics. They occupy, at least in one dimension, more space than most other apparatus, but this, particularly for research purposes, is rarely an insurmountable obstacle.

2. ANALYTICAL ZONE ELECTROPHORESIS

Paper electrophoresis [first practiced by König in **1937** (8) and first used for protein separation by König and von Klobusitzky in **1939** (9)] and the related electrophoresis on cellulose acetate strips (10) and on open-pored agar gels (see particularly ref. 11), have made the use of protein electrophoresis feasible and available to a myriad of laboratories (see the introduction to ref. **12**, part I). With this method, a narrow band of protein solution (e.g., whole serum) is deposited on one side of a support strip (which is soaked in buffer) and migrates in a constant voltage gradient applied through the two extremities of the strip, which separates the proteins into bands of different electrophoretic mobilities.

With paper electrophoresis, a not inconsiderable adsorption of protein by the paper generally occurs (**12**). This renders a complete separation of the various fractions difficult, as the slower-moving serum fractions migrate through the paper along a train of adsorbed albumin.

The use of cellulose acetate strips instead of filter paper has the advantage that there is virtually no adsorption. A second advantage is that cellulose acetate can be made completely transparent with paraffin oil or with treatment in 10–15% acetic acid in ethanol, except for the stained parts, after the electrophoresis run and the staining of the proteins. Thus optical density measurements of such stained and cleared strips, when plotted against distance of migration, will furnish serum electropherograms that are practically indistinguishable from schlieren curves obtained with a Tiselius apparatus. Some authors (11) prefer using an open-pored gel-like agar or agarose (at a concentration of ≈1%) for zone electrophoresis. Such gels, upon completion of the electrophoretic run, can be dried and stained and have then much the same aspect as stained and cleared cellulose acetate electropherograms. Agar gels, being rather strongly negatively charged, show a strong electroosmotic transport toward the cathode. Thus the starting sample should not be deposited too far from the middle of the gel plate. With the electrically practically neutral agarose gels, this problem does not arise. With the gel method, the sample is deposited by pipetting a measured amount of protein solution in a hole on the gel. Agar, being a polysaccharide sulfate, interacts more or less strongly with many proteins and particularly with lipoproteins, so that some proteins will not migrate as far in it as on paper strips (13). Agarose does not have this drawback.

a. Staining

Generally, the supporting medium (paper, cellulose acetate, or an open-pored gel, as agar or agarose) is "fixed" after the electrophoresis run, in ethanol, methanol, or acid, or by heating, to make the proteins insoluble. The strip is then stained with a protein stain, like bromophenol blue, Ponceau-S red, or amido black (12). Cellulose acetate strips are preferably stained with a Ponceau-S solution in 5% trichloracetic acid, thus combining the fixing and staining operations into one step. After a few washings with 4% acetic acid containing tap water, the strips can be kept and stored, or photographed, or scanned with a spectrophotometer. When amido black is used, the strips are washed for several hours in 10% acetic acid or methanol. When bromophenol blue (which tends to fade) has been used, treatment with ammonia fumes can largely restore the original sharpness and color. Ponceau-S also fades rather quickly. The pattern can at least partly be restored by rewetting the strips with 4% acetic acid containing tap water. When Ponceau-S is used on cellulose acetate strips, clearing with 10–15% acetic acid containing ethanol, which makes the strips transparent, also prevents further fading. Agar gels (see also Sect. I-A-3) should be dried while covered with a piece of wet filter paper, prior to the staining operation. Due to interactions between the strongly negatively charged agar and various proteins, the staining intensity, unless many precautions are taken, is only slightly proportional to the protein concentration (14). With agarose, that problem does not arise to any serious extent.

For lipoprotein staining, sudan black is used most. The strips are generally washed with 50% ethanol–water mixtures. Cellulose acetate is not very suitable for lipoprotein staining. Differences between lipids and phospholipids can be demonstrated by staining one strip directly with sudan black, while a second identical strip is washed in acetone before staining. The first strip will show *all* the lipids, the second the phospholipids only.

Glycoprotein staining is classically done with PAS, periodic acid–Schiff reagent containing fuchsin. One drawback is that the stock solution of fuchsin has to be made freshly quite frequently. When an isomer of fuchsin, pararosaniline, is used, this is no longer necessary (15). With the latter stain, stock solutions keep at least 6 months when conserved in the dark and refrigerated.

Apart from scanning the stained strips (see Sect. I-A-2-b), the stained spots can also be eluted and the intensity of the stain measured colorimetrically. In the case that all of the stained protein will not elute, even by heating the spots in $1N$ HCl at 100°C, the amount of uneluted stained protein can be calculated by assuming that it is adsorbed by the paper and by applying Langmuir's isotherm equation (16).

b. Scanning

A more recent and automated method of zone electrophoresis in very dilute agarose gels (Bausch & Lomb's Spectrophor I) does not involve fixing, staining, and washing, but scans the strips upon completion of the electrophoresis run with ultraviolet light of 205 mμ wavelength. 205 mμ is an absorption maximum of the peptide bond, so that scanning at this wavelength will not require any correction for the different sorts of protein scanned, the absorptivity of all proteins being about equal at this wavelength. Thus, a graph is obtained which comes very close to a Tiselius pattern; however, the price for the whole apparatus is also comparable to that of a Tiselius machine. Another new and automated method, Technicon Autophoresis, uses no solid or semisolid support at all, but contrives its electrophoretic separation by streaming a protein solution at constant speed through a density gradient, which is scanned after completion of the run at a wavelength of 280 mμ, the whole operation taking 20 min. The main fundamental drawback of this method is that the absorptivity of various proteins at 280 mμ differs widely, being mainly dependent on their tryptophan and tyrosine content (17). The Technicon apparatus contains a device to correct these deviations upon integration, but at this stage it is not known how successful these corrections are, particularly when sera with abnormal proteins are investigated. These investigations are frequently the very reason for the acquisition of such apparatus. The price of this machine is also of the same order of magnitude as that of moving boundary apparatus.

A last, recent, method of analytical zone electrophoresis also uses ultraviolet scanning for showing up the electrophoretic progress of proteins. Hjertén published this method of free zone electrophoresis in a long, horizontal, rotating quartz tube (18). The tube, filled with buffer, contains a small zone (not more than 0.1 ml of 0.1% protein is required) of protein solution which is electrophoresed through the length of the tube while the tube rotates slowly (15–60 rpm). The rotation stabilizes the proteins (which have a higher density than the buffer) in the liquid by preventing them from collecting in the lower parts of the tube, a low part in the tube in a very short while having become a high part again. The electrophoretic pattern is obtained by scanning the quartz tube over its whole length with ultraviolet light. The mobilities obtained are very similar to those obtained by moving boundary electrophoresis and the apparatus, when obtainable, is also not likely to cost less than a moving boundary machine.

In most cases, when electropherograms are stained, the scanning adds little new information, and no further precision to the stained strips. It is only when information on relative quantities of various proteins is required, that scanning, when followed by integration, has its uses. It must be emphasized, however, that the available automatic curve-integrating devices give the surfaces under the peaks from one minimum to the next, and do *not* first decompose the peaks into their constituent overlapping Gaussian curves. Particularly when the electrophoretic resolution is not total (as it rarely is), this type of integration will be fairly arbitrary, and the actual amount of information it furnishes will be less than that suggested by the hard figures it produces. Still, very much the same drawbacks obtain when the method of eluting the stained spots and measuring their color intensity is used; the main difference being that the latter method, although more tedious, obviates the necessity of acquiring rather expensive scanning equipment.

Apparatus

It is not feasible to describe even partly the various zone electrophoresis apparatus, accessories, and power supplies that are put out by over 100 different firms, and without prejudice to any of the other manufacturers, only a very few items will be mentioned which have some unusual or particularly useful feature. (See also above for Technicon's Autophoresis and Bausch & Lomb's Spectrophor I.)

Heath (Benton Harbor, Michigan) manufactures a direct current power supply, Heathkit 1P-32, giving up to 400 V constant voltage, with a maximum intensity of over 100 mA (which is high enough also to be used for gel electrophoresis and block electrophoresis, see Sects. I-A-3 and III-A), for well under 100 dollars. Gelman (Ann Arbor, Michigan) manufactures an extremely simple but quite versatile electrophoresis tank, the Rapid Electrophoresis Chamber, which can be used for paper and cellulose acetate strips as well as for immunoelectrophoresis and gel electrophoresis. The strips are quite easily attached to the tank with little magnets. The chamber's platinum electrodes, however, are quite rudimentary and are easily polarized and should, before the tank is used, be replaced with, for instance, graphite electrodes; but this, considering that the price of this tank is under 50 dollars, is no particular hardship. Gelman also furnishes cellulose acetate strips, Sepraphore II and III, with which quite good results can be consistently obtained; Sepraphore III, particularly is much less brittle than most other varieties. Finally, Gelman manufactures a sample applicator consisting of two taut parallel stainless steel wires attached to a holder, which is quite an improvement over a micropipet for depositing protein solutions in thin, straight, homogeneous lines on paper or cellulose acetate strips. Beckman Instruments (Fullerton, California) manufactures an interesting apparatus for cellulose acetate electrophoresis, Microzone, with which eight samples at a time are electrophoresed on small strips, stained with Ponceau-S, cleared, scanned, and integrated in one hour. This arrangement can also be used for immunoelectrophoresis (see Sect. I-A-6.)

3. IMMUNOELECTROPHORESIS

In principle, immunoelectrophoresis is nothing but a special variety of analytical zone electrophoresis, most generally in an open-pored agar gel, and with double diffusion immune precipitation (see Sect. V-A-1) in the same gel perpendicular to the direction of electrophoretic migration, as a method for marking a multitude of proteins extremely specifically by a fine, curved arch. This method, first published by Grabar and Williams in 1953 (19), has increased the number of plasma proteins that can be distinguished on electropherograms by about one order of magnitude! This has necessitated a further division of the α, β, and γ globulins into α_1, α_2, β_1, β_2, and γ_1, γ_2 globulins (β_2 and γ_1 overlap), which then are further subdivided according to decreasing mobility into, for instances, β_2A, β_2B, β_2C globulins. The notation M is reserved for globulins of exceptionally high molecular weight (M = 1,000,000): the macroglobulins α_2M and β_2M (20).

The most-used immunoelectrophoretic method probably is now the microtechnique on glass microscope slides, as developed by Scheidegger (21). The method is sensitive enough to maintain sufficient resolution of the fractions, shown by the various precipitable arches, even on this small scale, and it has the great advantage of needing only a minimum of rare or expensive antiserum for development. With the micro method, the electrophoresis part of the operation takes anywhere from $2\frac{1}{2}$ hr at 2 V/cm to 50 min at 6 V/cm. The author has consistently obtained the best results at the lower voltage. With immunoelectrophoresis, as with ordinary agar electrophoresis (see Sect. I-A-2), electroosmosis is an important factor, making it necessary that the sample to be electrophoresed be deposited fairly close to the middle of the slide. Most pattern cutters take this phenomenon into account (see below, under Apparatus). With agarose, this phenomenon does not arise, but this, unfortunately, makes the existing pattern cutters unsuitable for the use of this gel without modification. When the electrophoresis is finished, the antiserum is put in the channel with a syringe. The development is usually done at room temperature and in a humid chamber, that is, in a closed box in which with the help of a tray of water, an atmosphere saturated with water vapor is maintained. After development, the patterns can either be photographed directly or washed for 24 hr in physiological saline (to remove the excess soluble protein) and stained with any of the protein stains (see Sect. I-A-1-a), or with appropriate substrates, for showing up enzymatic and other specific activities (22). It is important that the slides be dried while covered with a piece of wet filter paper between the washing and the staining operations (23).

Immunoelectrophoresis has been the principal means of detection of a variety of new serum proteins, of subfraction of serum proteins and of hitherto unknown properties of certain serum proteins. The technique has become indispensable in the detection of abnormal amounts of certain immune globulins in a group of diseased states called dysproteinemias.

Apparatus

A practical tray which permits the use of the entire length of the microscope slides for electrophoretic migration, accommodating eight slides, is made by Shandon (Colab, Chicago Heights, Illinois). For the micro method practically any chamber can be used or adapted (see Gelman, under Apparatus in Sect. I-A-1-b.) One drawback is that not all brands of microscope slides easily fit into the tray and that some preliminary experiments are sometimes necessary.

If the Shandon microscope slide tray is used, the Shandon pattern cutter is a very practical auxiliary tool to have although its electrophoresis well cutters are on the small side and will afford a much improved performance if replaced with slightly wider cutting tubes. With most commercial agars no special purification is necessary.

Carrier materials. A good immunoelectrophoresis agar is Oxoid Ionagar No. 2 (Colab, Chicago Heights, Illinois). This can be directly and homogeneously dissolved in the electrophoresis buffer (at a concentration of approximately 1%) under continuous stirring, upon bringing it twice to the boiling point. Another carrier material that can be used is cellulose acetate (24). When this material is used, the precipitate lines will not show up until they are stained. With some practice great resolution can be attained with cellulose acetate, and finished stained strips can often yield even more information when inspected under a microscope. Other media have also been used for immunoelectrophoresis (25).

Antisera. With most commercial antihuman serum antisera (Antibodies, Inc., Davis, California; Behringwerke-Hoechst, Cincinnati, Ohio; Hyland, Los Angeles, California) development need not take more than 16 hr (overnight).

If it proves necessary or desirable to make one's own antisera, Kabat and Mayer's description of their preparation is quite useful (26).

4. BLOCK AND COLUMN ELECTROPHORESIS (27,28)

Electrophoresis in horizontal thick slabs of buffer soaked into potato starch or in vertical columns of some such medium, has for 15 years been the simplest method of accomplishing electrophoretic separations on a preparative scale. By this method, batches of 5–50 ml of serum (and more with cooled columns can be separated in 10–20 hr into its 6–8 main fractions.

a. Block Electrophoresis

This is by far the simplest preparative method, which nevertheless permits a quite fair separation of appreciable quantities of the various serum fractions in one step. For many purposes the separation obtainable with this method may already be sufficient. For practically all preparative purposes at the scale of 5–50 ml initial protein solutions, this method is probably the best starting point.

A simple but practical version of the method is the following:

A rectangular plastic tray (6–8 × 10–14 in. and ½–1 in. high) is filled with a slurry of the sedimented part of a suspension of the granular medium in the buffer used. With a knife or spatula a wedge-shaped slit is made in the block, in which the protein solution is slowly poured. When the solution is completely absorbed, the wedge is carefully pressed close again. The ends of the tray are then provided with wicks made out of several layers of thick filter paper or cotton towel cloth. The top surface of the block is closed off against evaporation by covering it with a thin plastic hydrophobic membrane (Saran wrap will do this quite well). The wicks dip into separate plastic boxes on either side of the tray which are filled with buffer and which contain the electrodes (graphite electrodes are quite satisfactory). The whole operation takes place best in a well-ventilated cold room, at 2–6°C. With no other cooling than this, it is generally advisable not to apply much more than 20 W to the block. This can be attained by making the block not too thick. Three to 10 V/cm, over 10–20 hr, is generally required to get good separations. The ionic strength of the buffer and the thickness of the block must in other words be such, for a 12-in. long and 6-in. wide block, at a total voltage of 90–300 V between the extremities of the block, that the current will not surpass 220 and 33 mA, respectively.

After the run, the block can be cut up and the fractions extracted or eluted from the various slices. In order to determine exactly where the fractions are situated in the block, it is desirable either to put a strip of filter paper on top of the block before starting, or to press such a strip on the block for a short time after the run and to stain the strip for protein. It will then be easy to see which slices of the block will correspond to the desired stained zone on the paper.

The first material used in this method as a medium was granular potato starch (29), and since then poly(vinyl chloride) (30), polystyrene, and other plastic powder or plastic latex beads were used, as well as glass powder (31) or beads, sand, cellulose flocs, Sephadex (Pharmacia, New York, N.Y.) and other gel grains. Instead of powders, blocks of sponge rubber or foam plastic have also been used. In the author's experience a latex of a copolymer of styrene and divinylbenzene (Dow Chemical Co., Midland, Michigan), 200–400 mesh, has the advantage over some other powders in that the separated fractions do not have to be eluted from the slices, which always dilutes the protein, but can be extracted from the intersticial spaces between the beads by simple suction on a fine or medium-fine fritted glass Buchner funnel without the necessity of adding any further solvent.

Some materials used as medium have a rather high negative electric charge, particularly sand and glass powder, giving rise to strong electroosmotic flow toward the cathode, which necessitates placing the starting slit near the middle of the block. For any given material and for any given dimensions of tray and ionic strength of buffer, the influence of electroosmotic counter flow will be different, and the optimal place of deposition of the sample is best found experimentally. When in doubt, a place at a distance of about one-third of the length of the tray from the cathode will be a reasonable starting point.

Besides a direct current constant voltage power supply, no apparatus is needed for this method other than two plastic boxes and a plastic tray of the desired dimensions. A quite extensive list of working conditions for block electrophoresis on the various media applied to the purification of a large variety of biopolymers will be found in Table II of Bloemendal's monograph (27). Avrameas and Uriel (32) have published a method for continuous elution of the fractions obtained by block electrophoresis in an agarose gel block. Here the block is provided with a slit, out of which the fractions (once they have migrated into it) are flushed by means of a continuous stream of buffer into a fraction collector.

b. Column Electrophoresis

Although at first sight, just another variety of block electrophoresis, this method, which is capable of treating greater quantities of protein solution due to greater cooling efficiency, is also much more complicated. Hydrostatic considerations dictate that the column always be one branch of a U-tube. Enclosure of the packing material in a column renders its being cut up in slices impractical. Introduction of the sample is a relatively intricate procedure.

While varieties of the column method abound, the most successful design is probably that of Porath (33), of which two versions are commercially available through LKB instruments (Washington, D.C.), the biggest of which is capable of coping with up to 300 ml of sample. Such columns cannot only be used with all the media already mentioned above under block electrophoresis (Sect. I-A-4-a), but the smaller of them can also be used for density gradient electrophoresis with only minor adjustments. For the latter purpose sucrose–water density gradients are the most generally used ones, although water–ethanol gradients as well as salt gradients have been described (4). Svensson has given a quite thorough treatment of density gradient electrophoresis (4). Fractions are generally obtained by eluting the column after the electrophoretic run into a fraction-collecting device, although Porath's larger column is also equipped to extract the fastest zones continuously by countercurrent elution during the electrophoretic run. Bloemendal, in Table XI of his monograph (27), gives a list of working conditions for column electrophoresis and the various media (including density gradients) used in the purification of a number of serum and other proteins.

5. CONTINUOUS ELECTROPHORESIS

This method, which is a bidimensional one combining simultaneously electrophoresis in a horizontal direction and continuous flow in a carrier in a vertical, downward direction, was independently developed by Haugaard and Kroner (34), by Svensson and Brattsten (35), and by Grassmann (36). The sample is generally continuously injected into the middle of the top of the carrier. The carrier can be any of the materials mentioned under block and column electrophoresis (Sect. I-A-4), or quite conveniently, a curtain of thick filter paper (37,38), or more recently, just a flowing band of liquid buffer (39).

The difficulty of the method lies in the necessity for thoroughly controlled synchronization and stabilization of the electric and hydrodynamic currents and the influx of sample. The advantage of the method is the relatively large amount of protein solution that can be purified: in free-flowing liquid or in a glass bead carrier up to 50 ml/hr and on paper curtains up to 5 ml/hr. Another advantage is that the slower-moving proteins can migrate over a path that has not already been passed by a faster protein, thus obviating a source of contamination that is very hard to avoid with the strip method. Peeters et al. (40) used a glass plate coated on one side with a mixture of cellulose powder and starch. They also described a method of "trickle-feeding," the downward buffer flow with buffers of varying pH, thus creating a horizontal pH gradient, which in many cases apparently can give rise to improved separation of the fractions.

Apparatus

Filter paper curtain apparatus are made by the Microchemical Specialties Co. (Berkeley, California) and by Beckman Spinco Model CP (Palo Alto, California). An apparatus using siliconized glass beads (10–15 μ diameter) as a carrier is made by JKM Instrument Co. (Durham, Pennsylvania), and a free-flowing liquid apparatus is marketed by Brinkman Instruments (Westbury, New York).

6. ELECTROCONVECTION METHODS

Electroconvection or electrodecantation is a simplified electrophoresis method by which use is made of the fact that when a given protein is attracted to one electrode during electrophoresis and accumulates there, the increased concentration of that protein gives rise to a locally increased density, which causes the protein concentrate in question to sag to the bottom of the vessel from where it can be collected (41). The method was first described by Wolfgang Pauli in 1924 (42), who has since introduced many improvements (43) and, in particular, worked on large-scale continuous purification of rubber latex (44). The method is particularly suitable to fairly large-scale separations, but due to its being based on convection in the liquid phase, the sharpness of separation leaves generally much to be desired. Other workers who have contributed to improvements in this method are Cann (45), Kirkwood (46), Gutfreund (47), and Timasheff (48). Polson has, at least partially, remedied the lack of sharpness of separation inherent to this method, by introducing a system of multiple membranes and by tilting the whole apparatus at an angle from the horizontal (49,50). Isliker (51) discussed the use of electroconvection in the fractionation of immune globulins. Bier (52) modified the method by superimposing a "forced-flow" liquid movement perpendicular to the direction of the electric field in a filter press, and the arrangement thus achieves a fairly large scale continuous separation. In practice, it would seem that even with this improvement, reasonable separations can only be expected of the very fast and the very slow plasma fractions, i.e., albumin and γ globulin.

Apparatus

Polson's multimembrane electrodecanter (50) is manufactured by Quickfit and Quartz Ltd. (Reeve Angel, Clifton, New Jersey) and is in the price range of the more expensive Tiselius moving boundary machines. A rather simple and fairly cheap electroconvection apparatus (under 500 dollars) designed by Raymond (53), the actual separation compartments of which consist of cellophane tubing, is made by E-C Apparatus Corp. (Philadelphia, Pennsylvania). Bier's device is made by Canalco (Rockville, Maryland).

7. OTHER ELECTROPHORETIC METHODS

a. Electrophoretic Fractionation with a Stepwise pH Gradient

Donnelly (54) devised a large scale method for electrophoretically purifying proteins in a multicompartment apparatus. Every compartment contains an electrode and is connected to a power supply and to its own stirrer, pH meter, set-point potentiometer, and pH controller unit. The first cell contains the cathode, the last one, the anode. A preliminary electrophoresis run, with just the buffer, creates the stepwise pH gradient, which, when it has attained its desired value, is maintained by the aid of the auxiliary electrodes, automatically controlled by the pH meters (or by a traveling pH meter). After introducing a protein mixture in one of the compartments, every given protein species will in due time reach a compartment with the pH of its isoelectric point and from then on stay there. A prototype of this system is capable of handling about 4 gallons of a protein solution and furnishing about 7 g of an enriched enzyme in one week (55). The method is a batch method and not, of course, very suitable for separating two proteins whose isoelectric points are fairly close, but when further developed, it may be of use as a step in fairly large scale purifications of proteins.

b. Electrophoresis of Adsorbed Proteins

The earliest data on electrophoretic mobilities were obtained by adsorbing the proteins in question onto tiny particles such as quartz or collodion, and by measuring the particles' mobility under the microscope (56). When data obtained with the Tiselius method became available, a comparison of the two methods showed that in many cases the electrophoretic mobility of free and adsorbed protein were the same (in particular in the case of horse serum albumin and of horse serum pseudoglobulin). In other cases, however, a slight shift of mobilities was noted upon adsorption of the protein (ovalbumin). The microscopic method is easy to use, and the electrophoretic mobilities of the particles are readily measured. The difficulty of the method lies in the fact that the electroosmotic back flow along the walls of the electrophoresis cell greatly complicates the interpretation of the results. According to whether the microscope is focused on particles farther from or closer to the glass walls, different velocities are noted, which are plotted against the depth of the cell. A parabola is thus obtained, and the exact place on that parabola which corresponds to the electrophoretic mobility only, is calculated.

Another method for measuring electrophoretic mobilities of protein-coated particles was published by the author (57). The proteins are adsorbed on polystyrene latex particles, and the latex migrates in a Tiselius cell, stabilized with a sucrose gradient.

No special optical system is necessary, as the latex–buffer boundary remains very visible to the naked eye. No electroosmotic flow complicates the experiments. No cooling baths are needed: placing the cell in a cold room or refrigerator suffices. The mobilities of free and adsorbed human γ globulins were found to be the same.

Nevertheless, measuring protein mobilities in the adsorbed state will at best remain an analytical method, applied to pure proteins. Desorption of the proteins is rarely possible. Adsorption of the proteins can in some cases alter their electrophoretic mobility (see above).

II. SEPARATION ACCORDING TO SIZE

The three methods here described are based on two totally different principles. The first one (ultracentrifugation) is a sedimentation method, using artificially increased gravitational fields. The second principle uses gels in both of its applications: in the form of membranes to keep molecules larger than given size outside one of the two compartments, the membrane being subjected to pressure (ultrafiltration), the molecules' own thermal motion (dialysis), or an electric field (electrodialysis). In the second class of applications of gels (gel filtration chromatography), gel grains are packed in columns, and their pores include small molecules and exclude big ones, over a given size. The two driving forces here are hydrodynamic flow and the thermal motion of the molecules themselves.

A. Ultracentrifugation

Sedimentation is one of the simplest ways of separating particles of different weights and sizes. But plasma proteins are so small that their thermal motion prevents their sedimenting in ordinary-sized vessels under the normal gravitational acceleration (g) of the earth. It was only since ultracentrifuges were constructed that are capable of subjecting protein solutions to gravitational fields of more than $10^5 g$, that plasma proteins could be separated according to their molecular weights, and their molecular weights determined by sedimentation.

1. MOVING BOUNDARY ULTRACENTRIFUGATION

Moving boundary ultracentrifugation, or ultracentrifugation of homogeneous protein solutions in tubes or cells, was first and is still mainly used for analytical purposes. Thus the first ultracentrifuge cells were already provided with windows so that the sedimentation of the various proteins could be observed and measured, with the aid of one of the usual optical systems. The optical system that was, and still is, most used for that purpose, is the astigmatic schlieren system, as developed by Philpott and Svensson (5), because it shows the *changes* in concentration versus distance, thus transforming a system of overlapping concentration gradients into a series of separate Gaussian peaks.

Moving boundary ultracentrifugation is the ideal method for determining the molecular weight distribution of *mixtures* of proteins, such as are present in serum. The sedimentation velocity method is here the most appropriate one. With this method, which makes use of schlieren optics (5), the rate at which the individual peaks (representing the different protein fractions) move towards the bottom of the cell at a given speed of rotation can be measured and is usually expressed in Svedbergs [one Svedberg (1S) = 10^{-13} sec]. The major nonlipoproteins in normal human serum have only three different sedimentation rates: about two-thirds of approximately 4.5S ($M \approx 70,000$), consisting mainly of albumin; less than one-third of 7S globulins ($M \approx 160,000$) containing most of the immune globulins; and small amounts (only a few per cent) of 19S macroglobulins ($M \approx 1,000,000$), consisting of approximately equal amounts of α_2 macroglobulins ($\alpha_2 M$) and β_2 macroglobulins ($\beta_2 M$ or IgM), the latter again being mainly immune globulin.

For serum protein studies it is useful to spin the serum, diluted with isotonic or slightly hypertonic saline to 1.0% protein concentration, at top speed (approximately 60,000 rpm with the current Spinco E analytical ultracentrifuge), taking 2 or 3 pictures at 8-min intervals (to get pictures of the 19S peak) and then at least 2 more pictures at 16-min intervals, to be able to measure the size of the 7S peak when it separates from the albumin (58).

To determine the sedimentation rates, the peaks on the photographic plate have to be measured to 0.01 mm, with an optical comparator provided with micrometers. The S values can then be calculated by using the formula

$$s = (\ln r_2 - \ln r_1)/\omega^2(t_2 - t_1) \tag{2}$$

In other words, the sedimentation constant s is the difference between the natural logarithms of the radii of a peak (i.e., of their distance to the center of centrifugation), divided by the square of the angular velocity (ω) and the time interval. Sedimentation rates are not by themselves sufficient to calculate molecular weights, but, as they are the only molecular weight parameter that can be determined in *mixtures,* serum proteins are frequently alluded to as 7S, 19S, etc. In order to calculate the molecular weight, both the sedimentation *and* the diffusion constants of the *pure isolated* protein must be known, according to the classical Svedberg equation:

$$M = RTs/D(1 - V\rho) \tag{3}$$

where R = the gas constant, T = the absolute temperature, s = the sedimentation constant, D = the diffusion constant, V the partial specific volume of the protein, and ρ the density of the solution. For determining the diffusion constant, a Tiselius apparatus (see Sect. I-A-1) is ideal, although a separate analytical ultracentrifugation at low speed (12,000 rpm, or less) may also serve (59).

The molecular weight of pure proteins can also be obtained while both the sedimentation and the diffusion constants remain unknown by means of the sedimentation equilibrium and the approach to equilibrium techniques. Here use is made of the equation

$$M = \frac{RT \, \partial c/\partial x}{(1 - V)\omega^2 xc} \tag{4}$$

where c is the protein concentration at the point x in the cell. When complete equilibrium is reached this equation is true at all points in the cell. Archibald (60) noted that eq. (4) *always* holds true, even before equilibrium is reached, for points at the *meniscus,* as well as at the *bottom* of the cell, hence the name *approach to equilibrium* for his technique. Determinations of the protein concentrations at these points are facilitated by Ehrenberg's technique (59). Trautman (61) has devised a novel way of plotting approach to equilibrium data which not only circumvents the necessity for determining the initial protein concentration, but makes it possible to calculate that concentration even if it was not previously known. The Archibald method (60) and the Trautman plot (61) are of importance in checking the purity of protein preparations as well as in the study of association and dissociation equilibria of proteins.

Next to schlieren optics, Rayleigh interference optics and ultraviolet absorption optics are also generally available (62). Interference optics furnish a repeating pattern of lines, representing the function of concentration versus distance (5). Ultraviolet absorption patterns have to be scanned in order to give a graph of concentration versus distance (63). Both these optical systems have their uses when concentration is the main desired datum and when the protein concentrations studied are very low. Since Trautman's indication (64) of the possibility to adjust the phase-plate train to and fro, perpendicularly to the optical axis of the Spinco E ultracentrifuge, so that the schlieren image (which tends to disappear upwards when the schlieren angle is lowered) can be brought back even when very low angles are used, schlieren optics can now be used to study proteins at concentrations down to 0.01%, at schlieren phase-plate angles down to 10°. Thus sedimentation rates of proteins can be studied at such low concentrations, that ultracentrifugations at several concentrations, with extrapolation to zero concentration in order to correct for concentration dependent sedimentation constants are superfluous. Particularly for protein work, schlieren optics are and remain the most indispensable method for visualizing sedimentation phenomena.

Moving boundary ultracentrifugation has only limited uses for preparative purposes. Only the lightest protein in a mixture of proteins can be obtained in the pure state by this method. The heavier fractions, which can be found at the bottom after a run, always also contain at least the initial amount of the lighter fractions. Still, for such separations as are desirable by this method, there exist separation cells with either a fixed porous partition or with a partition that gets into its place after the run, so that fractions, once they are wholly or partly separated, will not get remixed upon deceleration. The heaviest of the proteins present in a mixture (in serum, the macroglobulins) can be at least partly purified by spinning them very rapidly for a long time in tubes (7–9 hr at 40,000 rpm in a #40 Spinco preparative rotor), after which such macroglobulins can be found as a tight pellet at the bottom of the tubes, quite concentrated, but still mixed with some of the other proteins of course. Anderson (65) has presented some data on ultracentrifugation of proteins in packed potato starch and noticed an *increase* in sedimentation rate of bovine serum albumin in that medium. Further investigations into this type of phenomenon with, for instance, tubes packed with the more reproducible polystyrene particles of homogeneous particle size, are likely to lead to interesting results.

Apparatus

Analytical ultracentrifuges are made by Beckman Spinco Division (Palo Alto, California), who now have made more than 1000 such instruments in 20 years; Measuring and Scientific Equipment, Ltd. (London, England); Phywe (Göttingen, W. Germany), who make an airdriven ultracentrifuge; and Martin Christ (Osterode, W. Germany).

2. ZONE ULTRACENTRIFUGATION

A more efficient method of preparative protein purification than moving boundary ultracentrifugation is zone, or band, ultracentrifugation. It works by carefully layering a band of the protein mixture in solution on top of a column consisting of a solvent of higher density than the protein solution and preferably of a density gradient, in order to form a stable cushion for the solution, which must not be allowed to disintegrate into droplets shooting through the column, nor otherwise to cause convective disturbances during the layering or during the ultracentrifugation.

Density gradients (66,67) are formed of sucrose or of salt solutions. Glucose should be avoided: Aubel-Sadron (68) discovered that it has an irreversible polymerizing action on macroglobulins. Density gradients can be formed by machines, by simply layering a heavier solution under a lighter one and stirring lightly with a saw-toothed wire, or by even more simply layering a heavier solution under a lighter one and letting the gradient form itself by waiting the appropriate length of time. Budworth (69) gives ways of calculating the necessary conditions for the latter situation; see also Crank (70).

The actual layering by hand of the protein solution on top of the liquid column is always a source of convection. For analytical rotors there are cells that do this automatically after a given speed or a given temperature is reached (71). Now "band-forming caps" (72) are also available from Spinco for tubes of the swinging-bucket rotors.

In order to avoid remixing of the separated components, it is best to use a swinging-bucket rotor, unless extremely steep gradients are used. It is also advisable never to use a brake during deceleration (73).

After the rotor has stopped, the plastic tubes can be lifted out and have their bottoms pierced and be carefully emptied, drop by drop, into successive recipient tubes. Spinco has recently made devices available to make this operation more convenient and rigorous. Use of the older Spinco tube slicer is better avoided, because some remixing of the tubes' contents practically always occurs with it. Cowan and Trautman (74) have recently published a quite important fundamental improvement in the method. By combining a density gradient with an upside-down viscosity gradient (using two different salts), a stabilizing and zone-sharpening column was obtained through which immune globulins from undiluted serum could migrate at a constant (and not a constantly diminishing) velocity.

Theoretical considerations on zone ultracentrifugation have been given by Trautman and Breese (75), Schumaker and Rosenbloom (76), and Vinograd et al. (77).

Liquid traps for sedimenting proteins have been described by Trautman et al. (78), Koenig et al. (79), and Filitti-Wurmser and Hartmann (80).

Apparatus

Preparative ultracentrifuges are made by Beckman Spinco Division (Palo Alto, California), who make three types of preparative ultracentrifuges; International Equipment Co. (Needham Heights, Massachusetts); Arden Instruments (Rockville, Maryland), and the European Manufacturers already mentioned under Apparatus in the preceding section.

3. FLOTATION ULTRACENTRIFUGATION

Ultracentrifugation under normal circumstances, in isotonic saline and 1.0% protein and thus at a density of the solution of 1.008 g/cm, will not easily show up the serum lipoproteins because the specific densities of most of them (1.00–1.08) are too close to this value to float or to sediment with any noticeable speed. In order to study the lipoproteins they first have to be isolated from whole serum by flotation in the ultracentrifuge in a buffer of a density $\rho_{20} \approx 1.20$. The concentrate thus obtained, a mixture of low density ($\rho_{20} < 1.063$) and high density lipoproteins can then further be separated, or directly studied by ultracentrifugation in high density buffers. The peaks, which point downward, can then be seen migrating from the bottom of the cell to the meniscus. A negative sedimentaion rate or "flotation constant" will then be found, from which, again by using the Svedberg equation [eq. (3)], if the diffusion constant and the partial specific densities are known, the molecular weight can be calculated. The high density α lipoproteins have molecular weights of approximately 200,000 and 400,000. A very low density α_2 lipoprotein has a very high molecular weight of 5–20 $\times$ 10^6, while medium-low density β lipoprotein has a molecular weight of 3,200,000 (81).

4. LARGE-SCALE AND CONTINUOUS ULTRACENTRIFUGATION

Much work has recently been done on zonal and continuous-flow rotors (82) for the newest Spinco L4 preparative ultracentrifuge, and they are now commercially available. Although these rotors appear to have been designed with a view to large-scale separations of viruses and subcellular particles, their maximum speed of 40,000 rpm, giving them a maximum acceleration of 90,000g, seems even to promise possibilities for the larger-scale separations of proteins. As they work with density gradients and can function as zone or band ultracentrifuges, total serum protein separations according to molecular weight seem feasible. One of the available rotors, the B-IV, with a capacity of 1.7 liters, permits introduction of liquids and recovery of fractions during rotation, after introduction of a density gradient. The B-IX rotor permits introduction of a density gradient during rotation and accepts sample solution continuously at several liters per hour. For protein separations, this hourly capacity is, however, likely to be very much reduced, as fairly long rotation times are needed to separate, e.g., macroglobulins from the other proteins at 90,000g in a density gradient. Still, treatment of several 100 ml of protein solution per day would seem to be quite possible. A pump is needed with these rotors, which can provide a metered gradient supply and which can also continuously pump sample solution into the B-IX rotor. Measuring and Scientific Equipment, Ltd. (London, England) also manufactures preparative ultracentrifuges which can accommodate their zonal rotors B-XIV and B-XV.

B. Membrane Methods

The application of the three membrane methods for purifying protein solutions are classically: concentration by *ultrafiltration,* removal of small solutes and electrolytes by *dialysis* and desalination by *electrodialysis.*

1. ULTRAFILTRATION

Ultrafiltration is the forcing of solution and small solute molecules under pressure through membranes, which are impermeable to larger molecules. The latter are thus the only molecules to become more concentrated by this method. Of all protein concentration methods, ultrafiltration is doubtless the mildest one and the least inducive to protein denaturation, as it involves neither increases in temperature nor changes in phase, neither chemical aggression nor increases in ionic strength (83).

When a membrane has been found of a pore size that is sufficiently small to stop all molecules of the protein that is to be concentrated, the main remaining problem is to prevent the membrane from becoming clogged up by an accumulation of the very protein molecules its pores will not let pass. Vigorous agitation of the liquid with respect to the membrane is the solution to this problem, by oscillating or shaking the entire ultrafilter (84), by turbulent streaming (83), or by magnetic stirring (85).

By far the best type of ultrafilter now available is the one described by Blatt et al. (86). (Contrary to the statement in their reference 1, the device is manufactured by the Amicon Corporation, Cambridge, Mass., as are the membranes they describe.) Its useful volume is 400 ml; it is provided with a magnetic stirrer which can rotate just above the membrane without actually touching it, it can sustain pressures of 100 psi and over, and it is transparent, so that the progress of the ultrafiltration need not be deduced from the amounts put into the ultrafilter and actually ultrafiltered, but can be followed directly with the naked eye. With an appropriate membrane, 400 ml of a 0.2% protein solution can be concentrated 50 times in only a few hours. If further concentration is desired, the residual preliminary concentrate has to be either transferred to a smaller ultrafilter (e.g., the syringe-type), or continued in the initial ultrafilter after it has been provided with a washer-shaped metal mask, in order to reduce the surface of the membrane. (It is not feasible to agitate nor to recover about 1 ml of solution sitting on about 40 cm² of membrane.)

a. Membranes

For many years the standard membrane which stops all protein has been ordinary, thin, uncoated cellophane, of a dry thickness of 30μ. Under 100 psi surpressure, such a cellophane, membrane ultrafilters 15 ml water per hour per 40 cm² surface.

A more recently developed membrane, Amicon's (Cambridge, Mass.) Diaflo UM-1 (made of a complex of poly(vinylbenzyltrimethylammoniumchloride) and polysodium styrene sulfonate, a "polysalt" (87), which stops all plasma proteins, ultrafilters under the same circumstances $\approx$350 ml water/hr. When a 5% protein solution is ultrafiltered, this velocity can become reduced to anywhere between 30 and 100 ml/hr, but that is still 3–10 times as fast as with cellophane under the same circumstances. This membrane has a very thin, dense skin, which is the actual protein-stopping layer; it must always be used with that skin towards the liquid which is to be concentrated.

When dry cellophane is swollen for 15 min at 20°C in 64% (w/v) $ZnCl_2$ (density 1.83) and then quickly dipped in and thoroughly washed with cold tap water, a membrane is obtained that still stops all plasma proteins, but which will pass water at a rate five times as fast as untreated cellophane. These membranes can be conserved by drying them in air after impregnating them in a 50% glycerine solution in water.

Cellulose nitrate, or collodion, membranes have been used for various ultrafiltration purposes for more than a century. Zsigmondy (88) made the first commercially available collodion membranes in 1927. These membranes are manufactured by the Sartorius-Werke (Göttingen, Germany) and are available in the United States through Schleicher & Schull Co. (Keene, New Hampshire). Elford (89) did much to standardize the preparation methods for collodion membranes of graded porosity (Gradocol membranes), with the analytical aim of determining molecular weights of proteins. Unfortunately, too many parameters, like size, shape, pH, and concentration are involved (83), while at the same time more rigorous sedimentation and diffusion methods were developed for determining molecular weights, so that the utilization of these membranes has remained quite limited, coupled with the fact that it is very complicated to make collodion membranes of really reproducible pore sizes.

A more promising membrane material is cellulose acetate, which is the same type of material used in the membrane developed by Dr. Loeb (90) for reverse osmosis desalination of ocean water, except that his membrane has undergone a drastic heat treatment (necessary in the salt-stopping variety to provide the membrane with a dense, thin skin). This material will furnish a protein-stopping membrane of reasonably high flux. Gelman (Ann Arbor, Michigan) developed a protein-stopping membrane of cellulose triacetate, which is also "skinned": Type P.E.M., with a "pore size" of 75 Å; it has a flux of about 5 times that of untreated cellophane.

Other protein-stopping membranes, of medium flux, made of poly-(vinylbutyral), have been described by Gregor and Kantner (91).

Membranes that retain proteins of high molecular weight (macroglobulins; $M \approx 1,000,000$) but pass all the other serum proteins have been made (92) from 5% agarose, cast on a fritted glass Buchner funnel by the author and his collaborators. A short air treatment provides these membranes with a macroglobulin-stopping top layer. The membrane is cast in the shape of a cup, so that the serum that is being ultrafiltered touches nothing but membrane, thus excluding any leakage. Ultrafiltration can also be used for the measurement of the degree of nonequilibrium binding of small molecules or ions to proteins (93).

Apparatus

Apart from the earlier mentioned ultrafilter made by Amicon (Cambridge, Mass.), no other ultrafilter with a stirring mechanism is available. In the nonstirred variety there is a very simple one by LKB (Washington, D.C.) (LKB 6300 A) and a more complicated one by National Instrument Laboratories (Rockville, Maryland), both of which use cellophane tubing.

2. DIALYSIS

Dialysis is the process whereby with concentration difference as the sole driving force, equilibrium is ultimately established between two solutions separated by a membrane which is permeable to both solvents and to small solutes dissolved in them, but impermeable to macromolecules (generally proteins) present in solution on only one side of the membrane (or to different macromolecules present on either side of the membrane). Dialysis is most generally used as a means to remove small ions and other solutes from a protein solution, and also to equilibrate a protein solution with a given buffer, as a preliminary step in, for instance, electrophoresis, sedimentation, or diffusion measurements. Dialysis is also used for the osmotic concentration of protein solutions, with the help of very highly concentrated solutions of a polymer to which the membrane is also impermeable, in the other compartment. Rarer are the uses of dialysis for the determination of molecular sizes of solutes (94) and for the estimation of the degree of binding of small molecules or ions to proteins at equilibrium (95).

As the aim is most often to purify or desalinate a protein solution, cellophane tubing is by far the most commonly used type of membrane. Such tubes are best used in U-form. Quite liberal lengths of tubing should be used. After being partly filled with the solution to be dialyzed, the empty parts of the two branches of the U are best twisted together, care being taken not to trap too much air in them, and *knotted in at least three different places.* The author's experience has shown that *knots in cellophane are not impermeable to solvents nor to high molecular weight solutes.* The open ends of the tubing, beyond the knots, are best kept above the dialyzing solution. The solution inside the tubing generally has a much higher initial osmotic pressure than the outside solution (except when osmotic concentration is practiced) and will almost invariably eventually cross the first knot and sometimes even the second knot. But by the time the second compartment and the third knot are reached, this high osmotic energy is generally spent. More refined dialysis methods are described by Craig (96) and by Hjelm (97).

It is good practice to dialyze at refrigerator temperatures, to use large amounts of outer dialysis solution, to keep it stirred, and to renew the outer dialysis solution often.

As a rule it is more efficient to renew the outer dialysis solution more often, than to use vast volumes of that solution at any one time.

Desalting by dialysis always entails considerable dilution of the solution that is to be desalted.

a. Osmotic Concentration

For this use of dialysis it is also essential that the open ends of the dialysis tubing, above the knots, be kept out of the dialysis solution. The dialysis solution most often used for protein concentration by this method is a 20% solution of polyvinylpyrrolidone. Carbowax 20M (Union Carbide polyoxyethylene, a linear polymer of a molecular weight of about 20,000) is also frequently used in the dry, flaky form. When this material is used, the dry flakes are stuffed inside a length of dialysis tubing. Such a sausage, about 5 ft long, dipped in 300 ml of a dilute protein solution, will avidly soak up water and small molecular weight solutes, but, of course, no protein because it cannot pass the membrane. When almost completely soaked inside (which is easily noticeable by a change of aspect from white and flaky to transparent and soggy), the sausage is discarded and replaced with a new one, etc., until only a few milliliters of protein solution are left. A protein solution can be concentrated up to about 50–100 times by this method. One drawback is that a considerable proportion of the solution (up to 30%) is lost by adherence of solute to the outer surfaces of the multiple lengths of cellophane tubing.

A general drawback of these osmotic concentration methods is that a small proportion of both synthetic water-soluble polymers used have a molecular weight below 20,000 and can penetrate through the membrane into the protein solution and precipitate some of the plasma proteins (see Sect. IV-C-2).

Apparatus

Although generally no other apparatus is needed than some cellophane tubing, a beaker, and a magnetic stirrer, some more elaborate devices are available which can have their uses if greater efficiency is desired (obtainable with countercurrent methods) or if very small amounts of solution have to be dialyzed (97).

A Micro Dialyzer is made by Oxford Laboratories (San Mateo, California). Another small-volume dialyzer is made by the Chemical Rubber Company (Cleveland, Ohio) while other dialyzing devices are made by Buchler Instruments (New York, New York), and National Instrument Laboratories (Rockville, Maryland).

Membranes. Cellophane is almost invariably used. The best cellophane for the purpose is generally *not* the variety especially sold for dialysis, but rather the normal NoJax sausage casing, manufactured by the Visking Corporation (Chicago, Illinois). The thinnest cellophane (approximately 30 μ dry thickness and 50 μ swollen), which gives the quickest results, is found in the smaller sizes: $^{16}/_{32}$ to $^{28}/_{32}$ are the most satisfactory ones. This tubing is furnished in crinkled pieces of 30-ft extended length. After wetting, it is quite easy to handle and is so free of holes that it normally does not have to be tested prior to use. The special "dialysis" type of the same origin comes in flat tubing, rolled up, and is better avoided, because after some time of storage and slight desiccation, tiny cracks easily develop at the folds, causing leaks.

When very small amounts of solution have to be dialyzed, a smaller diameter cellophane tubing is required. Tubing of several millimeters to 1 cm diameter are obtainable from Union Carbide. These tubes come in rolls, but as this cellophane is considerably thicker than the above-mentioned NoJax variety (about 100 μ dry thickness), leakage is no particular problem.

Cellophane tubing is also used in the Kolff-type artificial kidney (98), where it serves to keep all the blood cells and plasma proteins inside the system and to dialyze out the excess salt, urea, creatin and creatinin.

Craig and Konigsberg have described a method to vary the pore size of cellophane by either unilateral (decrease in pore size) or bilateral (increase in pore size) stretching (99).

For other membranes see Section II-B-1.

3. ELECTRODIALYSIS

In electrodialysis, a method mainly used for desalination, the solution to be desalted is put between two membranes, while in two outer compartments, salt ions accumulate, due to an electric field applied through electrodes (100).

Electrodialysis is the quickest way to desalt protein solutions and, as an additional advantage, it does not dilute the protein solutions. It is quite a puzzling fact that the method is not more widely used.

Apparatus

There are several devices on the market—one is the Electric De-salter made by Research Specialties Co. (Richmond, California), and another the Torbal B.T.L. Desalter Model CD-1, made by the Torsion Balance Company (Clifton, New Jersey). The latter electrodialyzer utilizes ion-exchange membranes.

C. Molecular Sieve Chromatography

Molecular sieve chromatography or gel filtration is a method for separating solutes of different molecular weights by letting the smaller molecules penetrate into pores of the grains of a gel which are too small to give access to the bigger molecules. The driving force is simple diffusion. The method was developed by Porath (101).

1. COLUMN GEL FILTRATION

Most generally, gel filtration is practiced with columns packed with swollen grains of a gel (e.g., of crosslinked dextran). The biggest macromolecules, which cannot penetrate into the pores of the gel, will stay in the interstitial liquid between the grains and will thus upon elution be the *first* to emerge from the column. The molecules that are small enough to enter the pores of the gel grains will be much retarded through being continuously caught in these pores and will come off the column in the order of decreasing molecular weight. The method is particularly interesting for preparative separations (102). When used with dextran gel grains, columns wider than 6 in. in diameter and longer than 3–5 ft tend to clog up under the weight of their own gel grains, particularly with the more open-pored gels. For the separation of larger quantities of proteins in small columns, recycling is necessary (103). This method necessitates the inclusion of a recording ultraviolet absorption monitor in the circuit to keep track of the protein peaks.

Larger columns than those mentioned above, which nevertheless do not clog up, can be used, if they are first packed with some type of distillation column spacers, before the settling-in of gel grains. These spacers prevent the gel grains from exerting their own weight over more than about 1 in. height at any place in the column and thus will allow the use of much bigger columns (104). With tall columns of a smaller diameter than 6 in., clogging does not occur so easily.

Gels. The most commonly used gel grains are made of crosslinked dextran gels, which were originally developed for this purpose. They are manufactured by Pharmacia (New York, New York) and have become widely known as Sephadex. The existing types are known as Sephadex G-25, G-50, G-75, G-100, and G-200, with a molecular weight exclusion limit of 5000, 10,000, 50,000, 100,000, and 200,000 respectively. These gels are available as irregularly shaped particles and as spherical beads. The latter variety allows greater flow rates in columns. It must be noted that the Sephadex dextran particles, having a slight negative charge, can under conditions of low ionic strength, adsorb some of the more positively charged proteins (105).

Also available are polyacrylamide gel grains (106), manufactured by Bio-Rad Laboratories (Richmond, California) as Bio-Gel P-2 to P-300 with molecular weight exclusion limits of 1000–300,000. The higher porosity gel beads of this material are very hard to use and have a strong tendency to clog up even quite short columns by their own weight.

Much sturdier materials for gel filtration are agar (107) and a neutral polysaccharide derived from agar, agarose (108), both of which are particularly suitable in the very high molecular weight range. Although gel particles of these materials can be prepared by the user (108), the latter material is now also obtainable in a ready-made suspension of particles, from Mann Research Laboratories (New York, New York) in 10 down to 2% agarose, with molecular weight exclusion limits of from 400,000 to more than 10^8. A similar preparation is made by Seravac (Maidenhead, England) and obtainable in the United States through Gallard-Schlesinger (Carle Place, Long Island, New York) and by Pharmacia (New York, New York). Haller (109) describes the preparation and use of porous glass grains with pore sizes of 170–1700 Å diameter. No clogging of the column under its own weight is to be expected with this material.

2. THIN-LAYER GEL FILTRATION

For analytical purposes, gel grains, spread out in a thin layer on glass plates, can be used as a form of thin-layer chromatography (110), with the characteristic that the R_f's of the largest molecules have the highest value. Most of the manufacturers mentioned above make extra-small gel particles particularly suited for this technique.

3. BATCH GEL FILTRATION

The main use of batch gel filtration is for the concentration of proteins with the help of gel grains (111) or rods (112) of the denser variety. The *dry* grains or rods are put into the solution that is to be concentrated and are allowed to swell, thus taking up part of the water and small solutes, but the gel, being impermeable (in the swollen state) to the protein in question, will not take up any of the protein.

This method suffers of course from a rather important loss of protein due to adherence to the gel particles. The washing off of the particles with fresh solvent partly defeats the original purpose of concentrating the protein. The use of polyacrylamide rods (112), however, seem to have largely removed this inconvenience, due to the much smaller surface to volume ratio of the gel.

III. SEPARATION ACCORDING TO BOTH CHARGE AND SIZE

A number of methods are discussed here which give rise to separations of proteins by at least two mechanisms, acting *simultaneously,* and *in only one direction,* one of them separating proteins by size and shape, and the other by separating proteins by electrical charge. One must never lose sight of the fact that these methods, though quite useful in a variety of cases, can at no time by themselves furnish more information about the proteins that have been separated than the information that one algebraic equation with two unknown variables can furnish about two such variables.

Thus, of any protien isolated by this type of method, it is, in the absence of further tests, impossible to know if it has arrived where it is on account of its charge, its size, or both properties combined, nor is it proven (without additional data) that it is unaccompanied by other proteins which, although of different charges and sizes, can quite possibly have accumulated in the same fraction. Worse, one single homogeneous compound can, under certain circumstances, give rise to more than one band (113).

None of these arguments hold true for methods whereby proteins are separated according to charge and size *at different times, in different dimensions* or *by different mechanisms.* On the contrary, methods of the latter type, for instance electrophoresis, followed by analytical ultracentrifugation, or vice versa, furnish in fact almost the only solution for determining what the fractions separated by one of the above-mentioned hybrid methods actually are. This is admirably and strikingly illustrated in Figure 98 of Schultze and Heremans' work (114) (see also Fig. 58 of the same work), where a bidimensional electrophoresis starch-gel electrophoresis chart is compared with a bidimensional electrophoresis ultracentrifugal sedimentation chart. The hopelessness of extricating information from gel electropherograms without the aid of other methods becomes quite clear from that illustration, as well as the complication of interpreting immunoelectropherograms, although one should not lose sight of the fact that the latter method is really a bidimensional combination of two different methods, which has, unaided, increased our knowledge of serum proteins by almost one order of magnitude.

A. Molecular Sieve Electrophoresis

Dense gel electrophoresis of proteins was introduced by Smithies (115), who employed 1–15% (w/v) starch gels. Later the use of silica gel (116) (not actually in great use) and of polyacrylamide gels (117) were described; in the latter case, when it is done in small gel cylinders, it is often alluded to as "discelectrophoresis" (118). Agar is not much used for dense gel electrophoresis and dextran gels (Sephadex) are very rarely used.

1. STARCH-GEL ELECTROPHORESIS

The vertical "macro" method first described by Smithies (115) is still the one most in use. The preparation of bubble-free starch solution and the air-free filling of the apparatus require a considerable amount of practice. As the migration along the surface of the gel is less regular than in its center, the gel, upon completion of the electrophoresis run, has to be split laterally (with, for instance, a taut piano wire) over its entire surface into two halves prior to fixation and staining. Bromophenol blue or Nigrosine are the most satisfactory protein stains, while Sudan Black B or Oil Red O are suitable for staining lipoproteins. Owing to the thickness of the gel slabs, even after splitting, the staining and destaining processes take several hours.

Baur (119) and Lewis (120) have described methods for thin-layer starch-gel electrophoresis and for staining and plastifying such thin gels.

Smithies has demonstrated that, other conditions being equal, the mobility of a given protein is inversely proportional to the starch concentration (121). He also indicated that the degree of retardation of a given protein in a given gel (as compared to its migration in a very dilute gel) is proportional to the square root of its molecular weight.

The principal advance in our knowledge of serum proteins which we owe to this method is in the field of the haptoglobins. These are α_2 globulins, which serve as carriers for hemoglobin molecules when they accidentally find their way out of damaged erythrocytes into the plasma. The smallest haptoglobin molecules have a molecular weight of approximately 80,000 but they also occur in molecular weights of 160,000, 320,000, 640,000, 1,280,000, etc. The distribution of these various polymers in human sera is genetically determined (122).

Poulik (123) described a method of starch gel electrophoresis of proteins in the presence of high concentrations of urea ($5-8M$), which tends to depolymerize polymerized proteins. This technique is thus particularly useful for the study of subunits of proteins. Poulik (124) also described a discontinuous buffer system which can give better resolution of proteins. Poulik and Smithies (125) developed a method comprising paper electrophoresis in one dimension and starch-gel electrophoresis of the elctrophoresed proteins in a second dimension, perpendicular to the first. In this way they could differentiate between the influences of the size and charge of the proteins. Korngold (126) has developed a method comprising micro starch-gel electrophoresis followed by immunodiffusion in agar.

Apparatus

E-C Apparatus Company (Swarthmore, Pennsylvania) and Buchler Instruments (Fort Lee, New Jersey) manufacture apparatus for vertical starch-gel electrophoresis. E-C also furnish a destaining apparatus, the use of which materially accelerates the destaining time. The Spel Systems (Cleveland, Ohio) manufacture a thin-layer starch electrophoresis system. Micro starch-gel electrophoresis trays are manufactured (among others) by Shandon (Consolidated Laboratories, Chicago Heights, Illinois) and Gelman Instruments (Ann Arbor, Michigan).

2. POLYACRYLAMIDE ELECTROPHORESIS

Gels of even smaller pore sizes than are feasible with starch gels can be obtained with polyacrylamide (117). The latter gel also allows for carbonhydrate staining (for instance for the detection of glycoproteins (127)), which is of course impossible with starch gels.

These gels can be used for electrophoresis in tubes (discelectrophoresis) (118), which permits the application of the sample on top of the gel cylinder, giving rise to a very useful narrowing of the sample zone, due to the strong retardation of its mobility as soon as it migrates into the gel. Polyacrylamide gels can also be used with high concentrations of urea and serve otherwise very much the same purpose as starch gels and have the same limitations. The *monomer* acrylamide is *very toxic* before polymerization.

Apparatus

E-C and Buchler (see above) also make various apparatus for polyacrylamide-gel electrophoresis, including preparative apparatus. E-C make a device that uses electrophoresis in a direction perpendicular to the original gel electrophoresis, for eluting the various fractions after their separation. A polyacrylamide electrophoresis device for tubes (discelectrophoresis) is made by Canal Instrument Company, (Bethesda, Maryland).

B. Chromatography

Of the three mechanisms of liquid-phase chromatography: partition, adsorption, and ion exchange, only the latter two play a major role in protein separations, and the two mechanisms are hard to separate. Some charge phenomena always play a role even in adsorption chromatography, as it is quite rare to find an absorbent with no electric charge whatsoever. And some noncoulombic adsorption always plays a role even with strong ion-exchange materials. Some batch adsorption or ion-exchange methods may not perhaps be "chromatography" in the strictest sense, but for convenience sake they will still be treated here. For partition chromatography see Section IV-D-2.

1. ADSORPTION CHROMATOGRAPHY

Some inorganic gels or precipitates have been used for many years for purifying proteins (128,129). Alumina gels have been used for adsorbing enzymes since 1922 and have been used for selectively adsorbing serum prealbumin. Benzoic acid precipitates have been used for the adsorption of hormones. Calcium phosphate gels, silica gels, and diatomacious earth have also been used for purifying enzymes and for separating serum fractions. Magnesium hydroxide and barium sulfate have been used to purify prothrombin, the latter precipitate being particularly useful for the removal of this clotting factor from serum. Copper hydroxide and zinc sulfide have been used in enzyme purifications. Glass beads have been used for the separation of lipoproteins (130).

Still, apart from a number of specialized separations, with materials used for their specificity on an empirical basis, no particular inorganic adsorbent has any preponderant application in the field of serum protein purification.

Organic adsorbents, however, are in much wider use. (Gel grains used in molecular sieve chromatography and specific immunoadsorbents are treated elsewhere under these respective specific headings.) They fall mainly under the categories of synthetic crosslinked polymeric resins and the naturally occurring biopolymers of which cellulose is the main representative.

Hydrophobic resins. Polystyrene, a representative material of this class, has been much studied, in the form of very small beads or latex particles. It has been shown (131) that such particles adsorb the least hydrated serum proteins strongest, regardless of the electrical charges of either the oligo-hydrated proteins or the polystyrene particles. It has also been shown that a very large protein is more strongly adsorbed than a smaller but otherwise similar protein.

The adsorptive forces here at work are no other than "hydrophobic" van der Waals-London dispersion forces. The adsorbed proteins were very strongly bound to the polystyrene particles, but they could be eluted from them with the help of strong detergents. These properties are not peculiar to polystyrene, very much the same phenomena having been observed with polybutadiene, poly(vinyl chloride), and poly-(vinyl toluene). These properties are the same ones which prevent ion-exchange resins from being the ideal protein-fractionating materials which they at first sight would promise to be (see below). As Peterson and Sober (129) note, their main limitation is their irreversible binding of most proteins. They ascribe this to be principally due to strong electrostatic forces, the influence of which is probably less pronounced than they contend, in the light of the fact that we encountered the same phenomena under conditions where electrostatic forces played no or only a minor role.

Even the much more hydrophilic ion-exchange celluloses, which were specially developed to obviate these inconvenient nonelectrostatic interactions with proteins, show the same tendencies, although to a lesser degree. Thus, their general property of binding larger proteins more strongly than smaller ones (of otherwise largely the same charge density) is not due to stronger bonds with bigger proteins on account of a greater number of electrostatic bonds, as proposed by Peterson and Sober (132), because the charge density per unit surface is unchanged, but is rather caused by the fact that the attractive dispersion forces, being additive, are stronger for bigger and thicker molecules than for smaller and thinner ones (131).

2. ION-EXCHANGE CHROMATOGRAPHY

The ion-exchange celluloses, being much more hydrophilic than ion-exchange resins, are the materials of choice when protein separations are desired in which adsorption phenomena are minimized and in which charge plays an important role. Even here, where adsorption phenomena are never completely avoided, and other factors being equal, larger molecules are still more firmly bound to the cellulose and harder to elute than smaller molecules (see above).

To the extent where only coulombic bonds between protein and charged cellulose are taken into account, it will be obvious that ion-exchange chromatography of proteins is rather analogous to electrophoresis. At approximately neutral pH, all proteins are more or less negatively charged and will all become bound to a positively charged cellulose ion exchanger, for instance diethylamino ethyl (DEAE) cellulose. By eluting the column with a buffer of *increasing acidity*, the least negatively charged proteins will successively become neutral and then positively charged, upon which they will detach themselves from the ion-exchange material and emerge from the column. In this fashion the serum proteins will be collected in the order of increasing negative charge (barring some irregularities, already mentioned above, due to proteins being adsorbed by other mechanisms also): γ globulins first, then β and α globulins, and finally albumin. In practice, pH gradients (133) are used for elution, gradually decreasing in pH from approximately 7.5 to 4.5 and at the same time increasing in ionic strength from, say, 0.005 to 0.2 and higher. The addition of salt helps in detaching proteins from ion-exchange particles by actually "exchanging ions." If negatively charged ion exchangers are used, for instance carboxymethylcellulose (CM cellulose), at lower pH, the *more* negatively charged proteins at neutral pH, which are the least positively charged proteins at, for instance, pH 4.5 (like serum albumin), come off the column first and are followed by the α, β, and γ globulins in approximately that order. In reality, the order in which the proteins come off the columns is more complex, which is mainly due to the reasons already mentioned above. Still, for preparative work, and if the collection of the fractions is monitored by ultraviolet adsorption measurements at a wavelength of 280 mμ, by analytical zone electrophoresis or by other analytical methods, ion-exchange chromatography can be quite useful. Peterson and

Sober (129) give a most extensive table on the use of various ion-exchange materials for sundry protein purifications, with the elution system used and a multitude of references. They also have given a detailed description of a device for making elution gradients (133). (See also refs. 4, 134–136.) Moore and McGregor (137) have stressed the importance of the influence of the *shape* of the gradient used for elution on the efficiency of the fractionation (69,70). (See also under Sect. II-A-2.)

In some cases ion exchangers can be used batchwise for the isolation of a single protein species: DEAE cellulose powder, added to serum, under precise conditions of pH and ionic strength, can serve to isolate reasonably pure 7S γ globulin (138).

Ion-exchange resins are also used to remove the Ca^{2+} from whole blood to prevent clotting (139), to remove all salt ions for the elimination of fibrinogen and the euglobulins from plasma (140), and to remove Zn^{2+} from plasma (141) if salts of that metal have been added for the removal of these proteins.

Other cation-exchange celluloses are sulfo methyl (SM) cellulose and sulfo ethyl (SE) cellulose. Anion-exchange celluloses which are also used are epichloro triethanolamine (ECTEOLA) cellulose and triethylamino ethyl (TEAE) cellulose. Among the ion-exchange resins, the anion exchangers, Dowex 2 and 3 and Amberlite IRA 400 and the cation exchangers Dowex 50 and Amberlite IR120 and IRC50 are often used, generally in a finely powdered form. More recently, ion-exchanging, crosslinked dextrans have also come into use for protein purification, generally made with small-pored gels; DEAE and CM Sephadexes are the ones most used. These ion exchangers show little noncoulombic adsorption of proteins.

IV. SEPARATION ACCORDING TO SOLUBILITY

The most strongly charged protein, *albumin,* had long been known as the most soluble of the plasma proteins. A few generations ago it was even held to be only serum protein; "albumin" or "albumen" used to be synonymous with "protein." Then it became apparent that some of the serum proteins were less soluble in water of very low salt content. These proteins, insoluble in distilled water, were then called globulins, while the water-soluble proteins retained the name albumins. Later again, further fractionation attempts, particularly with the salting out method (see below), showed that only part of the globulins were really insoluble in distilled water, while the major part of the globulins (although less soluble in, for instance, 30% saturated $(NH_4)_2SO_4$ solutions than albumin) is after all quite soluble in water. The globulins that were still insoluble in distilled water were then baptized euglobulins (from Greek: ϵv = good) and the other, water-soluble globulins, pseudoglobulins (from Greek: $\psi\epsilon v\delta o$ meaning false, fake).

Although the electrical charge of proteins enhances their solubility in water, in many proteins this is not the only property that makes them soluble. Some proteins, in particular albumin, are very soluble even at their isoelectric pH, which is probably due to a comparatively high proportion of nonionic hydrophilic groups near the surface of their molecules, which contribute to their strong degree of hydration (131). Nevertheless, most proteins have their lowest solubility at a pH close to their isoelectric point. Thus control of pH is of great importance also in this class of separation methods.

Precipitation methods, of all the protein separation and concentration methods, are the most suitable for large-scale preparations.

A. Precipitation through Lack of Water (Salting Out)

When large amounts of salts are added to a protein solution the proteins will tend to precipitate through sheer lack of water: all their water of hydration necessary for staying dissolved is either captured as hydration water by the salt ions (142) or made unavailable for protein dissolution by unfavorable orientation of the water dipoles due to the presence of large amounts of highly charged salt ions of the same sign of charge as the proteins (143).

This method is generally called "salting out." As already noted by Hofmeister in 1889, the more negatively charged salt anions are, the more effective they are in salting out negatively charged proteins (144).

In the (approximate) order of decreasing salting-out effectiveness, the anions are (when they have the same cation): citrate^{3-}, tartrate^{2-}, $SO_4{}^{2-}$, F$^-$, IO$_3{}^-$, H$_2$PO$_4{}^-$, acetate$^-$, B$_2$O$_3{}^-$, Cl$^-$, ClO$_3{}^-$, B$_2{}^-$, NO$_3{}^-$, ClO$_4{}^-$, I$^-$, CNS$^-$ which is, roughly, in the order of decreasing hydration energies. When they have the same anion, the cations are in the (even more approximate) order of decreasing effectiveness: Th^{4+}, Al^{3+}, H$^+$, Ba^{2+}, Sr^{2+}, Ca^{2+}, Mg^{2+}, Cs$^+$, Rb$^+$, NH$_4{}^+$, K$^+$, Na$^+$, Li$^+$. These are generally called the Hofmeister, or lyotropic series. Of the plurivalent anions, sulfates are most generally used, preferably ammonium sulfate owing to its great solubility. Fibrinogen will precipitate first at fairly low salt concentrations, and then the euglobulins in 30% saturated ammonium sulfate, then the pseudoglobulins in 50% saturated ammonium sulfate, and finally albumin and α_1 globulins, which need even higher concentrations of ammonium sulfate to precipitate. It is advisable to add the appropriate amount of salt from a saturated solution, quite slowly and under constant stirring, in order to avoid precipitating undesirable proteins that are otherwise locally exposed to too high a degree of saturation of the added salt solution. When the desired protein is precipitated, the precipitate must be washed free of other, still-dissolved proteins, with a salt solution of the same degree of saturation as was necessary to attain its precipitation. The precipitate must be dialyzed against distilled water or physiological saline, or any other desirable buffer or solvent (see Sect. II-B-2), in order to remove excess salt and will redissolve. Owing to osmotic differences, the liquid volume inside the dialysis bag will usually expand considerably. The main drawback of the salting-out method is that the great reduction in volume of the initial precipitate cannot be maintained if the excess salt is to be removed. For this reason electrodialysis is a better way of removing excess salt than simple dialysis (see under electrodialysis, Sect. II-B-3). The protein can be reprecipitated a number of times, and greatly enhanced purity and reduction of volume can ultimately be attained in this manner. The method, though laborious, is a fairly mild one, and probably among those least liable to cause denaturation. If protein concentrations are to be measured by a chemical determination of nitrogen content, it is more practical to use Na_2SO_4 instead of $(NH_4)_2SO_4$. The concentration of this salt necessary for precipitating a given protein is generally less than that of $(NH_4)_2SO_4$, but on the other hand, Na_2SO_4 is rather less soluble in water than $(NH_4)_2SO_4$. Porath (144a) has published a method of salting out proteins on Sephadex columns, which he called zone precipitation.

Salting out methods are not now in use for the large-scale manufacture of plasma protein fractions, mainly owing to the tediousness of removing the excess salt, and they are now largely superseded by the cold alcohol method, treated in the following section. Still, for medium large-scale preparations in the laboratory, the salting-out methods are probably ideal as protein denaturation is no particular problem here, and no special cooling arrangements are needed.

B. Precipitation through Protein–Protein Interactions

The amphoteric character of proteins, that is, the fact that they possess a number of positive as well as negative charges, can be the cause of interactions between protein molecules, and thus of precipitation, under appropriate conditions of ionic strength, pH, and/or dielectric constant of their surroundings.

1. PRECIPITATION THROUGH LOW IONIC STRENGTH

Euglobulins are by definition, when not too far from their isoelectric pH, insoluble in distilled water. They' can be precipitated from a solution at the appropriate pH either by dilution with distilled water or by dialysis against distilled water that is weakly buffered at the desired pH. The cause of their precipitation by lack of salt is that these proteins do not have many charges of either sign, particularly at isoelectric pH, and are incapable of repelling one another electrically. Their hydration is also quite low, and they thus can get close enough together to form large complexes which are no longer soluble due to the mutual neutralization of their positively and negatively charged sites. Proteins thus precipitated are fairly easily denatured and should not be kept too long in the precipitated state.

When small salt ions are added to such precipitates, the strongly charged and hydrated salt ions will take over the neutralizing role of the other proteins, and with salt ions of both signs as counterions, most of the protein will redissolve. This is called "salting in."

This method is of course only of use for the isolation of euglobulins, among which is an important fraction of the β_2 macroglobulins (IgM), some factors of complement (C_1' and one of the C_3' fractions), as well as a small proportion of the γ globulins. The serum euglobulins and their propensity to precipitate in salt-poor media play an important role (145) in the agglomeration of erythrocytes in sugar solutions, which property is put to excellent use by Huggins (146), in his method of washing the excess glycerol out of thawed erythrocytes in the agglomerated state.

2. PRECIPITATION THROUGH LOW DIELECTRIC CONSTANT

Given that the attractive force F between charges Q and Q' of opposite signs is equal to the product of the charges, divided by the dielectric constant ϵ of the medium times the square of the distance between the molecules,

$$F = QQ'/\epsilon r^2 \tag{4}$$

it will be clear that if the dielectic constant ϵ of the solvent can be lowered sufficiently, the attractive force F between proteins should increase to such an extent that *all* plasma proteins (and not only the euglobulins) will neutralize each other by protein–protein interactions and precipitate. This can be done by adding organic solvents. The dielectric constant of water is 80, that of ethyl alcohol, 24.

By adding alcohol to watery solutions of proteins at low ionic strengths one can successively precipitate *all* the serum proteins. The drawback is, however, that proteins thus precipitated are easily denatured (particularly the lipoproteins), but this can be partly obviated by operating at low temperatures.

The cold ethanol fractionation methods are the ones most used for large-scale fractionation of human plasma and Cohn's (147) methods nos. 6, 9, and 10 are the most commonly used for that purpose. With these methods, most of the fibrinogen precipitates in fraction I, γ globulin in fraction II, and albumin in fraction V. Pennell (141) gives a description of these methods with diagrams (see also ref. 114). Methanol and acetone are not used much for this type of separation, but ether has been (141,148) and probably still is used in Great Britain.

C. Complex Coacervation Methods

Methods whereby the affinity between two solutes causes the redistribution inside the solution of these solutes into two layers of different concentration are called complex coacervation methods. Complex coacervation has been widely studied by Bungenberg de Jong (149). In this class the following precipitation methods can be grouped.

1. PRECIPITATION WITH POLYELECTROLYTES AND ASSOCIATION POLYELECTROLYTES

Positively charged proteins quite easily form precipitates with negatively charged polyelectrolytes. Alameri (150) has been able to separate γ globulin from albumin at pH 5.6 (where albumin is negatively and γ globulin positively charged) by selectively complexing and precipitating it with sodium methacrylate (see also ref. 151). The precipitate containing the γ globulin could then be redissolved by raising the pH, and the methacrylate could be removed from the γ globulin solution by precipitation with Ba^{2+} salts. Thus approximately 90% pure γ globulin was obtained. Instead of polyelectrolytes, association polyelectrolytes, that is, micelle-forming ionic detergents can also be used. After redissolution of the specific precipitate, the detergents can then either be precipitated (anionic detergents for instance with Ba^{2+} salts) or extracted by shaking with an appropriate organic solvent.

With these methods it is rather difficult to precipitate exactly one desired fraction out of a mixture without dragging along an important proportion of other fractions of neighboring isoelectric pH's. The complete removal of the complexing polyelectrolyte or association polyelectrolyte is also frequently quite difficult.

2. PRECIPITATION WITH OTHER POLYMERS

Water-soluble polymers need not have an electrostatic charge to form complexes with proteins, as long as the hydrophobic parts of protein and polymer can combine. For example polyoxyethylenes, or Carbowaxes (Union Carbide) preferentially coprecipitate with γ globulins at pH 7. The γ globulins can be recovered by shaking the precipitate with organic solvents (in the cold, in order to avoid denaturation), extracting the Carbowax, and leaving the protein. The higher the molecular weight of the Carbowax, the better it precipitates protein, though a molecular weight of 6000 seems optimal, because above that the solutions become very viscous (see ref. 152).

3. PRECIPITATION WITH HEAVY METAL IONS

Zinc ions in particular tend to precipitate with fibrinogen and the globulins. 20mM Zn^{2+} added to a protein solution at pH 7.2 will precipitate most globulins (141). The $ZnSO_4$ turbidity test (153) is based on this; it will give a first approximation of the ratio of albumin/globulin content of a serum. Cohn's zinc method No. 12 is a useful method for serum protein fractionation without addition of organic solvents (141).

Aluminum can also have its uses in plasma protein fractionation. With 0.05M Al^{3+}, all plasma proteins precipitate, except the γ globulins (141). Uranylacetate acts in a very similar fashion (154).

4. OTHER PROTEIN PRECIPITANTS

a. Tannic Acid

This compound will, at pH 4.7 and an ionic strength of 0.15, precipitate all, even very dilute proteins. Vigorous rubbing of the precipitate with powdered caffeine will keep the tannic acid insoluble in the form of a caffeine tannin precipitate and send the concentrated proteins back into solution. We have adapted this method to the approximately 500-fold concentration of urinary proteins (154a). With this method, not all the high molecular weight (macro) globulins go back into solution.

b. Rivanol (2-Ethoxy-6,9-Diamino-Acridine-Lactate)

This compound, an acridine dye, precipitates all serum proteins, except γ globulins (141). Rivanol can be used to precipitate β_2 macroglobulins (IgM) out of a solution containing other immune globulins (155).

c. Polyphosphates

Tetrametaphosphate and other polyphosphates precipitate all plasma proteins, at a pH between 4 and 5, except γ globulins (141).

d. Strong Organic Acids

Trichloroacetic acid, picric acid, or sulfo salicylic acid, in concentrations of 5 or 10% will precipitate all plasma proteins. These acids are mainly used to remove all proteins from a preparation or to test for the presence of proteins in liquids (e.g., urine). This type of precipitation irreversibly denatures most proteins.

D. Partition Methods

When several solutes, proteins for instance, are soluble to different extents in two solvents which are immiscible, or partially miscible, repeated shaking of the solvents, alternated by settling out and followed by decantation of each of the solvents into a different tube (each containing some more of the same mixture of solvents), etc., will finally redistribute the different solutes in different sets of tubes (156). This kind of liquid–liquid extraction can be performed in sets of multiple tubes, but also in a more continuous fashion in a single column (157), and even on a band of porous filter paper, in which latter case it has come to be called partition chromatography (158).

Although none of these methods is very commonly used for plasma protein fractionation, they have on occasion been so used and have yielded excellent purification of appreciable amounts of a variety of substances, so that their use is very much worth considering when a protein purification problem arises.

1. COUNTERCURRENT DISTRIBUTION

This method, operating with a train of a multitude of identical tubes, which can be manually or mechanically shaken and the contents of which can be automatically transferred to the next or to the preceding tube, was developed by Craig in 1949 (159). With this method, numerous proteins (160), peptides, enzymes, hormones, antibiotics, and other substances have been isolated. Detailed treatments of the method are given by Craig (156,161).

Kepes (157) has developed a column with many fine screens in which a solvent can be brought into multiple contacts with a lighter or a heavier solvent, thus ultimately redistributing initially mixed solutes which are retrieved in a fraction collector. Albertsson (162) has opened up a wholly new series of possibilities for countercurrent purification of proteins and other biopolymers with his work on aqueous polymer two-phase systems. As an example, a mixture of 5% dextran and 4% Carbowax 6000 in water will separate into two liquid layers with most of the Carbowax in the top layer and most of the dextran in the bottom layer; different proteins will tend to distribute themselves differently over such layers and can often be separated in relatively few countercurrent steps (163).

Apparatus

Craig-type installations are manufactured by E-C Apparatus Company (Swarthmore, Pennsylvania) who make fractionators of a variety of sizes, starting with a manual 20-tube machine for a few hundred dollars only, and Quickfit Reeve Angel (Clifton, New Jersey) who make a very large Steady State Distribution Machine.

2. PARTITION CHROMATOGRAPHY

More or less related to countercurrent distribution and to liquid–liquid extraction is partition chromatography, developed in 1941 by Martin and Synge (158). The stationary liquid is adsorbed on some porous material, which can be anything from rubber to filter paper. The sample is deposited and the chromatogram developed by letting another liquid, which is not completely miscible with the stationary solvent, creep along it, either ascending, or, more generally, descending. Partition chromatography has not been used much for protein separations, but it can be and has been so used on occasion, e.g., for the separation of rabbit γ globulins on a column of Celite, a diatomaceous earth (164).

V. SEPARATION ACCORDING TO IMMUNOCHEMICAL SPECIFICITY

Most mammals are capable of making antibodies against practically any known protein, provided that the protein in question is foreign to them, i.e., that the protein is not indigenous to their own species. For example, rabbits can make antisera against human serum albumin or against mouse serum albumin, etc., but not against rabbit serum albumin. A rabbit antiserum against human serum albumin will react strongly with human serum albumin but more weakly with serum albumin of related species (of, for example, other primates) and not at all with any of the other human serum proteins. The substances in antisera which react with the protein with which the animal was injected are generally called antibodies (AB), and the injected proteins themselves are called antigens (AG). The AB's made by animals and humans are serum proteins in the γ and the slow β range. They are also called immune globulins (Ig), of which three main varieties are known: IgG, ordinary γ globulin, of a molecular weight of $\approx$160,000 and present in normal serum to the extent of $\approx$10 g/liter; IgA, known as β_2A globulin, of the same weight as IgG but present in much smaller amounts; and IgM, known as β_2M globulin, a macroglobulin of a molecular weight of $\approx$1,000,000, present in small amounts in normal serum (59). Animals can make AB's against the Ig's of other species. AG–AB reactions are among the most specific reactions known. They play an increasingly important role in the analytical characterization of serum proteins as well as in their preparative isolation.

A. Immune Precipitation Methods

The manifestation of AG–AB reactions that is most pertinent to protein separations and purifications is their propensity to precipitate when brought into contact with one another. This precipitation reaction is highly specific for the AG and AB in question and can be put to great advantage for qualitative analytical identification as well as for quantitative preparative elimination or isolation.

1. QUALITATIVE COMPARISON OF IDENTITY AND NONIDENTITY BY THE IMMUNE DOUBLE DIFFUSION METHOD IN GELS

Double diffusion precipitation methods in flat bidimensional gel plates greatly facilitate the identification of an unknown protein by comparison with a known protein through the reaction of both with an AB.

a. Technique

Approximately 1% agar in physiological saline gellified in plastic flat-bottomed petri dishes are used as gel plates. Holes (about ¼ in. in diameter and about ¼ in. apart) for the reagents are punched in the gel with a cork borer (and the solid cylinders removed by suction). The reagents are deposited in the holes with the aid of a pipet. The gel plates are best kept in a humid chamber. Within 24 hr the lines formed by the precipitation reactions are clearly visible. The gel plates can then be either directly photographed or used as a negative in contact printing. The patterns are best viewed and photographed with side illumination and against a dark background.

b. Specifically Impermeable Barriers

The possibility of distinguishing between two or more AG–AB precipitating systems is based on the following law: Upon the meeting of two soluble substances which can precipitate one another when combined, their precipitate will form a specifically impermeable barrier to themselves, and to themselves only. This law holds true for inorganic (165), organic (166), and immunological (167,168) precipitates. It is limited by a second condition, which is that the precipitate barrier will be specifically impermeable to the substances that formed it only as long as there is some of the forming substances left in solution on either side of the precipitate barrier (165). In immunochemical systems and in all other systems where the precipitate is soluble in an excess of either of the reagents, this specific impermeability remains intact only as long as the forming substances on either side of the precipitate barrier are present in equivalent concentrations (166).

This last consideration is very important in *quantitative* interpretations of double diffusion precipitation patterns.

From the first condition of the law it follows that when two sub-

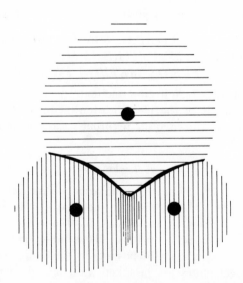

Fig. 1. Diagram of the fusing of precipitation bands formed when two substances, A and B, can both precipitate with C. (After ref. 58.)

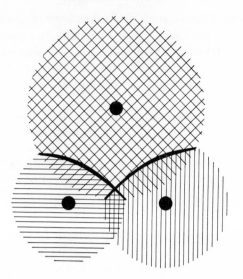

Fig. 2. Diagram of the crossing of precipitation bands formed between A and C and P and Q, the systems AC and PQ having nothing in common. (After ref. 58.)

stances A and B can both precipitate with C, their precipitate bands in double diffusion will fuse (Fig. 1). When two separate systems are such that A will precipitate with C and P with Q, the systems AC and PQ having nothing in common, the precipitate bands AC and PQ will cross one another unhindered (Fig. 2). In immunochemical systems, the first case (Fig. 1) is taken to be indicative of immunological identity. The second case (Fig. 2) is taken to be one of nonidentity. Spurs (Fig. 3a), often indicative of partial identity, are always composites of the first (Fig. 1) and the second (Fig. 2) case combined (Figs. 3b and 3c).

(a) (b) (c)

Fig. 3. (a) Diagram of spur formation in precipitation bands; this situation must always be considered as either the one depicted in (b) or the one drawn in (c), although it may not, in practice, always prove possible actually to decompose situation (a) into either (b) or (c), due to the fact that one and the same molecule may have several different antigenic properties. (After ref. 58.)

It will now be obvious that, as soon as enough (about 100 mg generally suffice) of a purified protein is available to induce AB formation against it in a laboratory animal, a reagent against it is obtained which far surpasses in specificity any known dye. As will be shown below, these properties of AG–AB precipitates also make quantitative titrations possible.

The distinction between the various fractions in immunoelectrophoresis (see Sect. I-A-3) is also based on these considerations.

2. QUANTITATIVE METHODS OF TITRATION BY IMMUNE PRECIPITATION

Many AG–AB combinations can occur in the most widely varying proportions, but they are *insoluble* in only a small range of AG–AB ratios. Soluble complexes are then formed in both the AB excess and the AG excess zones. Their equivalence point is held to be at the ratio of their maximum precipitate formation and upon this very definition their quantitative titration is based.

a. Immune Precipitation in Tubes

The technique most generally followed (169) utilizes a series of tubes which are all half filled with identical amounts of antiserum (at a given dilution), after which the other half of every tube is filled with the corresponding AG solution of increasing AG concentration. The tubes are best kept refrigerated for several days in order to obtain optimal amounts of precipitate. The protein or nitrogen content of the washed precipitate, or of the supernatant clear solution (or both) are then determined, and the highest found value gives the closest approximation to the equivalence point. If the antiserum has once been titrated with known amounts of the pure AG, further titrations with the same antiserum obviously will then enable one to obtain a quantitative estimation of the same AG, even in quite impure mixtures.

b. Immune Precipitation by Gel Double Diffusion

Double diffusion can now also be used for obtaining equivalence ratios of AG–AB systems. This method (166), which is at least as accurate as the precipitation technique in tubes, requires less reagents, is much less time consuming, and does not require any chemical nitrogen determinations.

It is based on the fact that the first formation of the precipitate lines, *regardless of the respective initial concentrations* of the reagents, always occurs in the same place (166,170): It divides the distance between the two reagent wells into Sections a and b proportionally to the square root of the ratios of the diffusion constants:

$$a/b = \sqrt{D_{AG}/D_{AB}} \qquad (5)$$

After some time has elapsed, however, the only place where the precipitate line has stayed immobile and has not thickened is the place where the reagents are still present in equivalent amounts (in the proximity of the precipitate line). This was first proven for non-immunological organic precipitates, the concentrations and valences of which were known, by van Oss and Heck (166), who also showed this to be true for immunological precipitates.

When one of the dissolved reagents is present in a higher concentration than the other, the precipitate band will do one of the following:

a. thicken;

b. multiply; [called periodic precipitation or Liesegang band formation (171)]; or

c. seem to move (but actually dissolve in excess reagent A and reprecipitate farther on), *in the direction of the well containing the most diluted reagent* (166).

One obviously can make use of the fact that the precipitate band stays thin and immobile only where the forming substances in solution on either side of it are present in equivalent concentrations to set up a quantitative titration (see Fig. 4). All top wells were filled with undiluted goat antihuman IgG serum (#602); the bottom wells were filled with 1/5, 1/10, 1/20, 1/40, and 1/80 dilutions of 1% human IgG (F II). As in this particular case both AG and AB are 7S γ globulins of identical diffusion constants, the site of equivalence obviously is the place where the precipitate band is exactly in the middle between the wells and where it has remained thinnest. In Figure 4, this corresponds to $\frac{1}{10}\%$ IgG, which means that 1 ml of antiserum #602 contains an amount of antibodies that will equivalently precipitate with 1 mg of antigen (human IgG). Parallel experiments with AG–AB precipitation in tubes have shown that the equivalence point of these titrations (the place of the thinnest, immobile, precipitation bands) corresponds to the point of most massive precipitation, when the variously diluted reagents, instead of being put into separate, oppositely situated holes in a gel, are mixed together in tubes (170). This holds equally true whether the gel titration and the precipitation in tubes are done at room temperature or at 4°C.

Fig. 4. Quantitative titration of goat antihuman γ globulin serum (#602) with 1/5, 1/10, 1/20, 1/40, and 1/80 dilutions of 1% human γ globulin (FII). The unit volume of antiserum here obviously corresponds to one volume of 1/10 of 1% IgG: Note the precipitation line between the second holes from the left, where it has stayed thinnest and remained immobile, in the exact middle between the wells. (In this particular case the diffusion constants of antibody and antigen are of course the same.) (From ref. 58. Courtesy J. B. Lippincott Company.)

B. Immune Adsorption Chromatography

In recent years the great specificity of the AG–AB combination has prompted research into the possibility of exploiting this property for the purification of proteins on a preparative scale.

Sutherland (172) has shown that AB protein may be coagulated in various manners without the complete destruction of the specific binding sites. Thus an AB against ovalbumin, rendered insoluble by a treatment with 90% ethanol, could repeatedly be used specifically to adsorb ovalbumin, which could then be desorbed (at pH 3); the method gave a 35-fold increase in purity in one step. AG or AB can also be covalently bound to many insoluble cellulose derivatives, and such adsorbents can still perfectly specifically combine with their immunological counterpart, which can then be eluted at a low pH (between 2 and 3), often with very high yields (173). This type of method is most used in the isolation of specific AB, with the help of AG bound to a solid material (174,175), but it definitely also has its uses in the reverse situation.

This method is potentially the most specific way of obtaining virtually any protein in a pure state, with a quite impure mixture as a starting source.

VI. OTHER PURIFICATION METHODS

When it is not possible to obtain plasma fractions in a totally sterile way, due to multiple manipulations, it may be necessary to sterilize such protein solutions in other ways. Also, in order to decrease the very small, but still finite, risk of viral hepatitis infections, particularly in plasma protein preparations from pools which originate from many donors, further measures are taken which decrease such a risk by several orders of magnitude.

A. Sterilization by Heating

Two types of heat treatment are used, to decrease, in the first case, and to abolish, in the second case, the incidence of serum hepatitis.

1. Whole plasma, pooled, with approximately 8 units (each originating from a different donation) per 2-liter bottle, with 5% glucose added (to decrease the precipitation of fibrin) must, according to the regulations of the National Institutes of Health, be kept for 6 months at 32°C. By this method the incidence of hepatitis is drastically decreased, and the majority of the plasma proteins are conserved without much alteration.

2. The only method of inactivating hepatitis virus possibly present in plasma protein solutions with virtually absolute certainty is by heating the solution at 60°C for 10 hr (176). Albumin and some of the globulins can stand this treatment with a minimum of denaturation in the presence of $0.004M$ sodium caprylate as a protective agent. The less soluble of the plasma proteins, which are at the same time the least heat-stable proteins, in particular fibrinogen and the euglobulins, are precipitated by this heat treatment and are best removed prior to it (140,177). This can be done by precipitation with Zn^{2+} salts (141) by desalination of the plasma with ion-exchange resins (140) or by the cold ethanol precipitation method (141).

B. Sterilization by Fine Filtration

Filtering solutions through membranes or porous masses with a maximum pore diameter of $0.5\,\mu$ or less, will by simple mechanical removal of all microorganisms, result in sterilization of the filtrate.

The classical filters of this type are asbestos pads (Seitz), filtration candles of the ceramic (Chamberland) and the compressed diatomaceous earth (Berkefeld) varieties, and the "ultrafine" grade of fritted glass filters (Zeiss G-5). The drawback of all these sterile filters is their thickness and, thus, frequently an appreciable loss of filtrate remaining adsorbed to the filter mass. All these filters are obtainable from most commercial supply houses.

A newer variety of filters, consisting of thin porous cellulose acetate membranes (and more recently also of membranes made of synthetic polymer materials) with a maximum pore size of $0.5\ \mu$, permits sterile filtration with much less loss of material due to adsorption (Millipore, Bedford, Mass., and Gelman, Ann Arbor, Michigan). It has been found recently (178) that these firms routinely add detergents to their membranes, which can be quite awkward for protein purification work. Filters free of detergent may, apparently be available from these manufacturers, upon special request. Similar membranes made of porous sheets of silver, Flotronics, are manufactured by Selas (Spring House, Pennsylvania). These heat-resistant membranes can of course be used repeatedly.

The most recent improvement in this field is the development of ultrathin membranes of pure, detergent-free polycarbonate, all the pores of which are the same size and cylindrical-shaped, with a diameter of exactly $0.5\ \mu$, with their axis essentially perpendicular to the membrane surface. These Nuclepore membranes are made by General Electric (Vallecitos Nuclear Center, Pleasanton, California) by bombardment of the membranes with collimated beams of fission fragments followed by chemical etching of the ionized paths. This is the only membrane of this type of which the pore size is exactly known and directly measurable by microscopic and electron microscopic methods and of which all the pores have the optimal size.

C. Sterilization by Irradiation

As an extra precaution, pooled plasma which has been kept at 32°C for 6 months (see above) is generally, prior to bottling and transfusion, subjected to controlled ultraviolet irradiation of 250–260 μ wavelength in very thin layers (of the order of 10–15 μ thickness), during a period of $\frac{1}{10}$ sec. It can then be stored for up to $2\frac{1}{2}$ years without undergoing further significant changes.

VII. OTHER CONCENTRATION METHODS

A number of drying methods are considered here, as well as a method for protein enrichment by freezing. Various large-scale drying methods such as the heated cylinder methods, although used for drying animal blood proteins of lower quality, will not be further discussed.

A. Evaporation and Pervaporation

Although vacuum evaporation of dilute protein solution at room temperature has been used occasionally (179) on a small scale, the method is fairly tedious and, as it concentrates the salts with the proteins, is generally accompanied by a certain degree of denaturation.

Pervaporation, that is evaporation of water through a membrane which is impermeable to proteins (generally cellophane), is also sometimes practiced (mostly at room temperature) on diluted protein solutions. As here too the salts are concentrated with the protein, this method also leads to some unavoidable denaturation, unless the protein solution has been subjected to exhaustive dialysis against distilled water prior to the pervaporation. The lengthiness of this method also generally entails some alteration of the proteins.

Most other protein concentration methods are preferable to these two.

B. Spray Drying

Animal plasma proteins are usually spray dried when a high quality animal protein is to be obtained (to be used as a glue, in the textile printing industry, etc.). In this process, the protein solution is sprayed in very finely pulverized droplets against a stream of hot air, close to the top of a big conical container and at its widest part. The protein very quickly solidifies into tiny grains which fall to the narrow bottom of the cone where they are collected as a dry powder. Although dried by hot air, the drying of every individual droplet is so fast and its downward travel towards colder regions of the cone so quick that the heat denaturation is kept to a minimum. This is one of the more expensive drying methods.

C. Freeze Drying

Freeze drying, the most expensive method of water removal by evaporation, if done properly, is the method for concentrating proteins by evaporation that will give rise to a minimum of denaturation.

It consists of freezing the protein solution very quickly (generally to at least $-70°C$, with the help of Dry Ice in alcohol), and then, while still in the frozen state, subjecting it to a high vacuum until all the water has been removed by sublimation. If long-term conservation is desired, this should still be done at low temperatures: freeze drying is in itself no substitute for refrigeration.

The freeze-dried protein can be redissolved to any desired concentration in *distilled water*.

D. Freezing

Castro and Ehrlich (180) have made use of the fact that solutes often freeze later than the solvent (when they are frozen slowly) to concentrate serum proteins severalfold in the bottom layers of tall 2 liter bottles of serum, which they stored away in a freezer at $-11°C$. The exact mechanism of this is fairly complex, and salting-out phenomena, caused by locally increasingly high salt concentrations due to initial freezing of mainly distilled water, probably play an important role here.

The important implication of this phenomenon is that serum samples in test tubes or bottles, when stored frozen, have to be thawed *completely* and must be *well shaken* after thawing before any aliquot of the contents of such a tube may be considered representative of the whole. Particularly the $\beta_2 M$ macroglobulins sometimes have, in the author's experience, the tendency to collect in the bottom of a tube upon freezing, and to remain there as a pellet if the contents of the tube are not thoroughly agitated.

References

1. D. C. Henry, *Proc. Roy. Soc. (London), Ser. A,* **133,** 106 (1931).
2. J. T. G. Overbeek, Thesis, University of Utrecht (1941); *Advan. Colloid Sci.,* **3,** 97 (1950).
3. A. Tiselius, *Trans. Faraday Soc.,* **33,** 524 (1937).
4. H. Svensson, "Zonal Density Gradient Electrophoresis," in *Analytical Methods of Protein Chemistry,* Vol. I, P. Alexander and R. J. Block, Eds., Pergamon, London, 1960, p. 195.
5. H. Svensson and T. E. Thompson, "Translational Diffusion Methods in Protein Chemistry," in *Analytical Methods of Protein Chemistry,* Vol. 3, P. Alexander and R. J. Block, Eds., Pergamon, London, 1961, p. 58.
6. A. Tiselius, *Biochem. J.,* **31,** 1464 (1937).
7. R. A. Brown and S. N. Timasheff, "Applications of Moving Boundary Electrophoresis to Protein Systems," in *Electrophoresis,* M. Bier, Ed., Academic Press, New York, 1959, p. 317.
8. P. König, *Actas III Congr. Sul-Am. Chim.,* **2,** 334 (1937).
9. D. von Klobusitzky and P. König, *Arch. Exp. Pathol. Pharmakol.,* **192,** 271 (1939).
10. J. Kohn, *Nature,* **181,** 839 (1958).
11. R. J. Wieme, *Agar Gel Electrophoresis,* Elsevier, Amsterdam, 1965.
12. R. J. Block, E. L. Durrum, and G. Zweig, *Paper Chromatography and Electrophoresis,* Academic Press, New York, 1955.
13. R. J. Wieme, *Protides Biol. Fluids, Proc. Colloq., Bruges, 1959,* **7,** 18 (1960).

14. H. J. H. Kreutzer, Thesis, University of Amsterdam, 1965.
15. C. J. van Oss, D. Annicolas, and C. Labie, *Bull. Soc. Chim. Biol.*, **41**, 1711 (1959).
16. A. P. Gaunce and P. A. Anastassiadis, *Anal. Biochem.*, **17**, 357 (1966).
17. R. A. Phelps and F. W. Putnam, "Chemical Composition and Molecular Parameters of Purified Plasma Proteins," in *The Plasma Proteins*, Vol. 1, F. W. Putnam, Ed., Academic Press, New York, 1960, Table I, p. 146.
18. S. Hjertén, *Protides Biol. Fluids, Proc. Colloq., Bruges, 1959*, **7**, 28 (1960).
19. P. Grabar and C. A. Williams, *Biochim. Biophys. Acta*, **10**, 193 (1953); **17**, 67 (1955).
20. P. Burtin, "The Proteins of Normal Human Plasma," in *Immunoelectrophoretic Analysis*, P. Grabar and P. Burtin, Eds., Elsevier, Amsterdam, 1964, p. 94.
21. J. J. Scheidegger, *Intern. Arch. Allergy Appl. Immunol.*, **7**, 103 (1955).
22. J. Uriel, "The Characterization Reactions of the Protein Constituents Following Electrophoresis or Immunoelectrophoresis in Agar," in *Immunoelectrophoretic Analysis*, P. Grabar and P. Burtin, Eds., Elsevier, Amsterdam, 1964, p. 30.
23. F. Peetoom, *The Agar Precipitation Technique*, Stenfert-Kroese, Leiden, 1963.
24. J. Kohn, *Protides Biol. Fluids, Proc. Colloq., Bruges, 1961*, **9**, 120 (1962).
25. M. D. Poulik, *Can. J. Med. Sci.*, **30**, 417 (1952); *J. Immunol.*, **82**, 502 (1959); *Nature*, **198**, 752 (1963).
26. E. A. Kabat and M. M. Mayer, *Experimental Immunochemistry*, 2nd ed., C. C Thomas, Springfield, Ill., 1961, appendices F, G, H, p. 811.
27. H. Bloemendal, *Zone Electrophoresis in Blocks and Columns*, Elsevier, Amsterdam, 1963.
28. H. G. Kunkel and R. Trautman, "Zone Electrophoresis in Various Types of Supporting Media," in *Electrophoresis*, M. Bier, Ed., Academic Press, New York, 1959, p. 225.
29. H. G. Kunkel, in *Methods of Biochemical Analysis*, Vol. 1, D. Glick, Ed., Interscience, New York, 1954, p. 141.
30. H. J. Müller-Eberhard, *Scand. J. Clin. Lab. Invest.*, **12**, 33 (1960).
31. C. J. Bradish and N. V. Smart, *Nature*, **174**, 272 (1954).
32. S. Avrameas and J. Uriel, *Nature*, **202**, 1005 (1964).
33. J. Porath, *Sci. Tools*, **11**, 21 (1964).
34. G. Haugaard and T. D. Kroner, U. S. Pat., 2,555,487 (1948).
35. H. Svensson and I. Brattsten, *Arkiv Kemi*, **1**, 401 (1949).
36. W. Grassmann, *Angew. Chem.*, **62**, 170 (1950); *Ber. Ges. Physiol. Exptl. Pharmakol.*, **139**, 220 (1950).
37. W. Grassmann and K. Hannig, *Z. Angew. Chem.*, **62**, 170 (1950); *Naturwiss.*, **37**, 397 (1950).
38. E. L. Durrum, *J. Am. Chem. Soc.*, **73**, 4875 (1951).
39. K. Hannig, *Z. Anal. Chem.*, **181**, 244 (1961).
40. H. Peeters, P. Vuylsteke, and R. Noë, *J. Chromatog.*, **2**, 308 (1959).
41. F. Blank and E. Valko, *Biochem. Z.*, **195**, 220 (1928).
42. W. Pauli, *Biochem. Z.*, **152**, 355, 360 (1924).
43. W. Pauli and E. Valko, *Elektrochemie der Kolloide*, Springer, Vienna, 1929.
44. W. Pauli, P. Stamberger, and E. Schmidt, Brit. Pat., 492,030; *Chem. Abstr.*, **33**, 19897 (1939).

45. J. R. Cann, J. G. Kirkwood, R. A. Brown, and O. J. Plescia, *J. Am. Chem. Soc.*, **71**, 1603 (1949); *J. Biol. Chem.*, **181**, 161 (1949).
46. J. G. Kirkwood, *J. Chem. Phys.*, **9**, 878 (1941).
47. H. Gutfreund, *Biochem. J.*, **37**, 186 (1943).
48. S. N. Timasheff, R. A. Brown, and J. G. Kirkwood, *J. Am. Chem. Soc.*, **75**, 3121, 3124 (1953).
49. A. Polson, *Biochim. Biophys. Acta*, **11**, 315 (1953).
50. A. Polson and J. F. Largier, "Multi-Membrane Decantation," in *Analytical Methods of Protein Chemistry*, Vol. 1, P. Alexander and R. J. Block, Eds., Pergamon, London, 1960, p. 161.
51. H. C. Isliker, *Advan. Protein Chem.*, **12**, 416 (1957).
52. M. Bier, "Preparative Electrophoresis Without Supporting Media," in *Electrophoresis*, M. Bier, Ed., Academic Press, New York, 1959, p. 263; *Chem. Eng. News*, 60 (Oct. 17, 1960).
53. S. Raymond, *Proc. Soc. Exptl. Biol. Med.*, **81**, 278 (1952); U.S. Pat. 2,758,966 (1956).
54. T. H. Donnelly, D. B. Badgley, and L. M. Galvez, 59th Annual Meeting A. Ch. I. E., Proc. Symp. Biochem. Eng., Detroit, 1966, 14B.
55. *Chem. Eng. News*, 57 (May 24, 1965).
56. H. A. Abramson, L. S. Moyer, and M. H. Gorin, *Electrophoresis of Proteins*, Reinhold, New York, 1942; reprinted by Hafner, New York, 1964.
57. C. J. van Oss and J. M. Singer, *J. Colloid Interface Sci.*, **21**, 118 (1966).
58. C. J. van Oss, "Methods Used in the Characterization of Immune Globulins," in *Proceedings Immunogenetics Symposium*, T. J. Greenwalt, Ed., Wood, 1966, Lippincott, Philadelphia, 1967, p. 1.
59. A. Ehrenberg, *Acta Chem. Scand.*, **11**, 1257 (1957).
60. W. J. Archibald, *J. Phys. and Colloid Chem.*, **51**, 1204 (1947).
61. R. Trautman and C. F. Crampton, *J. Am. Chem. Soc.*, **31**, 4036 (1959).
62. S. Claesson and I. Morning-Claesson, "Ultracentrifugation," in *Analytical Methods of Protein Chemistry*, P. Alexander and R. J. Block, Eds., Pergamon, London, 1961, p. 119.
63. H. Schachman, *Biochem.*, **2**, 887 (1963); *Ultracentrifugation in Biochemistry*, Academic Press, New York, 1959.
64. R. Trautman, *J. Phys. Chem.*, **60**, 1211 (1956).
65. N. G. Anderson, *Nature*, **181**, 45 (1958).
66. H. G. Kunkel, "Macroglobulins and High Molecular Weight Antibodies," in *The Plasma Proteins*, Vol. I, F. W. Putnam, Ed., Academic Press, New York, 1960, p. 279.
67. G. M. Edelman, H. G. Kunkel, and E. C. Franklin, *J. Exptl. Med.*, **108**, 105 (1958).
68. G. Aubel-Sadron, *Bull. Soc. Chim. Biol.*, **41**, 1361 (1959).
69. D. W. Budworth, *J. Sci. Instr.*, **39**, 377 (1962).
70. J. Crank, *The Mathematics of Diffusion*, Clarendon Press, Oxford, 1964.
71. J. H. Fessler and J. Vinograd, *Biochim. Biophys. Acta*, **103**, 160 (1965).
72. L. Gropper and O. Griffith, *Anal. Biochem.*, **16**, 171 (1966).
73. "An Introduction to Density Gradient Centrifugation," Tech. Rev. No. 1, Beckman Instruments, Spinco Division, Palo Alto, California, 1960.
74. K. M. Cowan and R. Trautman, *J. Immunol.*, **94**, 858 (1965).
75. R. Trautman and S. S. Breese, *J. Phys. Chem.*, **63**, 41 (1959).

76. V. N. Schumaker and J. Rosenbloom, *Biochemistry*, **4**, 1005 (1965).
77. J. Vinograd, R. Bruner, R. Kent, and J. Weigle, *Proc. Natl. Acad. Sci. U.S.*, **49**, 902 (1963).
78. R. Trautman, S. S. Breese, and H. L. Bachrach, Natl. Colloid Symp. 36th Preprints **1962**, 178.
79. F. Koenig, J. K. Palmer, and C. J. Likes, *Science*, **128**, 533 (1958).
80. S. Filitti-Wurmser and L. Hartmann, *Bull. Soc. Chim. Biol.*, **44**, 725 (1962).
81. K. Jahnke and W. Scholtan, *Die Bluteiweisskörper in der Ultrazentrifuge*, Georg Thieme Verlag, Stuttgart, 1960.
82. N. G. Anderson, *Fractions*, **1**, 2 (1965); *New Scientist*, **421**, 732 (Dec. 1964).
83. C. J. van Oss, *L'Ultrafiltration, Thesis*, University of Paris, 1955; de Bussy, Amsterdam, 1955.
84. L. Ambard and S. Trautmann, *Ultrafiltration*, Thomas, Springfield, Ill., 1960.
85. C. J. van Oss and N. Beyrard, *J. Chim. Phys.*, **60**, 451 (1963).
86. W. F. Blatt, M. P. Feinberg, H. B. Hopfenberg, and C. A. Sorovis, *Science*, **150**, 224 (1965).
87. A. S. Michaels, *Ind. Eng. Chem.*, **57**, 32 (1965).
88. R. Zsigmondy, *Biochem. Z.*, **171**, 198 (1926); *Z. Angew, Chem.*, **39**, 398 (1926).
89. W. J. Elford, *J. Pathol. Bacteriol*, **34**, 505 (1931); *Proc. Roy. Soc. (London)*, *Ser. B*, **106**, 216 (1930); **112**, 384 (1933); *Trans. Faraday Soc.*, **33**, 1094 (1937); with J. D. Ferry, *Biochem. J.*, **28**, 650 (1934); **30**, 84 (1936).
90. S. Loeb, *Desalination*, **1**, 35 (1966); with S. Sourirajan, Report 60-60, Dept. of Engineering, UCLA, Los Angeles, 1960.
91. H. P. Gregor and E. Kantner, *J. Phys. Chem.*, **61**, 1169 (1957).
92. C. J. van Oss, A. Scheinman, and J. E. Lord, *Nature*, **215**, 639 (1967).
93. C. J. van Oss, *Rec. Trav. Chim.*, **77**, 479 (1958); with H. Simonnet and D. Annicolas, *ibid.*, **78**, 425 (1959).
94. L. C. Craig, "Fractionation and Characterization by Dialysis," in *Analytical Methods of Protein Chemistry*, Vol. 1, P. Alexander and R. J. Block, Eds., Pergamon, London, 1960, p. 103; *Science*, **144**, 1093 (1964).
95. I. M. Klotz, "Protein Interactions," in *The Proteins*, Vol. 1B, H. Neurath and K. Bailey, Eds., Academic Press, New York, 1953, p. 727.
96. L. C. Craig and K. Stewart, *Biochemistry*, **4**, 2712 (1965).
97. K. K. Hjelm, *Compt. Rend. Trav. Lab. Carlsberg*, **32**, 409 (1962).
98. J. P. Merrill, *Sci. Am.*, **205**(1), 56 (1961); *New Engl. J. Med.*, **246**, 17 (1952).
99. L. C. Craig and W. Konigsberg, *J. Phys. Chem.*, **65**, 166 (1961).
100. T. Wood, *Biochem. J.*, **62**, 611 (1956).
101. J. Porath, *Adv. Protein Chem.*, **17**, 209 (1962).
102. P. Flodin and J. Killander, *Biochem. Biophys. Acta*, **63**, 403 (1962).
103. J. Porath and H. Bennich, *Arch. Biochem. Biophys., Suppl.*, **1**, 152 (1962).
104. J. Porath, private communication (1965).
105. A. N. Glazer and D. Wellner, *Nature*, **194**, 862 (1962).
106. S. Hjertén and R. Mosbach, *Anal. Biochem.*, **3**, 109 (1962).
107. R. L. Steere and G. K. Ackers, *Nature*, **194**, 114 (1962); **196**, 475 (1962).
108. S. Hjertén, *Arch. Biochem. Biophys.*, **99**, 466 (1962).
109. W. Haller, *Nature*, **206**, 693 (1965).

244 C. J. VAN OSS

110. B. G. Johansson and L. Rymo, *Acta Chem. Scand.*, **18**, 217 (1964).
111. P. Flodin, B. Gelotte, and J. Porath, *Nature*, **188**, 493 (1960).
112. C. Curtain, *Nature*, **203**, 1380 (1964).
113. G. Franglen and C. Gosselin, *Nature*, **181**, 1152 (1958).
114. H. E. Schultze and J. F. Heremans, *Molecular Biology of Human Proteins*, Vol. 1, Elsevier, New York, 1966.
115. O. Smithies, *Biochem. J.*, **61**, 629 (1955).
116. F. F. Davis, *Biochim. Biophys. Acta*, **42**, 1 (1960).
117. S. Raymond and L. Weintraub, *Science*, **130**, 711 (1959).
118. L. Ornstein and B. J. Davis, *Disc Electrophoresis*, Distillation Products Industries, Rochester, N.Y., 1962.
119. E. W. Baur, *J. Lab. Clin. Med.*, **61**, 166 (1963).
120. L. A. Lewis, *Clin. Chem.*, **12**, 596 (1966).
121. O. Smithies, *Arch. Biochem. Biophys., Suppl.*, **1**, 125 (1962).
122. Ref. 114, pp. 384–402.
123. M. D. Poulik and G. M. Edelman, *Protides Biol. Fluids, Proc. Colloq., Bruges, 1961*, **9**, 126 (1962); M. D. Poulik, *ibid., Bruges, 1964*, **12**, 400 (1965).
124. M. D. Poulik, *Nature*, **180**, 1477 (1957).
125. M. D. Poulik and O. Smithies, *Biochem. J.*, **68**, 636 (1958).
126. L. Korngold, *Int. Arch. Allergy Appl. Immunol.*, **23**, 268 (1963).
127. J. W. Keyser, *Anal. Biochem.*, **9**, 249 (1964).
128. S. Keller and R. J. Block, "Fractionation of Proteins by Adsorption and Ion Exchange," in *Analytical Methods of Protein Chemistry*, Vol. 1, P. Alexander and R. J. Block, Eds., Pergamon, London, 1960, p. 67.
129. E. A. Peterson and H. A. Sober, *Chromatography of the Plasma Proteins*, in *The Plasma Proteins*, F. W. Putnam, Ed., Academic Press, New York, 1960, p. 105.
130. L. A. Carlson, *Acta Chem. Scand.*, **9**, 1046 (1955).
131. C. J. van Oss and J. M. Singer, *J. Reticuloendothelial Soc.*, **3**, 29 (1966).
132. Ref. 129, p. 109.
133. E. A. Peterson and H. A. Sober, "A Variable Gradient Device for Chromatography," in *Analytical Methods of Protein Chemistry*, P. Alexander and R. J. Block, Eds., Vol. 1, Pergamon, London, 1960, p. 88.
134. C. W. Parr, *Biochem. J., Proc.*, **56**, xxvii (1954).
135. R. M. Bock and N. S. Ling, *Anal. Chem.*, **26**, 1543 (1954).
136. G. Oster, *Sci. Am.*, **213**, 70 (August 1965).
137. B. W. Moore and D. McGregor, *J. Biol. Chem.*, **240**, 1647 (1965).
138. D. R. Stanworth, *Nature*, **188**, 156 (1960).
139. M. Stefanini and S. I. Magalini, "Applications of Ion Exchangers to Blood," in *Ion Exchangers in Organic and Biochemistry*, C. Calmon and T. R. E. Kressman, Eds., Interscience, New York, 1957, p. 432.
140. H. Nitschmann, P. Kistler, H. R. Renfer, A. Hässig, and A. Joss, *Vox Sanguinis*, **1**, 183 (1956).
141. R. B. Pennell, "Fractionation and Isolation of Purified Components by Precipitation Methods," in *The Plasma Proteins*, Vol. 1, F. W. Putnam, Ed., Academic Press, New York, 1960, p. 9.
142. P. Debye, *Z. Physik. Chem. (Leipzig)*, **A130**, 56 (1927); with J. McAulay, *Physik. Z.*, **26**, 22 (1925).
143. J. Lyklema, "Adsorption of Gegenions," Thesis, University of Utrecht, 1957.

144. F. Hofmeister, *Arch. Exptl. Pathol. Pharmakol.*, **25**, 1 (1889); **27**, 395 (1890); **28**, 210 (1891).
144a. J. Porath, *Nature*, **196**, 47 (1962).
145. C. J. van Oss and S. Buenting, *Transfusion*, **7**, 77 (1967).
146. C. E. Huggins, *Science*, **139**, 504 (1963); *J. Am. Med. Assoc.*, **193**, 941 (1965).
147. C. J. Cohn, W. L. Hughes, D. J. Mulford, J. N. Ashworth, and M. Melin, *J. Am. Chem. Soc.*, **68**, 459 (1946).
148. R. A. Keckwick and M. E. Mackay, *Med. Res. Council Spec. Rept. Ser.* **286** (1954).
149. H. G. Bungenberg de Jong, "Complex Colloid Systems" and "Morphology of Coacervates," in *Colloid Science*, Vol. 2, H. R. Kruyt, Ed., Elsevier, Amsterdam, 1949, pp. 335, 433.
150. E. Alameri, *Suomen Kemistiehti*, **28B**, No. 1, 28 (1955).
151. H. Morawetz and W. L. Hughes, *J. Phys. Chem.*, **56**, 64 (1952).
152. A. Polson, G. M. Potgieter, J. F. Largier, G. E. F. Mears, and F. J. Jouberth, *Biochim. Biophys. Acta*, **82**, 463 (1964).
153. H. G. Kunkel, *Proc. Soc. Exptl. Biol. Med.*, **66**, 217 (1947).
154. K. Dirr, P. Decker, and M. Becker, *Z. Physiol. Chem.*, **307**, 97 (1957).
154a. T. J. Greenwalt, E. A. Steane, and C. J. van Oss, *Federation Proc.*, **25**, 612 (1966).
155. A. Saifer, *J. Lab. Clin. Med.*, **63**, 1054 (1964).
156. L. C. Craig, *Partition*, in *Analytical Methods of Protein Chemistry*, Vol. 1, P. Alexander and R. J. Block, Eds., Pergamon, London, 1960, p. 122.
157. A. Kepes, *Bull. Soc. Chim. Biol.*, **35**, 1243 (1953).
158. A. J. P. Martin and R. L. M. Synge, *Biochem. J.*, **35**, 1358 (1941).
159. L. C. Craig and O. Post, *Anal. Chem.*, **21**, 500 (1949).
160. W. Haussmann and L. C. Craig, *J. Am. Chem. Soc.*, **80**, 2703 (1958).
161. L. C. Craig and D. Craig, "Separation and Purification," in *Technique of Organic Chemistry*, Part I, Vol. 3, 2nd ed., A. Weissberger, Ed., Interscience, New York, 1956, p. 149; see also E. C. Scheibel, *ibid.*, p. 332.
162. P. A. Albertsson, *Partition of Cell Particles and Macromolecules*, Wiley, New York, 1960.
163. P. A. Albertsson and E. J. Nyns, *Nature*, **184**, 1465 (1959).
164. B. A. Askonas, J. H. Humphrey and R. R. Porter, *Biochem. J.*, **63**, 412 (1956).
165. P. Hirsch-Ayalon, *Rec. Trav. Chim.*, **75**, 1065 (1965); **79**, 382 (1960); **80**, 365, 376 (1961); *J. Polymer Sci.*, **23**, 697 (1957); *Electrochim. Acta*, **10**, 773 (1963).
166. C. J. van Oss and Y. S. L. Heck, *Z. Immunitätsforsch.*, **122**, 44 (1961) (article in English).
167. Ö. Ouchterlony, *Acta Pathol. Microbiol. Scand.*, **32**, 231 (1953).
168. A. C. Allison and J. H. Humphrey, *Nature*, **183**, 1590 (1959); *Immunology*, **3**, 95 (1960).
169. E. A. Kabat, "Precipitin Reaction," in *Experimental Immunochemistry*, 2nd ed., E. A. Kabat and M. M. Mayer, Eds., Thomas, Springfield, Illinois, 1961, p. 22.
170. C. J. van Oss, in preparation.
171. C. J. van Oss and P. Hirsch-Ayalon, *Science*, **129**, 1365 (1959).
172. G. B. Sutherland, *Science*, **132**, 1252 (1960).

173. N. Weliky and H. H. Weetall, *Immunochemistry*, **2**, 293 (1965); N. Weliky, H. H. Weetall, R. V. Gilden, and D. H. Campbell, *Immunochemistry*, **1**, 219 (1964).
174. K. Onoue, Y. Yogi, and D. Pressman, *Immunochemistry*, **2**, 181 (1965).
175. A. A. Hirata and D. Campbell, *Immunochemistry*, **2**, 195 (1965).
176. S. S. Gellis, J. R. Neefe, J. Stokes, L. E. Strong, C. A. Janeway, and G. Scotchord, *J. Clin. Invest.*, **27**, 239 (1948).
177. J. H. Hink, J. Hidalgo, V. P. Seeberg, and F. F. Johnson, *Vox Sanguinis*, **2**, 174 (1957).
178. R. D. Cahn, *Science*, **155**, 195 (1967).
179. C. de Vaux St. Cyr and J. C. Patte, "Immuno-Electrophoretic Study of Proteinurias," in *Immuno-Electrophoretic Analysis*, P. Grabar and P. Burtin, Eds., Elsevier, Amsterdam, 1964, p. 230.
180. A. Castro and A. Ehrlich, *Transfusion* (*Paris*), **6**, 594 (1966).

Purification of Zirconium by a Dry Process. Chemical Separation of Hafnium from Zirconium*

ORVILLE D. FRAMPTON AND JULIAN FELDMAN

*Research Department U.S. Industrial Chemicals Co.
Division of National Distillers and Chemical Corporation
Cincinnati, Ohio*

* Major features of this manuscript were presented before the Division of Industrial and Engineering Chemistry of the American Chemical Society at St. Louis, Missouri, during March, 1961.

I. INTRODUCTION

A. Zirconium for Atomic Reactors

Zirconium is especially desirable as a construction material for liquid-cooled atomic reactors of the thermal type because of its requisite properties of low cross section for thermal neutron capture [0.18 barns (1)], resistance to corrosion (2), high temperature strength (3), and relatively high melting point, 1855 ± 15°C (4).

Naturally occurring zirconium ore contains hafnium to the extent of 2–3% based on zirconium. In contrast to zirconium, hafnium has a high cross section for thermal neutron capture [115 barns (5)]. Thus, in order to provide reactor grade zirconium, hafnium must be substantially removed. The AEC specifications for hafnium content in reactor grade zirconium is 0.01% or less.

B. Separation Methods

1. GENERAL

Until Newnham's discovery (6) of the separation of hafnium from zirconium by a differential reduction of their tetrachlorides, few practical chemical separations of hafnium from zirconium had been devised because in most cases their compounds have almost identical properties. Steidler (7), for example, found not the slightest difference in several reactions of compounds of these two elements. Larsen (8) ascribed the similarity to the configurations of the valence electrons, $4d^25s^2$ and $5d^26s^2$ for zirconium and hafnium, respectively, and to the nearly equal ionic radii of the M^{4+} ions, i.e., $Zr^{4+} = 0.74\text{Å}$ and $Hf^{4+} = 0.75\text{ Å}$.

A wide variety of schemes have been proposed for the separation of compounds of hafnium and zirconium based on slight differences in physical and/or chemical properties. The most important of these have been discussed in detail by Herndon (9) and Miller (10). They include fractional crystallization and precipitation, fractional distillation and sublimation, ion migration, column chromatography, vapor-phase dechlorination, thermal decomposition, and liquid–liquid extraction.

2. COMMERCIAL PROCEDURES

Liquid–liquid extraction methods have proven to be practical, and two procedures have been developed into commercial operations (11). One method (12), which has been used for a number of years by the U.S. Bureau of Mines, the Carborundum Corporation, and U.S. Industrial Chemicals Company, is a modification (13) of a process devised by W. Fischer et al. (14), and depends upon the distribution of the thiocyanates of zirconium and hafnium between aqueous acid and methyl isobutyl ketone. Many stages are required in this extraction process to achieve the required low concentration of hafnium. In addition, the complications of converting the metallic ion to the solid tetrachloride required as feed for reduction by the Kroll (15) or U.S.I. process, with resulting problems of corrosion, add to the expense of this procedure.

The NRC Metals Company, a subsidiary of National Research Corporation, uses an extraction procedure based on the work of H. A. Wilhelm (16). In this method, tributyl phosphate preferentially extracts zirconium from hafnium when they are present as a solution of nitrates in aqueous nitric acid.

3. DRY CHEMICAL METHODS

Dry chemical methods have been devised by Newnham (6) and Chandler (17). These have a major advantage over the commercial extraction procedures in that the $ZrCl_4$ required for sponge preparation is obtained directly. Thus, reconstitution of dissolved zirconium salts to the anhydrous tetrachloride state is avoided.

Chandler's method is based on a vapor–solids system and depends on a metathetical reaction between $HfCl_4$ impurity in the feed and ZrO_2 in the bed to yield nonvolatile HfO_2 and $ZrCl_4$ vapor [eq. (1)]

$$HfCl_4(g) + ZrO_2(s) \rightarrow ZrCl_4(g) + HfO_2(s) \qquad (1)$$

Using 15 in. of bed, he obtained a decrease in hafnium content of the tetrachlorides of from 3 to 1%.

Newnham's method developed from an observation that different temperatures had been reported for the reduction of $ZrBr_4$ (18) and $HfBr_4$ (19) with aluminum to the respective nonvolatile tribromides. Using this observation as a clue together with established chemistry of the lower valence zirconium and hafnium halides (described in Sect. II-A), he also developed a vapor–solids system of separation. Employing static, closed systems *in vacuo*, principally small glass bulbs, he demonstrated the feasibility of separation of $HfCl_4$ from $ZrCl_4$ by reduction of the latter to nonvolatile $ZrCl_3$ with subsequent sublimation of unchanged $HfCl_4$.

It was our purpose to develop a laboratory method into a practical semicontinuous bench-scale process operating at atmospheric pressure. The U.S. Industrial Chemical Company could then develop this further toward a commercial operation.

II. CHEMISTRY OF ZIRCONIUM AND HAFNIUM HALIDES

A. Lower Valence State Halides

Ruff and Wallstein (20) first observed the reduction *in vacuo* of $ZrCl_4$ to $ZrCl_3$, a brown powder, by aluminum at 300°C [eq. (2)], the thermal disproportionation of the $ZrCl_3$ *in vacuo* at 330°C to the tetrachloride and the nonvolatile $ZrCl_2$ [eq. (3)], and the thermal disproportionation *in vacuo* of the dichloride to the tetrachloride and metallic zirconium at 600°C [eq. (4)].

$$3ZrCl_4(g) + Al(s) \rightarrow 3ZrCl_3(s) + AlCl_3(g) \tag{2}$$

$$2ZrCl_3(s) \rightarrow ZrCl_4(g) + ZrCl_2(s) \tag{3}$$

$$2ZrCl_2(s) \rightarrow ZrCl_4(g) + Zr(s) \tag{4}$$

In eq. (2), aluminum trichloride and unreacted zirconium tetrachloride were removed from the reaction mixture by volatilization.

deBoer and Fast (21) observed that zirconium metal was attacked by hot (1200°C) vapors of $ZrCl_4$ or $ZrBr_4$. The former yielded a black, shiny compound with properties similar to those of the $ZrCl_2$ described by Ruff and Wallstein, and the latter yielded a compound having the composition $ZrBr_2$. This reaction may be expressed by eq. (5):

$$ZrBr_4(g) + Zr(s) \rightarrow 2ZrBr_2(s) \tag{5}$$

Young (18) prepared $ZrBr_3$ by the reduction of $ZrBr_4$ vapor with aluminum at 450°C in a stream of hydrogen gas. By-product $AlBr_3$ was removed by sublimation *in vacuo* at 140–200°C. The sublimate was condensed on a cooled surface which was later sealed off. Then the removal of unreacted $ZrBr_4$ at 290°C in a similar fashion left the $ZrBr_3$. However, careful investigation of this product indicated the presence of small amounts of metallic zirconium and zirconium dibromide in the sample. To eliminate these substances the preparation was altered so that at the point following removal of the $AlBr_3$, the mixture was heated to 300°C for 8 hr in the presence of the excess $ZrBr_4$. This insured conversions as indicated by eqs. (6) and (7):

$$3ZrBr_4(g) + Zr(s) \rightarrow 4ZrBr_3(s) \tag{6}$$

$$ZrBr_4(g) + ZrBr_2(s) \rightarrow 2ZrBr_3(s) \tag{7}$$

Young demonstrated the reaction shown by eq. (6) and postulated eq. (7) in analogy to similar reactions for titanium bromides. The remaining unreacted $ZrBr_4$ was then removed as already described above.

Young also prepared $ZrBr_2$ by thermal disproportionation *in vacuo* of zirconium tribromide at 350–390°C, similar to the disproportionation of $ZrCl_3$ as shown by eq. (3).

Fast (22) demonstrated the reduction of ZrI_4 to ZrI_3 with zirconium metal at 400°C *in vacuo*, analogous to that shown in eq. (6) for the bromides. The disproportionation reactions of the iodides were analogous to those shown in eqs. (3) and (4).

Schumb and Morehouse (19) reduced $HfBr_4$ to the blue-black tribromide with aluminum in a stream of hydrogen gas at 600°C, employing procedures used by Young (18) for preparing $ZrBr_3$ from $ZrBr_4$. They found that heating the tribromide *in vacuo* to 300–360°C caused it to disproportionate, liberating $HfBr_4$ and leaving a jet-black residue, presumably $HfBr_2$ in a reaction similar to that shown by eq. (3). On further heating to above 400°C, they found the $HfBr_2$ yielded $HfBr_4$ and presumably hafnium metal.

Larsen and Leddy (23) studied the reduction of $ZrBr_4$ with Zr [eq. (6)] along with similar reactions of $ZrCl_4$, ZrI_4, and ZrF_4 and corresponding reactions of hafnium metal with hafnium tetrahalides. Reductions in evacuated glass ampules were conducted at 400, 500, and 700°C, using stoichiometric amounts of reactants. The autogeneous pressure developed inside the ampules was balanced by controlled pressure applied externally to the ampule. Tetrahalide was collected in an air-cooled portion of the ampule at the conclusion of the reaction period, and this portion was then sealed off. The reaction mixture was added to water, and after the resulting hydrogen evolution subsided, metallic zirconium was filtered off and the aqueous solution was analyzed for metal and halide ions. The analysis showed the soluble materials to have the composition of trihalide, except the fluoride which failed to react even at 700°C for 72 hr. The stoichiometry was further substantiated by determining the amount of hydrogen evolved on solution of the trihalide.

Larsen and Leddy found the rate of disproportionation, corresponding to eq. (3), to be appreciable only at a temperature above 450°C, a temperature higher than that observed by others. They failed to detect the presence of pure dichlorides over the temperature range 250–500°C, and suggested that the residues were mixtures of trichloride and dichloride.

B. Newnham Separation

In their reduction studies of $HfCl_4$ with Hf metal and $ZrCl_4$ with Zr metal, Larsen and Leddy observed that the extent of reaction of $HfCl_4$ was slightly lower than that for $ZrCl_4$. They concluded that the difference in ease of reduction did not appear to be great enough to be utilized in a separation procedure. Schumb and Morehouse (19) found a temperature of 600°C to be necessary for reduction of $HfBr_4$ to $HfBr_3$ with aluminum. Young found the lower temperature of 450°C to be sufficient for a similar reduction of $ZrBr_4$ to $ZrBr_3$ (18). These findings suggested to Newnham that reaction of a mixture of the tetrachlorides of zirconium and hafnium at *selected* temperatures with *selected* reducing agents might result in a preferential reduction of $ZrCl_4$. Separation might then be achieved because the resulting zirconium trichloride would not be volatile at the sublimation temperature of $HfCl_4$.

Newnham selected zirconium powder prepared by dissociation of zirconium hydride at 900°C as the reducing agent. Using small (50-ml) evacuated glass bulbs containing zirconium powder and solid crude $ZrCl_4$ containing 1.5% $HfCl_4$, he demonstrated the differential reduction at 420°C. Newnham also separated $HfCl_4$ from the reaction mixture by sublimation at 200°C *in vacuo*. He obtained a greenish-black, solid residue of $ZrCl_3$, mixed with unreacted Zr, which had a hafnium content as low as 0.01%.

Newnham outlined steps for two separation schemes each free of solution chemistry. In one (6,24), $ZrCl_4$ in crude feed is reduced by $ZrCl_2$ to the trichloride and $HfCl_4$ is separated by sublimation. The purified $ZrCl_4$ product is produced by the disproportionation of $ZrCl_3$ [eq. (3)], and the $ZrCl_2$ reducing agent is simultaneously regenerated. The $ZrCl_2$ used for the initial reduction is prepared by reacting $ZrCl_4$ and metallic Zr. The resulting $ZrCl_3$ is disproportionated to produce the $ZrCl_2$ and residual $ZrCl_4$ is volatilized out.

In the other scheme (6), aluminum is used as the reducing agent. This is advantageous in that a lower temperature is permissable, thus minimizing disproportionation of $ZrCl_3$ formed. The $ZrCl_3$ dissolves in the $AlCl_3$ melt, thus eliminating the problem of caking encountered when Zr is the reducing agent.

On the basis of thermodynamic calculations and the general behavior of transition elements, Prakash and Sundaram (25) have indicated the feasibility of the selective reduction at 300–400°C of $ZrCl_4$ with aluminum to comparatively nonvolatile di- and trihalides, leaving $HfCl_4$ mostly unchanged. This was suggested as an effective scheme for obtaining reactor grade zirconium.

Alternative schemes suggested by Newnham include solution of the subhalide in water from which resulting oxychloride may be extracted by various methods, or dissociation of the subhalide at 600°C to yield Zr metal itself [eqs. (3) and (4)].

We employed Newnham's scheme (6) in our studies and developed a dynamic bench-scale procedure operating at atmospheric pressure which yields $ZrCl_4$ in purity suitable for preparation of reactor grade zirconium.

C. The Nature of $ZrCl_3$

There is disagreement among some investigators on the crystal structure of $ZrCl_3$ as determined by x-ray structural analysis (26–29). Watts (27) suggested that the differences were due to the methods of preparation. He reported two forms; a β form prepared at moderate or low temperatures, and an α form probably formed at higher temperatures. He also suggested other forms which were similar in structure to the γ and δ forms of $TiCl_3$. The $ZrCl_3$ employed by Watts in his studies was prepared by a method developed by Newnham and Watts in which $ZrCl_4$ was reduced at 200°C with atomic hydrogen. The latter was made by passing a mixture of hydrogen gas and the $ZrCl_4$ vapor, at a pressure of 3–4 mm, through an electric glow discharge.

D. Zr_3Cl_8

In part of our studies detailed later, crude $ZrCl_4$ containing about 2.5% Hf was reduced with preformed $ZrCl_2$ powder to the nonvolatile zirconium subchlorides. The experiments were carried out in a stirred reactor at about 330°C under an argon atmosphere. Unreacted tetra-

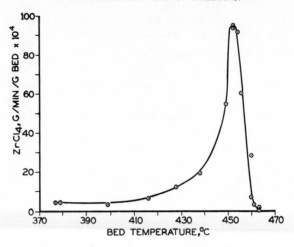

Fig. 1. Dissociation rate of zirconium subhalide.

chloride vapors (including the $HfCl_4$) were swept out of the reactor with argon. Then the temperature was raised and the purified $ZrCl_4$ vapor was swept out. All effluent vapors were collected in consecutive fractions for analysis. During the temperature-rise period the effluent rate was relatively low, between 350 and 420°C. The rate became high in the range 420–460°C due to disproportionation of $ZrCl_3$ in the bed. Figure 1 shows an example of the effect of temperature on dissociation rate. Purified $ZrCl_4$, which we refer to as "product," was collected over the temperature range 420–460°C.

The formation of the $ZrCl_4$ product can best be understood by the following considerations. The production of the $ZrCl_3$ at low temperature takes place according to eq. (8).

$$ZrCl_4(g) + ZrCl_2(s) \xrightarrow{350-420°C} 2ZrCl_3(s) \tag{8}$$

At temperatures above 420°C, $ZrCl_3$ disproportionates to $ZrCl_4(g)$ and $ZrCl_2(s)$ according to eq. (3). However, above 420°C, the composition of the solids in the reactor approximates a formula represented by Zr_3Cl_8. It is conceivable that this composition is fortuitous and that it is only a result of disproportionation of some of the $ZrCl_3$. However, this means that the bed composition would have to be a mixture comprising 2 moles of $ZrCl_3$ and 1 mole of $ZrCl_2$, regardless of the composition of the reactants, $ZrCl_4$ and $ZrCl_2$. This does not appear to be plausible to us. Table I relates the amount of $ZrCl_4$ product to the $ZrCl_4$ feed and $ZrCl_2$ present in the reactor bed initially. The values are given in moles of material. The last column

TABLE I
Molar Stoichiometry in Newnham Process

Preformed ZrCl$_2$	ZrCl$_4$ product of:		ZrCl$_2$/product ZrCl$_4$
	Feed	Product	
3.2	3.6	1.7	1.9
	1.8	1.6	2.0
	2.5	1.6	2.0
6.9	5.0	2.9	2.4
	3.9	2.9	2.4
10.7	6.7	4.9	2.2
8.3	6.2	3.5	2.3
	4.9	4.1	2.0
	4.9	4.4	1.9
		Av.	2.1

in the table indicates that when ZrCl$_4$ is charged into the reactor at a molar ratio $\leqq$ ZrCl$_4$/2ZrCl$_2$, no more than 1 mole of ZrCl$_4$ is obtained, i.e.,

$$2ZrCl_2(s) + (1 + x)ZrCl_4(g) \xrightarrow{350\text{--}420°C} \text{complex}(s) + xZrCl_4(g)$$

$$\text{Complex}(s) \xrightarrow{420\text{--}460°C} 2ZrCl_2(s) + 1ZrCl_4(g)$$

On the basis of eq. (7) (applied to chlorides) and eq. (3), one should expect to recover 1 mole of ZrCl$_4$ per mole of ZrCl$_2$. The complexity of factors which determine the dissociation rate make it unlikely that the extent of dissociation of ZrCl$_3$ would result in the identical final bed stoichiometries despite variations in conditions. Such factors include time, rate of increase in temperature, flow rates

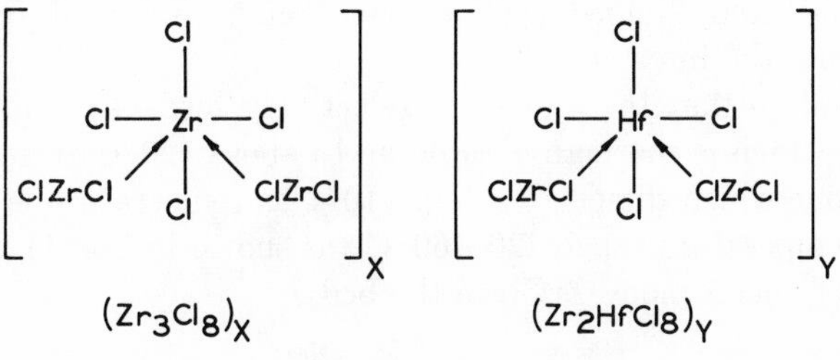

Fig. 2. Structure of subhalides.

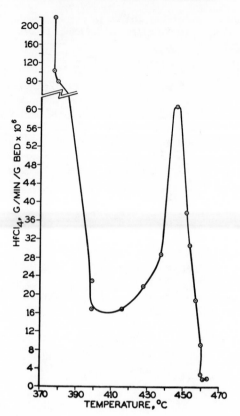

Fig. 3.   Dissociation rate of hafnium subhalide.

of argon, and particle size distribution, as well as rate of formation of nuclei, and relative rates at which the interface between gas and solid advances in different directions, crystal type, previous history of the solid, and "impedance" effects such as rates of removal of product from the interface (30).

We postulate that there exists a selectively formed complex, Zr_3Cl_8 [eq. (9)], which is thermally stable up to about 420°C at atmospheric pressure, but which dissociates [eq. (10)] as temperature is increased over the temperature range 420–460°C (as shown in Fig. 1) to liberate 1 mole $ZrCl_4$ per 2 moles $ZrCl_2$ in the bed.

$$ZrCl_4(g) + 2ZrCl_2(s) \xrightarrow{330–420°C} Zr_3Cl_8(s) \tag{9}$$

$$Zr_3Cl_8(s) \xrightarrow{420–460°C} ZrCl_4(g) + 2ZrCl_2(s) \tag{10}$$

Figure 2 shows a proposed structure for the "selective" zirconium subhalide Zr_3Cl_8 formed between 2 moles of $ZrCl_2$ and 1 mole of $ZrCl_4$. As will be discussed later, a hafnium subhalide Zr_2HfCl_8 analogous to Zr_3Cl_8 is also formed in minor quantities. It dissociates in a manner similar to Zr_3Cl_8 [see (eq.) 11 and Fig. 3].

$$Zr_2HfCl_8(s) \xrightarrow{\text{420–460°C}} HfCl_4(g) + 2ZrCl_2(s) \qquad (11)$$

The formulas Zr_3Cl_8 and Zr_2HfCl_8 are based on the fact that, as Blumenthal (31) pointed out, although zirconium has a valence of 4, valences of 5, 6, 7, and 8 can be realized by coordination. Zirconium apparently does not form valence bonds with itself, but can accept electrons from electronegative elements such as chlorine. $ZrCl_4$ thus forms addition compounds with molecules having available pairs of unshared electrons such as H_2O, NH_3, halogen ions, amines, and alcohols. One may postulate, therefore, that either Zr_3Cl_8 or Zr_2HfCl_8 could be formed by coordination of one molecule of $ZrCl_4$ (or $HfCl_4$) with a chlorine atom from each of two $ZrCl_2$ molecules. The actual structure may be polymeric since the subhalides are nonvolatile.

E. Nonselective Complexes

During reduction of crude $ZrCl_4$ containing about 2.5% $HfCl_4$, part of the $HfCl_4$ and some $ZrCl_4$ as well appeared to be bound nonselectively.

1. Although a large proportion of the $HfCl_4$ (which exists as vapor above 317°C) was swept out of the reactor at temperatures near 350°C, a considerable amount mixed with $ZrCl_4$ continued to flow out as the bed temperature was increased over the range of 350–420°C. We refer to this as the "intermediate" sublimate. Table II shows that for several experiments, large volumes of argon failed to sweep out all the $HfCl_4$ as the temperature was increased over the range 340–400°C, although this should be achieved readily because both $HfCl_4$ and $ZrCl_4$ are vapors above 331°C.

As a further example, in the experiment represented by Figure 4, in which the feed contained 2.3% Hf, the sublimate may be divided into three parts as follows: 9% at 332–334°C containing 13% Hf, 9% ("intermediate" sublimate) at 334–417°C containing 3.8% Hf, and 81% at 417–470°C containing 0.54% Hf. The "intermediate" sublimate had a higher hafnium content than that of the feed, suggesting that it arose from dissociation of a nonselectively formed complex in the bed. [See eqs. (12) and (13).]

TABLE II

Argon Purge of Reactor

Expt.	Bed temperature,[a] °C	Argon vol./ reactor vol.	Hf in terminal sublimate, %
1	340	4.5	2.2
	379	9.5	2.3
	432	11.0	1.0
2	390	2.1	9.4
	404	3.9	2.8
	430	4.8	1.6
3	387	3.7	1.5
	404	5.2	1.8
	440	6.8	0.7

[a] Reaction temperature was below 334°C; feed $ZrCl_4$ contained 2.2% Hf, product $ZrCl_4$ about 0.3% Hf.

2. The dissociation pressures developed by reaction mixtures containing more than 0.5 mole crude $ZrCl_4$ per mole $ZrCl_2$ were greater, below 420°C, than those for mixtures containing less than 0.5 mole (Fig. 5). The former mixtures contained hafnium and zirconium complexes which were less stable than the Zr_3Cl_8 present in the latter

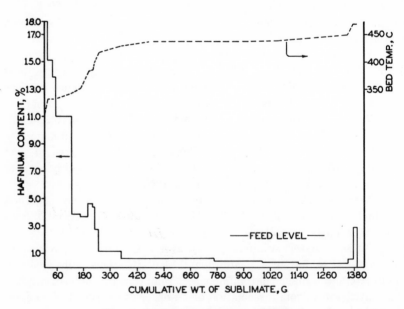

Fig. 4. Hafnium separation—solid feed.

mixture.   The curves in Figure 5 merge in the region near 420–430°C.
At this point the nonselective complexes had substantially dissociated,
whereas dissociation of Zr_3Cl_8 had just begun.

3. Much lower separation factors were obtained in experiments in
which reduction temperature of crude $ZrCl_4$ with $ZrCl_2$ was relatively
low (330–350°C) as compared with those obtained when the reduction
temperature was relatively high, near 400°C.   (See Table III; explanation appears later.)

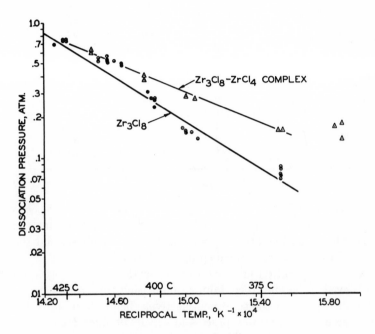

Fig. 5.   Dissociation pressures of Zr_3Cl_8 and $Zr_3Cl_8 \cdot ZrCl_4$ complex.

The low reduction temperatures in the experiments referred to were
achieved by adding solid crude $ZrCl_4$ ("solid feed" experiment) to
the $ZrCl_2$ bed, then raising the temperature.   Reaction is substantial
in the temperature range 317–331°C, the atmospheric sublimation
temperatures of $HfCl_4$ and $ZrCl_4$, respectively.

Reduction at higher temperatures was achieved by vaporizing the
crude $ZrCl_4$ feed from a sublimer ("vapor feed" experiment), heating
the vapors to 400°C and passing them into the stirred powdered $ZrCl_2$
bed at 400°C.

Figures 4 and 6 show typical runs for low reaction temperature

TABLE III
Stage Dependence on Temperature for Reactor Grade Zirconium

Reaction temperature, °C	% yield of $ZrCl_4$ from one stage suitable as feed for		
	Two-stage process	Three-stage process	Four-stage process
Below 334	0	22	79
	0	24	71
	0	36	57
	0	69	92
	0	70	80
	0	52	58
	0	46	73 Av.
370–375	29	63	69
390	41	64	85
400	54	77	87
415	39	46	51
	41	62	73 Av.

(solid feed) and high reaction temperature (vapor feed), respectively. The curves show the increase above feed level in hafnium content of the effluent tetrachloride during the absorption phase, and the decrease below feed level in the disproportionation phase.

The lower separation factors observed for the solid feed experiments are believed to be due to contamination arising from a higher concentration in the bed of the solid nonselectively bound hafnium. During that stage of the separation process in which "product $ZrCl_4$" is recovered as vapor from the solid bed by disproportionation over the temperature range 420–460°C, dissociation of the solid nonselective hafnium complex yields $HfCl_4$ vapor which mixes with, and contaminates the "product $ZrCl_4$." However, high concentrations of this solid nonselective complex are not developed in the vapor-feed experiment as explained below, and contamination is reduced.

The concentration of nonselective hafnium complex depends on the composition of the vapors above the bed, the reaction time, and temperature. In the vapor-feed experiments, $HfCl_4$ vapors were continually swept out of the reactor whereas they were retained during the absorption phase in the solid-feed runs. Vapors above the bed were generally higher in hafnium content in the solid feed experiments (7 times that of the feed) than in the vapor feed experiments (3.5

times that of the feed), allowing for a buildup of the solid nonselective hafnium complex in the bed because of mass action effect. Furthermore, contact times were longer in the solid feed experiments which favored an increase in concentration of nonselective hafnium complex in the bed.

Temperatures obtained during the reduction phase are also important factors affecting concentration of the nonselective hafnium complex. In the vapor-feed experiments where reduction temperature was high, the dissociation of the nonselective complex was rapid with respect to its formation, preventing an increase in the concentration of the complex. On the other hand, at low reduction temperatures (330°C) in solid-feed experiments, the dissociation rate was low relative to the rate of formation of the complex.

The data taken from experiments presented in Figures 4 and 6 were used to compare solid and vapor feed with respect to the relative amounts of hafnium subliming from the bed with increasing temperature. These appear in Figure 7, which demonstrates the liberation of bound hafnium tetrachloride through dissociation over the temperature range 334–430°C of the nonselective complex in the bed of the solid feed run.

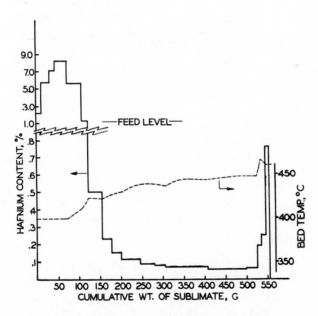

Fig. 6. Hafnium separation—vapor feed.

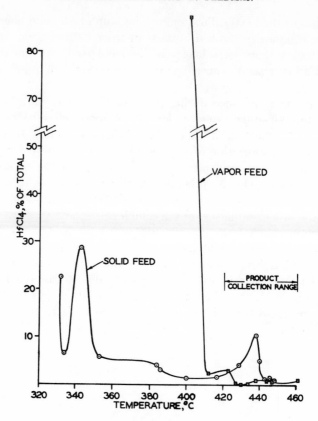

Fig. 7.   Sublimation of hafnium.

4. A dark-green colored product was formed when excess $ZrCl_4$ reacted with reducing agent. However, when the reducing agent was present in excess, this green color was absent. Zirconium trichloride appears to be black like the dichloride.

We postulate that the selectively formed solid "stable" Zr_3Cl_8 reacts nonselectively with either $HfCl_4$ or $ZrCl_4$ in the vapor, as suggested by eqs. (12) and (13), to form the nonselective, less-stable complexes discussed above.

$$Zr_3Cl_8(s) + HfCl_4(g) \underset{350\text{--}430°C}{\overset{350\ C}{\rightleftharpoons}} Zr_3Cl_8 \cdot HfCl_4(s) \tag{12}$$

$$Zr_3Cl_8(s) + ZrCl_4(g) \underset{350\text{--}430°C}{\overset{350°C}{\rightleftharpoons}} Zr_3Cl_8 \cdot ZrCl_4(s) \tag{13}$$

F. Hafnium in the Bed

Equation (11) and structure 2 of Figure 2 referred to a hafnium subhalide Zr_2HfCl_8 which dissociates to liberate $HfCl_4$ as the temperature is increased over the range 420–460°C. Liberation of $HfCl_4$ over this temperature range is shown by the curve labeled "Solid Feed" in Figure 7.

Apparently $HfCl_2$ also was formed in the process. Hafnium was found in the bed even after prolonged heating at temperatures above 470°C, and, in addition, the hafnium content of the bed rose perceptibly on reuse. For example, a bed prepared from reagents containing less than 350 parts per million Hf, contained 800 parts per million after several uses. Furthermore, hafnium in this state can be worked out of the bed with $ZrCl_4$. Thus, a bed containing more hafnium than the feed (0.35% as against 0.027%) transferred hafnium to the product (0.06% Hf).

$HfCl_4$, similarly to $ZrCl_4$, can be reduced to the $HfCl_2$ by zirconium metal. A sample of the fine zirconium sponge (325 mesh) containing 0.007% hafnium was reacted with 98% $HfCl_4$ for 24 hr at 540°C in an evacuated sealed bulb. After exhaustive sublimation of this reaction mixture for 6 hr at 360°C, then 21 hr at 540°C, the residue was found to contain 49% hafnium, probably in the form of $HfCl_2$.

The following explanations are suggested for (1) the formation of hafnium subhalide Zr_2HfCl_8, (2) the increase in hafnium content of the $ZrCl_2$ bed on repeated use, and (3) the removal of hafnium from the bed with $ZrCl_4$:

The nonselective complex formed between $HfCl_4$ and Zr_3Cl_8 can liberate either $HfCl_4$ or $ZrCl_4$ on dissociation. This is shown below where the nonselective hafnium complex is written in three resonating forms. Liberation of $ZrCl_4$ gives rise to the hafnium subhalide. $ZrHfCl_4$ formed by dissociation of the hafnium subhalide is probably the hafnium contaminant that appears in the bed on repeated use. This, however, can be removed by reaction with $ZrCl_4$ to again form the subhalide which can dissociate by an alternate path to $ZrCl_2$ and $HfCl_4$,

$$Zr_3Cl_8(s) + HfCl_4(g) \rightleftarrows Zr_3Cl_8 \cdot HfCl_4(s)$$
$$\updownarrow$$
$$Zr_3HfCl_{12}(s)$$
$$\updownarrow$$
$$Zr_2HfCl_8(s) + ZrCl_4(g) \rightleftarrows Zr_2HfCl_8 \cdot ZrCl_4(s)$$
$$\downarrow\uparrow$$
$$ZrHfCl_4(s) + ZrCl_4(g)$$
$$\text{and/or}$$
$$2ZrCl_2(s) + HfCl_4(g)$$

The postulations presented serve to explain the phenomena observed during the laboratory and bench-scale development of a practical process for separating hafnium from zirconium. In order to substantiate these hypotheses it would be desirable to study some of the more fundamental properties of the intermediates and complexes. x-ray diffraction diagrams and more precise kinetic and equilibrium data are needed.

III. BENCH-SCALE PROCESS

A. Purpose

The Newnham separation scheme which was demonstrated in evacuated small glass systems merited development on a bench scale because of important advantages it offered over present commercial extraction methods. These advantages are: (1) chemical requirements are smaller; (2) it is a cyclic process adaptable to either batch or continuous operation; (3) only a few stages are needed; (4) the chemical nature of the feed to the reducer for sponge preparation is not altered; (5) a hafnium concentrate is obtained as a by-product, in the form of the tetrachloride.

The scheme shares with the extraction process the benefit of removing certain impurities such as iron.

Briefly, the process is divided into the following steps:

1. Preparation of $ZrCl_2$ bed.
2. Selective reaction of $ZrCl_4$ with $ZrCl_2$ to yield $ZrCl_3$.
3. Sweeping out of unreduced $HfCl_4$.
4. Disproportionation of $ZrCl_3$ to regenerate solid $ZrCl_2$ and liberate purified $ZrCl_4$ product as vapor.

B. Procedure

1. REACTANTS

a. Zirconium Metal

Zirconium metal was obtained in the form of sponge from U.S.I. pilot plant operations. It was prepared by reduction of hafnium-free sublimed $ZrCl_4$ with metallic sodium by a process similar to that described by Schott and Hansley (32). The material was sieved to obtain the powdered zirconium.

b. Zirconium Dichloride

Solid $ZrCl_2$ used as the reducing agent was prepared in two steps. In the first step, fine zirconium sponge was heated without stirring with an excess of low hafnium $ZrCl_4$ at 430°C for several days in a 1-liter autoclave, all operations being conducted under inert gas (argon). The mixture of zirconium chlorides was transferred from the autoclave to a flask in an argon-filled polyethylene bag, and, after weighing, was transferred under argon at room temperature to the reactor described below.

In the second step, the mixture was heated (with stirring when the tin was molten) to 460°C and kept at that temperature for 4 hr to disproportionate the $ZrCl_3$. During this time a stream of argon removed the subliming $ZrCl_4$. The number of moles of $ZrCl_2$ was equal to twice the number of moles of $ZrCl_4$ which had reacted [see eq. (5) for the analogous reaction with $ZrBr_4$], calculated from the increase in weight of the solid remaining after the extended disproportionation at 430–460°C.

c. Zirconium Tetrachloride

The crude zirconium tetrachloride was purchased from Stauffer Chemical Company. This material contained 2.3–2.5% hafnium.

d. Hafnium Tetrachloride

A sample of hafnium tetrachloride was obtained from the U.S. Bureau of Mines. Analysis showed it to contain 2% Zr.

2. APPARATUS

The operation was conducted at atmospheric pressure under argon in a stirred reactor immersed in molten tin (Figs. 8 and 9). The reactor was a 5-liter cylindrical vessel, 8 in. high and 7 in. in diameter, fabricated from stainless steel and containing an anchor-type stirrer (A in Fig. 8) in the center, having a blade which scraped the bottom and the lower 4 in. of the sides.

The reducing bed (B in Fig. 8) in the reactor was either powdered zirconium sponge or powdered zirconium dichloride. Crude $ZrCl_4$ was fed into the reactor as a solid in one set of experiments and as a vapor in another. The apparatus differed in the two instances and is described in Sections III-D-1 and III-D-2.

a. Materials of Construction

Since the apparatus was to be immersed in molten tin to obtain good heat transfer, it was necessary to choose materials of construction which would not react with molten tin within the operating temperature range of 300–500°C. Precipitation of chromium carbide takes place in this temperature range; thus certain stainless steels such as 316 or 304 are ruled out. However, columbium-stabilized stainless

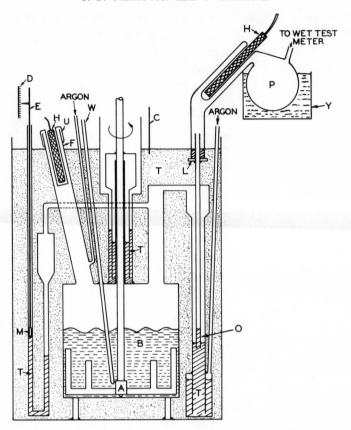

Fig. 8. Stirred reactor (solid feed): (A) stirrer, (B) ZrCl₂ bed. (C) controlling thermocouple, (D) scale, (E) pennant, (F) feed port, solid ZrCl₄, (H) cartridge heater, (L) ledge, (M) manometer, (P) product collector (glass), (T) molten tin, (U) recessed pipe cap, (W) thermowell, (Y) coolant bath, (O) exit valve.

steels such as 347 or low-carbon stainless steels such as 304 ELC may be used. Type 316 stainless steel apparently does not alloy with tin but ordinary steel dissolves.

Our equipment was fabricated initially of 316 stainless steel which provided prolonged use before leaks developed. Equipment used later was fabricated of 347 stainless steel.

b. Solid Feed

(1) **Equipment.** Figure 8 is a drawing of the system used in the solid-feed studies. The port (F) for solid ZrCl₄ feed consists of a

17-in. threaded pipe, 2 in. in diameter, placed at an angle of 5° from the vertical into the top of the reactor. A portion (8 in.) of this pipe protruded out of the molten tin bath. This port was covered by a recessed threaded pipe cap (U) carrying a 300 W stainless steel cartridge heater (H). Heating of the recessed portion of the cap prevented condensation of $ZrCl_4$ and $HfCl_4$ in that part of the port protruding from the molten tin bath. A thermocouple was inserted in the recessed cap.

Argon entered through a ⅜-in. diameter stainless steel tube at a point on the feed port below the level of the molten tin bath (T). Flow was controlled by a needle valve and its rate was measured with a calibrated orifice flowmeter after it left the cylinder through a pressure-reducing valve.

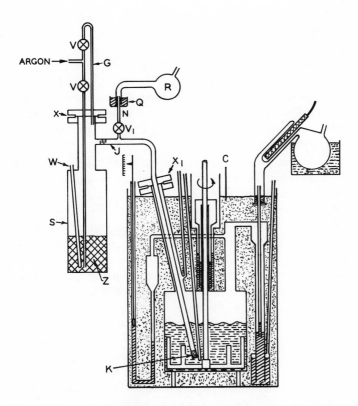

Fig. 9. Stirred reactor (vapor feed): (G) argon purge assembly, (J) side arm delivery tube, (K) sintered metal cap, (N) feed sample port, (Q) annular trough, (R) sample collector (glass), (S) sublimer, (V) valve, (V_1) sample valve, (X) flange, (X_1) recessed flange, (Z) solid crude $ZrCl_4$.

Product $ZrCl_4$ flowed from the reactor via an exit valve (O) into collecting vessels. This valve was closed by the application of argon gas pressure to molten tin in a reservoir causing it to flow up a tube and block an opening into the reactor. Reduction of pressure allowed the tin level to drop, opening the reactor to the exit. The valve tubes were of such length that a pressure of 80 mm Hg could be maintained in the reactor.

Pressure in the reactor was measured with a manometer carrying a float (M). The manometer was a U-tube with one end open to the reactor. The pressure was indicated by a pennant (E) (attached to M) on scale D. The other end of the U-tube led out of the tin bath and opened to the atmosphere. The open end should be blanketed with nitrogen. It was partly filled with tin and operated when the tin was molten. A second device (not shown) for measuring pressure was an open-end mercury manometer connected to the argon inlet line through a T. The bed temperature was measured with an ammeter-calibrated Brown Pyr-o-vane Controller (not shown) with a thermocouple inserted into the well (W) extending into the reactor almost to the impeller blade. The Pyr-o-vane Controller was calibrated against melting tin. The bath temperature was measured and controlled in a similar manner using a controlling thermocouple (C) which was inserted in a well (not shown) in the tin bath in which the reactor was immersed.

The stirrer motor (not shown) was a $\frac{1}{3}$-hp 220-V motor, reduced to turn at 40 rpm. In some of the earlier experiments, the stirring rate was 220 rpm, which caused excessive dusting of the bed. The stirrer shaft entered the top of the reactor through molten tin seals, as shown in Figure 8, similar to the well-known mercury-sealed stirrer. Although this seal gave satisfactory service for an extended time, eventually solid material, possibly tin oxide, appeared in the seal and jammed the stirrer. In a later modification, asbestos packing was successfully used as a seal. Two types of product collector were used; one (P) fabricated of glass for collecting small fractions of product at low rates, and the other (not shown), a metal "Stauffer" condenser for collecting larger fractions at high sublimation rates.

As shown in Figure 8, the entry duct to the glass product collector (P) was inserted to a depth of 1 in. under the melt surface around the mouth of the exit tube from the exit valve, which projected an inch above the tin surface, and rested on a ledge (L).

The glass product collector (P) consisted of a glass tube, $1\frac{1}{2}$ in. in diameter, which conducted the tetrachloride vapors out of the reactor to a large glass bulb (made from a 500 ml Kjeldahl flask), placed at an angle to the tube so that any solid formed from condensing vapors would not fall back into the reactor. Another glass tube 1 in. in diameter and 10 in. long, attached to the bulb, vented the collector. A reentrant well, 6 in. long and ¾ in. in diameter, carrying a 300-W cartridge heater (H), was placed inside and concentric with the tube leading to the reactor. Maintaining the well at a suitable temperature insured that all tetrachloride was vaporized out of the tube into the collector bulb. The effluent tetrachloride vapors were condensed in the collector bulb which was cooled by immersion in a liquid. Several such collectors which could be removed and replaced quickly and easily were used to collect consecutive fractions of the effluent tetrachloride.

The Stauffer condenser operated in a fashion similar to the glass product collector. Its entry duct was also immersed in the molten tin and rested on the ledge (L). However, this condenser was a jacketed 316 stainless steel vessel with a 3-in. o.d. and 20-in. length with a conical bottom to which a glass jar could be attached by means of a rubber stopper for collection of product. The vessel contained a motor-driven (60 rpm) horizontal arm at the top, carrying hanging chains which served to dislodge the sublimate from the sides. A vibrator attached to the side at the bottom aided in the dislodgement. Product vapor carried by a stream of argon into the condenser at a port located near its center sublimed on the cylinder wall which was cooled by air circulating through the concentric jacket. On being dislodged, the product fell into the jar collector. Argon exited through an opening at the top.

The Stauffer condenser, when used, was employed to collect the purified product fractions only, because of possible contamination from previous samples which are high in hafnium and which may cling to the sides of the condenser and on the chains.

The container for the molten tin used as the heat transfer medium was a vessel 12 × 12 × 24 in., fabricated of ⅛-in. 316SS plate. The sides and bottom were covered with 3 in. of high-temperature insulation. The tin was heated with six 1000-W Watlow monel-sheathed cartridge heaters (not shown), ⅞ in. in diameter and 22 in. long, placed in vertical stainless steel wells in the tin. Four of the heaters were connected in series and parallel combinations to a range switch (3-heat), and received current from a 230-V source through an Allen Bradley No. 702 contactor controlled by a Brown Pyr-o-vane. In this system, it was possible to obtain heat inputs in 500-W increments up to 6000 W.

(2) **Method of Operation.** Solid crude $ZrCl_4$ was charged into the argon-purged reactor from a 2-liter Erlenmeyer flask through a 12-in. length of glass tubing 1 in. in diameter attached at the mouth

with rubber tubing. The glass tube was inserted into the charging port below the argon inlet. The solid $ZrCl_4$ was added over a period of 1–2 min to the stirred $ZrCl_2$ bed which was at a temperature of about 300°C. The addition of the charge immediately cooled the bed to about 200°C. The charging port was then capped, the exit valve closed, and the bath heated rapidly to raise the mixture to reaction temperature. When the pressure exceeded 50–70 mm above atmospheric, it was relieved through the exit valve (O) into the product collector (P).

The temperature of the stirred mixture rose in about 45 min. to the sublimation temperature of $ZrCl_4$, namely, 331°C. The bath was maintained at this temperature to avoid pressures in excess of atmospheric. From the results of rate studies plotted in Figure 10, it can be estimated by extrapolation of the ascending portion of the

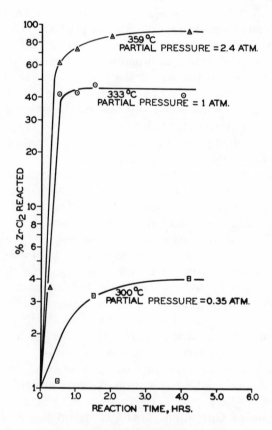

Fig. 10. Reaction of $ZrCl_2$ with $ZrCl_4$.

curve for 333°C that reaction should be complete in about 1 hr. The plateaus in the curves of Figure 10 may be caused by deposition of reaction products on the surface of the $ZrCl_2$ since the beds were not stirred in these experiments. After 2–4 hr from the time of addition of $ZrCl_4$, the exit valve was opened to allow the effluent tetrachloride vapors to pass into the cool argon-purged glass product collector where condensation to solid tetrachloride took place. The reactor was now purged with argon to sweep out unreacted $HfCl_4$. The bed temperature was then raised to disproportionate the $ZrCl_3$ [eq. (3)]. All of the sublimate was collected in consecutive fractions. During this period of fraction collection the pressure in the reactor was atmospheric.

c. Vapor Feed

(1) Equipment. The apparatus shown in Figure 8 was modified to admit $ZrCl_4$ feed vapors from a sublimer through the feed port to the reactor. Figure 9 is a drawing of the modified system. Labels are shown only for the modified parts, the other parts being the same as in Figure 8. Figure 11 is a photograph of the bench-scale system using the vapor feed method.

The modifications were as follows:

The threaded closure of the recessed cap (U in Fig. 8) was changed to a flange fitting sealed with a Flexitallic gasket as shown by X_1 in Figure 9. The reentrant well for carrying the $\frac{1}{2} \times 6$ in. 300-W cartridge heater was retained but is not shown in the drawing.

A $\frac{3}{4}$-in. stainless steel tube (J) served to lead the $ZrCl_4$ vapor feed from the sublimer (S) into the reactor to a point below the bed surface. A sintered metal cap (K) attached to the end of the tube (J) prevented plugging of the tube. The other end of the tube (J) was joined to the sublimer assembly with small flanges (not shown) and sealed by a Flexitallic gasket.

Figure 9 shows the sublimer assembly consisting of a stainless steel cylindrical vessel, 10 in. deep and $3\frac{1}{4}$ in. in diameter, an electrically lagged delivery tube (J), and a valved, electrically lagged sampling port (N). The sublimer fitted into a cylindrical aluminum sheath (not shown) $\frac{1}{2}$ in. thick, which was wrapped with two 1000-W Calrod heaters, 6 ft long and $\frac{1}{2}$ in. in diameter, each controlled by a variable voltage transformer. The sublimation temperature was measured by a thermocouple inserted in well W. The sublimer sheath temperature also was measured by a thermocouple placed in a 2-in. deep hole drilled into the wall of the sheath. Other thermocouples were welded to the electrically lagged neck of the sublimer, to points along the

Fig. 11.   Separation system (vapor feed).

delivery tube (J), and to the valved sampling port (N). A thermocouple was placed in the reentrant well in flange X_1. A controlling thermocouple (C) was inserted in a well in the tin bath in which the reactor was immersed. The thermocouple temperatures were recorded on a Brown Multipoint Recorder. Heat was applied to the tin bath as described in Section III-B-2-b.

During sublimation of the $ZrCl_4$ feed, metered argon gas was introduced at the point indicated in the argon gas purge assembly (G). Part of this gas was directed to the bottom of the sublimer, and part to the top by means of valves (V). Baffles in the delivery tube (J) aided in mixing of the argon–$ZrCl_4$ vapor mixture. The entering argon gas was metered through a wet test meter containing transformer oil. Flow rates were observed by means of Micromite flowmeters.

An annular trough (Q) at the exit of the sampling port contained molten tin. To sample the feed vapor, the mouth of a small glass sample collector (R) was placed over the sampling port (N) and under the molten tin surface. When the sampler valve (V_1) was opened, part of the argon–$ZrCl_4$ vapor mixture was diverted into the glass sample collector, where the ZrCl₄ vapor was condensed and collected. Argon gas leaving the sample collector was cooled, passed through aqueous alkali, and metered through a wet test meter filled with water. Another wet test meter measured the argon gas leaving the reactor.

(2) **Method of Operation.** The sublimer (S) was charged with a weighed quantity of solid crude ZrCl₄ through the neck of the sublimer, after which the argon purge assembly (G) was attached and argon gas made to flow through the lower tube. The ZrCl₂ bed in the reactor was brought to the desired reaction temperature. The temperatures of the delivery tube (J), the sampling port (N), the re-entrant well in flange X₁, and the neck of the sublimer were all raised well above the sublimation temperature of ZrCl₄ (331°C). The sublimer was heated to vaporize the ZrCl₄ which was carried into the reactor by the argon gas flowing at a rate of 40–60 cc/min. At intervals during the sublimation period, the sampler valve (V₁) was opened for short, measured periods of time. The emerging ZrCl₄ vapor was collected in the sample collector (R) and weighed, and the accompanying effluent argon gas was metered. The data showed when sublimation of feed ZrCl₄ was complete and also served to measure the sublimation rate. This rate could be changed by varying the heat input to the sublimer jacket.

The ZrCl₄ effluent was collected in several successive fractions during the reaction, sweep-out, and disproportionation phases of the operation. This effluent was collected in the glass product collectors (P), then weighed and analyzed to determine which fractions should be retained as "product." Several small fractions were collected in order to zero in on the cut point for product collection. Fractions containing hafnium in lower concentration than the feed could be combined later, as desired. The effluent rate was calculated from these data and from the time of collection. When addition of vapor feed to the reactor was complete, the argon rate was greatly increased, generally to 600–1000 cc/min, to sweep out unreacted tetrachloride. The bed temperature was then raised to disproportionate the subchloride at 450°C as described in solid feed experiments and as shown in eq. (3).

C. Analysis

1. METHOD

The per cent Hf/(Hf + Zr) was determined by an emission spectrographic method using a Baird 3-meter spectrograph. The densitometer was also a Baird instrument. A modification of the oxide technique of Mortimore and Noble (33) was used for ratios in the range 0.007–0.1%. For water (or dilute HCl) soluble compounds ($ZrCl_4$,$HfCl_4$) having a ratio in the range 0.1–30%, a new rapid procedure was developed in our laboratory (34) because conversion of the samples to the oxide form was tedious. In the new method the sample was dissolved to make a solution containing approximately 4 g of metal per 100 cc. Standards were prepared by mixing aliquots of stock solution of pure $HfCl_4$ and $ZrCl_4$.

A 0.1-ml portion of a solution, sample or standard, was pipetted into the cavity of a graphite electrode and evaporated to dryness. A special drier was constructed to facilitate rapid drying of the samples in the electrode without splattering. The sample in the electrode was volatized in an ac arc.

2. DEFINITIONS

a. *Separation factor.* The separation factor is defined as the ratio of hafnium concentration in the feed to hafnium concentration in the product. The factors needed to decrease the hafnium content from 2.5 to 0.01% required for reactor grade zirconium, in 1, 2, 3, and 4 stages are 250, 15.6, 6.3, and 4.0, respectively.

b. *Product.* The product is defined as that portion of disproportionated material having a hafnium content less than that of the original charge.

c. *Yield.* The yield in per cent is the weight of $ZrCl_4$ obtained as product multiplied by 100 and divided by the stoichiometric weight of $ZrCl_4$ calculated to react on an equimolar basis with the $ZrCl_2$ in the bed. However, when $ZrCl_2$ was in excess, the yield was based on $ZrCl_4$ charged.

D. Experimental

1. SOLID FEED

Table IV and Figures 1, 3, and 4 present typical data obtained from three separate "solid-feed" experiments; Figures 1 and 3 are from the same run. As shown by Table IV, fractions of effluent tetrachloride were taken in sequence (first column). Primary data included elapsed time (after charging $ZrCl_4$) for each fraction (second column), bed and bath temperatures when the fraction was taken (third and fourth columns, respectively), the weight of the fraction (fifth column), its analysis (seventh column) and the argon rate (thirteenth column). Values in the other columns were calculated from the primary data. As indicated in Table IV, the hafnium content of the crude $ZrCl_4$ was 2.3%, and that in the product was 0.56% corresponding to a separation factor of 2.3%/0.56% or 4.

The "product $ZrCl_4$" includes fractions 8–19 (Table IV) since the hafnium contents in each of these were less than that of the feed. It is obvious that fractions could be selected which have a lower hafnium content with a sacrifice in yield.

In this particular experiment, during the reaction phase, the ranges of temperature increase over the first and second hours were 299–343°C and 343–380°C, respectively. The temperature was maintained at 380°C during the third hour. The mole ratio of $ZrCl_4$ to $ZrCl_2$ was 0.57 for the feed and 0.49 for the product.

Table V gives selected typical examples of separation factors for runs in which feed $ZrCl_4$ had different hafnium contents. The results show that the separation factor is essentially independent of the hafnium concentration in the feed when the bed is free of hafnium.

As indicated in Section III-C-2-a, separation factors greater than 15.8 are required for a two-stage separation procedure for reducing hafnium content from 2.5 to 0.01%. Such separation factors were never observed in solid-feed runs. However, as shown in Table VI, a three-stage separation was achieved, the net yield being 27% of theory.

Table III demonstrates the yield of $ZrCl_4$ produced from one stage which has the required hafnium content to obtain reactor grade $ZrCl_4$ in two-, three-, or four-stage operations. In the case of the solid-feed (low-temperature) experiments, no $ZrCl_4$ product was obtained from the first stage which, if used in the second stage, would yield $ZrCl_4$ having 0.01% Hf or less. At least a third stage would be required for minimal yields and a fourth stage for practical yields.

Figure 4 depicts data from a solid-feed experiment relating the hafnium content of each successive effluent tetrachloride fraction (as in the seventh column, Table IV) to the individual and cumulative weights (as in fifth and sixth columns, Table IV). The figure demonstrates the change in composition of the effluent tetrachloride as it emerges from the reactor and relates this to the bed temperature. The mole ratio of feed $ZrCl_4$ to $ZrCl_2$ in the bed was 0.57. The figure shows that effluent tetrachloride obtained in the first stages of the run over the bed temperature range 332–417°C was much higher in hafnium than was the feed. This represented only about 245 g of tetrachloride compared to 1163 g obtained as product where each successive fraction had a hafnium content below feed level. The product contained 0.60% Hf compared to 2.3% in the feed. A slight rise

TABLE IV
Separation Process in Stirred Reactor with Solid $ZrCl_4$ Feed

Fraction no.	Collection time from startup, min	Temperature, °C Bed	Temperature, °C Bath	Sample weight Individual, g	Sample weight Cumulative, g	Hf/(Hf + Zr) in sample, %	Weight $HfCl_4$ in sample, g	Weight $ZrCl_4$ in sample, g	Sublimation rate (g sublimate per min per g bed × 10⁴)	Relative rate $ZrCl_4/HfCl_4$	Cumulative weight $HfCl_4$, g	Argon rate, cc/min
0	0	290	—	—	—	—	—	—	—	—	—	0
1	185	380	377	1.7	1.7	14.0	0.238	1.5	0.17	6.30	0.238	105
2	193	377	377	4.3	6.0	23.5	1.01	3.3	10.5	3.27	1.25	105
3	213	377	—	4.8	10.8	25.2	1.21	3.6	4.7	2.98	2.46	105
4	249	379	—	8.9	19.7	19.0	1.69	7.2	4.9	4.26	4.15	320
5	268	399	399	3.8	23.5	6.6	0.251	3.6	3.9	14.3	4.40	375
6	282	399	399	3.1	26.6	4.5	0.139	3.0	4.3	21.6	4.54	375
7	297	416	—	5.5	32.1	2.7	0.149	5.3	7.2	35.6	4.69	245
8	312	428	435	10.6	42.7	1.8	0.191	10.4	13.9	54.5	4.88	60
9	327	438	446	16.7	59.4	1.5	0.251	16.5	21.7	65.8	5.13	30
10	336	449	460	28.4	87.8	1.1	0.313	28.1	61.2	90.0	5.44	30
11	345	452	460	36.5	124.3	0.53	0.194	36.3	79	187	5.64	30
12	354	452	460	48.6	172.9	0.42	0.205	48.2	105	235	5.84	60
13	367	452	460	70.9	243.8	0.38	0.269	70.6	106	262	6.11	60
14	378	454	460	57.8	301.6	0.34	0.197	57.6	103	293	6.31	60
15	391	457	460	45.2	346.8	0.31	0.14	45.0	67.7	321	6.45	60

16	406	460	—	24.5	371.3	0.33	0.081	24.4	31.8	301	6.53	60
17	426	460	—	7.9	379.2	0.41	0.0324	7.9	7.8	247	6.56	60
18	452	461	—	5.0	384.2	0.59	0.0295	5.0	3.7	175	6.59	60
19	502	463	—	5.1	389.3	0.98	0.05	5.1	1.9	102	6.64	60
20	562	463	—	3.9	393.2	2.1	0.082	3.9	1.2	47.5	6.72	60
21	695	463	—	3.7	396.9	1.7	0.062	3.7	0.6	59.6	6.78	60

Bed weight ($ZrCl_2$): 512 g

$ZrCl_4$ charge: 423.3 g, 2.3% Hf
Product: 85% theory, 0.558% Hf

Product yield: 360.9 g
Recovery $ZrCl_4$: 93.8%
HfCl$_4$: 69.7%

TABLE V

Separation Factors in Solid Feed Experiments

Hafnium content in ZrCl$_4$, %		Product	
Feed	Product	Yield, %	Separation factor
2.3	0.558	85	4.1
0.83	0.12	48	7.1
0.29	0.046	52	6.3
0.048	0.012	80	4.0

in hafnium content in fraction 7 (cumulative wt 217 g) resulted from a brief increase in the rate of argon sweep (not shown) from about 100 cc/min to about 300 cc/min. The increase in hafnium content of the terminal sample (Hf = 3.0%) is presumed to be due to continued dissociation of Zr$_2$HfCl$_8$ postulated to be present in the bed.

Figure 1 illustrates the effect of temperature on the rate at which effluent zirconium tetrachloride vapor sublimed from the reactor. Figure 3 shows the corresponding curve for HfCl$_4$. The marked increase in rate (Fig. 1), beginning at about 420°C is due to dissociation of a zirconium subhalide in the bed—the postulated Zr$_3$Cl$_8$—the nonselective complex [see eq. (13)] having dissociated at a lower temperature. The increase in the rate of sublimation of HfCl$_4$ in the region 420–460°C (Fig. 3) is associated with the disproportionation of Zr$_2$HfCl$_8$ (2, Fig. 2). The sublimation of HfCl$_4$ in the region 380–420°C is believed to be due to dissociation of the nonselective complex Zr$_3$Cl$_8$·HfCl$_4$ [eq. (12)]. In examining these curves the difference in scale of the ordinates should be noted.

TABLE VI

Three-Stage Separation of HfCl$_4$; Solid Feed

Reaction conditions		Hf content, %		Yield, %		Separation factor/ stage
Temp., °C	Time, hr	Feed	Product	Per stage	Net	
301–328	1.1	2.4	0.31	65	65	7.7
302–319	0.9	0.29	0.05	54	35	5.5
318–338	3.2	0.05	0.01	77	27	4.8

2. VAPOR FEED

Larger separation factors than those obtained from solid-feed experiments resulted when the reaction temperature was above 340°C and approached 400°C as in the vapor-feed experiments.

Table VII shows yields and corresponding separation factors realized at higher reaction temperatures. As in the solid-feed experiment, effluent tetrachloride was collected in consecutive fractions and analyzed. In Table VII the yields shown are obtained by combining selected contiguous fractions of the effluent. The higher yields show poorer separation factors because fractions with higher hafnium content were included.

Table III, discussed previously, shows that a two-stage process was possible only if the reaction temperature was above 340°C. No material suitable as feed for the second stage of a two-stage procedure was obtained when the original feed was charged as a solid. On the other hand, yields up to 54% of suitable feed material for a second stage of a two-stage operation were obtained using vapor feed and a reaction temperature of 400°C. Also, feed suitable for three-stage procedures was obtained in higher yield, on the average, when the original feed was charged as a vapor.

This is interpreted to mean that the main hafnium contaminating source, $Zr_3Cl_8 \cdot HfCl_4$, fails to form to as great an extent at the higher temperatures since these are close to its dissociation temperature. The complex Zr_2HfCl_8 which we believe derives from $Zr_3Cl_8 \cdot HfCl_4$ also fails to form to as great an extent.

TABLE VII
Yields for Newnham Separation Using Vapor Feed

Mole ratio, charge $ZrCl_4/ZrCl_2$	0.33	0.21	0.31	0.37
Hafnium content of charge $ZrCl_4$, %	1.20	1.37	1.15	1.32
Reaction temperature, °C	370–5	390	400	415
Product				
Yield, % (on $ZrCl_4$ charge)	58	63	71	42
Hafnium content, %	0.11	0.21	0.10	0.11
Separation factor—total product	11	6.5	11.5	12
$ZrCl_4$ suitable for 3-stage operation				
Yield, %	63	64	77	46
Hafnium content, %	0.19	0.22	0.18	0.21
$ZrCl_4$ suitable for 2-stage operation				
Yield, %	29	41	54	39
Hafnium content, %	0.076	0.087	0.073	0.083

Figure 6 represents a vapor-feed experiment plot similar to that in Figure 4. In this experiment the greatest decrease in hafnium content was from 1.15 (in the feed) to 0.052%, for a separation factor of 22. In contrast, the greatest decrease observed in the experiment represented by Figure 4 was from 2.3% in the feed to 0.35%, corresponding to a separation factor of 6.6.

Figure 7 gives a further comparison of the effect of reaction temperature on the separation. It shows the difference between solid feed and vapor feed in the relative amounts of hafnium reacted in the temperature range 420–450°C using data from Figures 4 and 6. A large portion of hafnium is evidently in the form of a subhalide in the solid-feed experiment. Figure 7 also demonstrates that, in the solid-feed run, hafnium is bound by the nonselective complex over the temperature range 334–420°C.

3. PRODUCT QUALITY

The separation procedure effectively removed iron from feed $ZrCl_4$ (35), even when the concentration of iron in the reducing bed was very high. Table VIII shows a comparison of iron concentration

TABLE VIII
Distribution of Impurities, ppm

Element	Feed	Hf concentrate	Product	Bed
Al	100	50 est.	55	—
B	8 est.	4 est.	1	—
Cd	<0.5	Interference	0.7	<0.5
Co	<5	Interference	<5	—
Cr	45	>600	40	—
Fe	980	2200 est.	60	70,000
Mg	22	1,000–10,000 est.	42	—
Mn	<5	230 est.	<5	280
Mo	20	50 est.	<10	—
Ni	16	>200	11	>200
Pb	15	>200	<10	—
Si	105	400 est.	40	260
Sn	17	50,000–100,000	380	1,000
Ti	26	<20 est.	<20	—
V	<10	<10	<10	—
Hf/(Hf + Zr)	23,000	180,000	6,900	800[a]
H$_2$O insoluble as %	—	—	—	37.18

[a] Soluble portion of bed.

(980 ppm) and other metals in the feed $ZrCl_4$ with that in the effluent $ZrCl_4$ and in the bed. The initial sample of effluent containing the bulk of the hafnium had a high concentration of iron (2200 ppm). However, the product $ZrCl_4$ obtained by disproportionation of $ZrCl_3$ was very low in iron (60 ppm), even though the bed contained over 1000 times that concentration (70,000 ppm). The original $ZrCl_2$ bed was prepared from low hafnium–zirconium sponge (70 ppm Hf) and zirconium tetrachloride (140 ppm Hf). Silicon and molybdenum were also removed. Zirconium and hafnium oxides were left in the bed.

4. HAFNIUM CONCENTRATION

The procedure can be utilized to produce high hafnium concentrates (36). Using a low reaction temperature in a closed reactor, fractions containing up to 43% Hf have been obtained from feed containing 2.3% Hf. With vapor feed, hafnium has been concentrated from 15 to 42% in a single stage.

Hafnium was concentrated from 77 to 99% (37) in a 13-in. fixed-bed reactor using $ZrCl_2$ as the reducing bed as follows:

The reactor was a 4 ft long pyrex tube, ⅝ in. in diameter, placed horizontally in a cylindrical furnace built to heat 32 in. of the tube, allowing 8 in. of the tube to protrude on each side. The protruding tube on one end was charged with $ZrCl_4$, and the tube on the other end served to collect samples of effluent $ZrCl_4$.

The reactor could be slipped back and forth axially through the furnace as desired. The charge was volatilized through the bed by sliding the charged end into the furnace. Argon flowing through the tube prevented back diffusion of the $ZrCl_4$. Temperature measurements were made by thermocouples in wells placed radially into the cylindrical furnace. Fourteen grams of a mixture containing 77% $HfCl_4$ and 23% $ZrCl_4$ were vaporized over a bed of $ZrCl_2$ at 325–330°C during a period of 2 hr. The effluent collected from the reactor during this time weighed 8.4 g and contained 99% hafnium. After reaction the bed was heated to 480°C, and 2.38 g of sublimate containing 74% hafnium was collected. $ZrCl_4$ (12.39 g) containing only 0.0088% Hf was next passed over the bed at 325–330°C for 2 hr. During this time 2.5 g of effluent containing 6.1% hafnium was collected. Disproportionation of the residue for 1 hr at 480°C, using an argon sweep, yielded 7.7 g of sublimate analyzing 2.3% hafnium.

In addition to demonstrating the preparation of purified $HfCl_4$, these experiments showed that the Newnham separation was more dependent on relative rates of reaction of the tetrachlorides with the reducing bed, than on equilibria. The appreciable hafnium concentra-

tion in the disproportionated product as well as in the second disproportionated product obtained after reaction of the bed with low hafnium $ZrCl_4$ was evidence for reaction of $HfCl_4$ with $ZrCl_2$. Further
work demonstrated that hafnium in the bed could be removed gradually by repeated successive reaction and disproportionation with low
hafnium $ZrCl_4$ feed.

5. PHYSICAL CHEMISTRY

a. Reaction Rates

(1) $ZrCl_4$ with $ZrCl_2$. The rates of reaction of $ZrCl_4$ with $ZrCl_2$
were studied by means of evacuated, sealed, glass U-tubes. The
U-tubes were inverted, with a known weight of $ZrCl_4$ charged to
one arm and a known weight of $ZrCl_2$ charged to the other. The
tubes were immersed simultaneously in molten tin and removed
successively at time intervals. After immersion $ZrCl_4$ sublimed into
the $ZrCl_2$ bed at a rate determined by the reaction rate. On removing
a tube, the arms were separated at the U-bend by means of a gas flame,
and the weight of unreacted $ZrCl_4$ determined. The increase in
weight of the $ZrCl_2$ was calculated as a function of time at 300, 330,
and 360°C, and at partial pressures of $ZrCl_4$ of 0.35, 1.0, and 2.6 atm,
respectively.

Figure 10 shows results obtained at different temperatures. Two
rates are observed, a rapid initial rate followed by a slower one. The
initial rate may depend on reaction of $ZrCl_4$ with $ZrCl_2$ surface and
with the resulting subhalides. The final rate may depend on diffusion
of $ZrCl_4$ vapor through these subhalides to the underlying $ZrCl_2$. The
data are presented on the basis of 1 mole of $ZrCl_4$ reacting with
1 mole of $ZrCl_2$.

It was important for process purposes to determine the rate of
absorption (RA) of $ZrCl_4$ by a stirred $ZrCl_2$ bed at higher temperatures, i.e., 402–404°C and at atmospheric pressure. The highest rates
observed at these temperatures were 2.8 and 3.2×10^{-3} moles of $ZrCl_4$
per minute per mole of $ZrCl_2$. The absorption rate was apparently
a linear function of the feed rate (FR) over most of the reaction
time, the slope constants k [eq. (14)] being 0.73 for 402–404°C and
0.55 for 415°C.

$$RA = kFR \qquad (14)$$

The feed rates ranged from 2.5×10^{-4} to 50×10^{-4} moles of $ZrCl_4$
per minute per mole of $ZrCl_2$. These observations indicate that the
factor limiting the absorption rate is the mechanical contacting of
$ZrCl_2$ powder with $ZrCl_4$ vapor.

(2) ZrCl₄ with Zr. The rate of reduction of $ZrCl_4$ with Zr was measured by heating premixed charges of excess $ZrCl_4$ with zirconium sponge in sealed, evacuated, glass tubes. Several tubes were immersed simultaneously in molten tin at the desired temperature, then successively removed at chosen time intervals. The extent of the reaction was determined by measuring the water-insoluble material

TABLE IX
Reduction of $ZrCl_4$ with Zr

Particle size Zr mesh	Bath temp., °C	Reaction time, min	Metal remaining, %
30–40	290	6	99.3
	315	20	94.7
	290	35	98.0
	290	50	93.3
	318	85	90.7
100–250	325	30	96.1
	335	90	95.6
	328	150	93.7
	333	210	93.5
	330	270	94.1
100–250	361	30	87.6
	369	90	82.7
	365	270	76.8
	362	450	71.0
100–250	390	30	80.5
	385	270	66.5
	390	1860	43.5
	370	4440	30.1
<325	380	900	1.3
100–250	380	900	48.3
40–60	380	900	57.6
>10	380	900	77.6
<325	380	880	8.6

remaining. Since zirconium halides are water soluble, the water-insoluble material is presumed to be unreacted zirconium. The data are presented in Table IX. There appear to be two rates, a high initial rate followed by a lower one (Fig. 12). The explanation of the rates is similar to that for $ZrCl_4$ reacting with $ZrCl_2$, except that the underlying surface is Zr metal. Important factors relating to reaction rate are the particle size of the zirconium sponge and reaction temperature. Thus 99% reaction was obtained at 380°C after

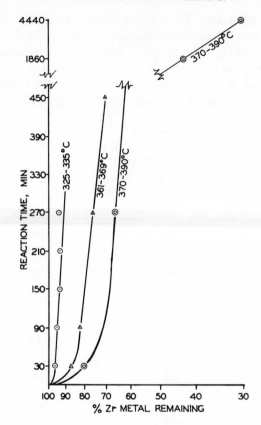

Fig. 12. Reaction of Zr sponge (100–250 mesh) with ZrCl₄.

15 hr when the particles were less than 325 mesh, compared to 52% with 100–250 mesh or 22% with larger than 10 mesh. The reaction is particularly rapid above 420°C where disproportionation of ZrCl₃ can take place. Thus, ZrCl₃ formed at the Zr surface is immediately disproportionated, causing spalling of the surface coat with exposure of fresh metal to the ZrCl₄ vapors.

b. Dissociation Rates

Dissociation rates for the "trichloride" were measured at atmospheric pressure for several temperatures in the stirred reactor using sufficiently high rates of argon flow to sweep the vapors from the reactor as soon as they formed. In addition to the temperature, the rates depended on whether the mole ratio of ZrCl₄ to ZrCl₂ was less or more than 0.5. In the former case the rate is represented by the symbol

$R_{Zr_3Cl_8}$ and in the latter by $R_{complex}$. The rates expressed in moles of $ZrCl_4$ liberated per minute per mole of charged $ZrCl_4$ remaining in the bed are given by eqs. (15) and (16) where T is the absolute temperature.

$$\log R_{Zr_3Cl_8} = 13.1828 - 10{,}869/T \tag{15}$$

$$\log R_{complex} = 6.4363 - 6707/T \tag{16}$$

The value for the dissociation rate at 400°C from eq. (16), 3.2×10^{-4} moles of $ZrCl_4$ per minute per mole of subhalide remaining, is in agreement with that estimated from the data of Larsen and Leddy (23).

c. Dissociation Pressure

The partial pressure in atmospheres of $ZrCl_4$ vapor over the bed, P_{ZrCl_4}, was calculated from the effluent rate of $ZrCl_4$ vapors, R_{ZrCl_4}, transported by argon gas under atmospheric pressure flowing over the bed at known low velocities, R_A, according to eq. (17), where the rates are expressed in moles per minute.

$$R_{ZrCl_4}/(R_{ZrCl_4} + R_A) = P_{ZrCl_4} \tag{17}$$

The dissociation pressure was assumed to be the partial pressure under conditions such that the partial pressure of $ZrCl_4$ was independent of argon flow rate. Thus the dissociation rate was greater than the rate of removal of vapors and the partial pressure was maintained constant. The dissociation pressure depended on the temperature and also on whether the mole ratio of $ZrCl_4$ to $ZrCl_2$ in the bed was more or less than 0.5. Pressure–temperature curves of the former lie above those of the latter as shown in Figure 5.

IV. SCALEUP

A. Operational Technology

1. PROCESS FLOW SHEET

A process flow sheet is given in Figure 13. The $ZrCl_4$ feed is sublimed through a $CaCl_2$–steel wool bed for preliminary purification and then is introduced as vapor into reactors R1-AZ and R1-BZ which contain preformed beds. The selective reduction of $ZrCl_4$ to $ZrCl_3$ (Step 1) takes place in these reactors at 400°C with the simultaneous purge of unreduced $HfCl_4$ as vapor (step 2). The hafnium-enriched fraction (ca. 5.4% Hf) is fed to a recovery stage (R1-AH) containing a preformed bed of $ZrCl_2$, where steps 1 and 2 are repeated. The first effluent from reactor R1-AH contains about 40% $HfCl_4$, while

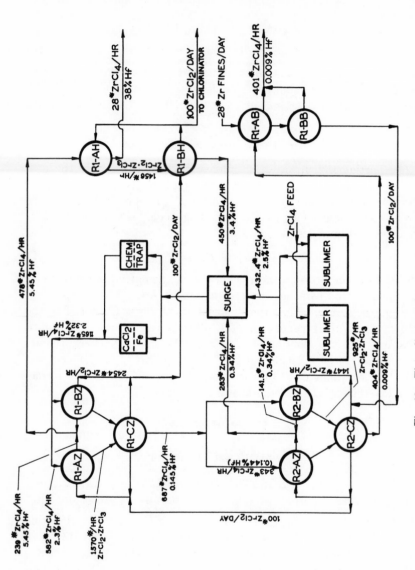

Fig. 13. Flow diagram for two-stage plant process.

the $ZrCl_4$ disproportionated from the bed and moved to R1-BH contains 3.4% (or feed-level) hafnium.

In order to carry out the disproportionation of $ZrCl_3$ to recover low-hafnium $ZrCl_4$ (step 3), the $ZrCl_3$ beds in R1-AZ and R1-BZ are introduced into reactor R1-CZ. The temperature in this reactor is held at 460°C where the reaction in eq. (3) takes place. The regenerated $ZrCl_2$ bed formed in this step is recirculated to reactors R1-AZ and R1-BZ and the cycle is repeated. The disproportionated $ZrCl_4$ from reactor R1-BH is recirculated to R1-AZ and R1-BZ via the surge since its hafnium content approximates that of the feed.

Since $ZrCl_4$ from R1-CZ contains about 0.14% hafnium, another stage is required to reduce the hafnium content to 0.005–0.010%. This stage takes place on the dichloride bed in reactor R2-AZ and R2-BZ with disproportionation in reactor R2-CZ. The final $ZrCl_4$ product is below the AEC specifications of 0.10% hafnium. The $ZrCl_2$ bed (which is prepared in reactors R1-AB and R1-BB) is introduced into this final stage and moves counter-current to the $ZrCl_4$ toward the high hafnium stage from which it is discharged. At this point it may be converted to tetrachloride for feed material by chlorination.

2. PREPARATION OF $ZrCl_2$ BED

The $ZrCl_2$ bed is produced in 90% yield by heating a mixture of 325-mesh zirconium sponge and $ZrCl_4$ at 450°C in a sealed vessel for several days. The $ZrCl_2$ can also be made by passing $ZrCl_4$ through a moving bed of zirconium sponge maintained at 450°C. The latter method may be preferred for preparing large quantities of $ZrCl_2$.

The operation does not require removal of old bed and replacement. Bed impurities arise from raw $ZrCl_4$ feed consisting of reducible and nonvolatile compounds, e.g., $FeCl_3$, sulfides, oxygenated compounds, etc. These are removed via a purge stream of $ZrCl_2$. This stream is formed in a reactor (R1-AB in Fig. 13) from low hafnium-zirconium fines and low hafnium $ZrCl_4$, and may average about 1 lb of zirconium fines per 270 lb of $ZrCl_4$ feed, with an equivalent amount purged from the hafnium concentration stage (R1-BH in Fig. 13). Removal of iron and aluminum halides impurities from crude $ZrCl_4$ feed by scrubbing with molten $CaCl_2$ and iron wool (38) permits the use of low purge rates.

3. ABSORPTION STEP

In the preferred absorption step, $ZrCl_4$ is fed as vapor, since under these conditions a separation factor of about 16 may be achieved.

The preferred temperature is 400°C. Contact times of 15 min are sufficient to give complexing of $ZrCl_4$ with the $ZrCl_2$ bed. Longer times reduce separation factors, since hafnium appears to work into the resulting $ZrCl_3$ bed. The partial pressure of the $ZrCl_4$ should be near the total pressure of the system to prevent disproportionation of the $ZrCl_3$ at 400–415°C. Consequently, the inert gas purge employed for removal of the $HfCl_4$ should be held to only that necessary to sweep out the $HfCl_4$. Close control of temperature, partial pressure, and time is necessary.

4. DISSOCIATION STEP

The recovery of purified $ZrCl_4$ from the $ZrCl_3$ involves heating the bed to approximately 450°C in a rapid argon flow and with sufficient contact time.

Figure 13 shows that product $ZrCl_4$ containing 0.009% Hf and representing 93% of the feed flows out of the system from reactor R1-AB. Most of the hafnium flows out (from reactor R1-AH) in a $ZrCl_4$ stream which is 6% of the feed and which has a hafnium concentration of 38%.

The yields and separation factors in the two separation stages are 59% and 16, respectively. In the first stage, 38.6% of the $ZrCl_4$ (3.4% Hf) is recycled via the hafnium concentration stage (R1-AH and R1-BH), the remaining 2.4% corresponding to the 38% Hf concentrate mentioned above. The separation factor in the hafnium concentration stage is 1.6.

In the second separation stage, 41% of the $ZrCl_4$ (0.34% Hf) is recycled to the first separation stage.

B. Engineering

The Newnham procedure, because of its simplicity, introduces fewer engineering operational problems than other hafnium separation processes. In this operation, zirconium is introduced as the tetrachloride and removed as such, ready for reduction to metal by sodium or magnesium without further processing.

The separation process requires about an 8-hr time cycle, and the labor requirement is modest.

1. MATERIALS OF CONSTRUCTION

The metals suitable for construction of the reactors have been discussed (Sect. III-B-2-a) where the heat transfer medium was molten tin. Molten sodium is suggested for plant operation. Suitable metals for operation within the temperature range 300–460°C are iron or steel.

2. MATERIALS HANDLING

a. Zirconium Trichloride

In our small-scale studies in glass bulbs with no stirring, caking of the trichloride preparations was observed on cooling to room temperature. This is believed to be due to condensed unreacted $ZrCl_4$ acting as a cementing agent for the particles. However, the trichloride could be maintained as a free-flowing powder if formed (and cooled) with stirring. The cold material is not pyrophoric.

b. Zirconium Dichloride

No caking was observed for the hot or cold dichloride. The powder was always free flowing. This material is not immediately pyrophoric in air at room temperature, but is pyrophoric at slightly elevated temperatures.

c. Zirconium Tetrachloride

This tetrachloride reacts as a vapor with the solid dichloride. The tetrachloride may be fed into the reactor as a gas or as a solid. The latter is vaporized in the reactor. In the former case nonvolatile materials are not introduced into the bed, but this is not true with solid feed unless the tetrachloride is presublimed. Certain nonvolatile and adsorbed materials, such as moist $ZrCl_4$, $FeCl_2$, Fe_2O_3, phosgene, etc., may react with the bed, causing its effectiveness to decrease, and others such as ZrO_2 may unnecessarily dilute the bed.

3. OPERATIONAL FACTORS

a. Free Board of the Reactor

A reactor should minimize the possibility of contamination of the product with unreacted $HfCl_4$ in the gas phase above the bed. Thus, the free board of the reactor should be held to a minimum, and arrangements made to sweep out unreacted vapor with argon.

b. Cut Points on Fractionation

A monitoring system may be necessary to detect the rapid change in the hafnium content of the sublimate which occurs just prior to product collection. In our studies we have found this change to occur at about 430°C.   This change is accompanied by a rapid rise in sublimation rate.

c. Heat Transfer

Adequate control of bed temperature is a critical factor for successful operation of the Newnham procedure.   A narrow temperature interval (50°C) exists between formation and dissociation of Zr_3Cl_8. The dissociation rate changes very rapidly with temperature in a 15°C interval (410–425°C).

These factors make it necessary to have good heat transfer to and through the solid in the bed.   In our work, molten tin was used to transfer heat to the reactor.   Heat input was controlled by the bath temperature.

Fluidized bed techniques are particulary suitable for the heat-transfer requirements outlined above, especially in view of the powdery, noncohesive nature of $ZrCl_2$ and $ZrCl_3$.

d. Cold Spots

Particular care must be taken to eliminate cold spots in any reactor designed.   Under atmospheric pressure, $ZrCl_4$ will condense into very hard deposits on any metal surface below 331°C.   These deposits may interfere with moving parts of the reactor or may act as a source of contamination during product collection.   Elimination of cold spots in the laboratory was accomplished by immersion of the entire reactor and valving system in molten tin.

C. Physical Properties of Reactants

1. VAPOR PRESSURE

The vapor pressure of solid $ZrCl_4$ according to Kuhn, Ryan, and Polko (39) is given by eq. (18).

$$\log P_{mm} = 11.76 - 5400/T \quad (T = 500\text{–}710°K) \tag{18}$$

An approximate relationship between vapor pressure of liquid $ZrCl_4$ and temperature is also given by these authors.

$$\log P_{mm} = 9.11 - 3400/T \tag{19}$$

Howell, Sommer, and Kellogg (40) have determined portions of the phase diagrams and vapor pressures of the systems $NaCl\text{–}ZrCl_4$, $KCl\text{–}ZrCl_4$, and 1:1 molar $NaCl\text{–}KCl$ with $ZrCl_4$.

2. THERMODYNAMIC PROPERTIES

Rossini et al. (41) list the thermodynamic properties for sublimation of $ZrCl_4$ and $HfCl_4$ as, respectively, sublimation temperature $= 331$ and $317°C$, $\Delta H = 25.3$ and 24 cal/deg-mole, and $\Delta S = 41.9$ and 40 cal/deg-mole. In addition, the melting points are 437 and 432°C.

Gross et al. (42) have obtained a value of -234.7 kcal/mole for the standard heat of formation of $ZrCl_4$ (solid).

3. VAPOR DENSITY OF $ZrCl_4$

The vapor density of $ZrCl_4$ at various temperatures has been studied by Friend, Colley, and Hayes (43), who found no evidence of association at 400°C or above. They believe that above 500°C, some decomposition takes place according to eq. (20).

$$2ZrCl_4 \rightleftarrows 2ZrCl_3 + Cl_2 \tag{20}$$

Their data may be expressed in the form of eq. (21).

$$D_v = 10.41 - 7.62 \times 10^{-3}T \tag{21}$$

where D_v is the vapor density relative to air and T is the temperature in °C.

Deville and Troost (44) obtained values of the vapor density of $ZrCl_4$ of 8.10 and 8.21 as the monomer at the boiling point of sulfur (444°C) which is in close agreement with the calculated value of 8.09 for unassociated $ZrCl_4$.

D. Economics

A preliminary cost estimate of the Newnham process was made, based on the laboratory findings, for comparison with hexone extraction. An overall production rate of 1.2 million pounds of zirconium as zirconium tetrachloride was assumed with a recovery of 90% in processing crude zirconium tetrachloride to a reducible product. Three stages were assumed using horizontal tube reactors. A total of 10 units were required. A material balance and schematic flow diagram for the process are included in Figure 13. Double streams are included for the reaction of $ZrCl_4$ with $ZrCl_2$ in two of the stages since residence time is important in this phase of the process. Only one reactor is included for treatment of the hafnium-rich $ZrCl_4$ stream since amounts at this stage are lower.

A net manufacturing cost was obtained which was $1.50 less per

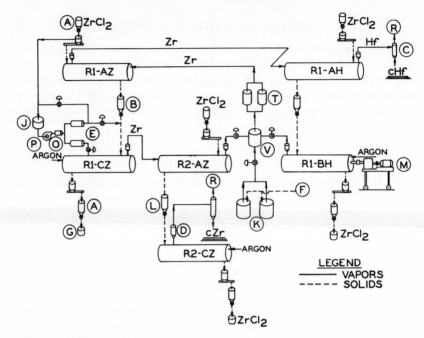

Fig. 14. Pilot plant flow diagram: (A) surge hopper, (B) seal hopper, (C) condenser, (cHf) condensed product HfCl₄, (cZr) condensed product ZrCl₄, (D) dust filter, (E) heat exchangers, (F) solid crude ZrCl₄ feed, (G) tote can, (Hf) HfCl₄, (J) sodium heating system, (K) sublimers, (L) seal, (M) screw drive assembly, (O) filter, (P) pump, (R) to argon recycle (T) chemical trap, (V) vapor surge, (Zr) ZrCl₄.

pound of zirconium content than the comparable figure for the hexone (methylisobutyl ketone) extraction process at this stage as of the period 1957–1958.

E. Pilot Plant

A three-stage pilot plant process design is shown in Figure 14.   The design assumes recovery of 25 lb/hr of hafnium-free ZrCl₄.   Horizontal tube screw-fed heated reactors, as assumed for the plant, are employed.

V. SUMMARY

A practical procedure has been demonstrated for separating HfCl₄ from ZrCl₄ based on selective reduction. Yields of ZrCl₄ of 90% having less than 100 parts per million of hafnium can be obtained in a two-stage operation with heads recycle.

We have shown that:

1. Adequate separations are obtained at reduction temperatures in the range of 320–420°C. Solid feed limits reaction temperature to about 330°C. At these temperatures separation factors are lower. Thus, 3–4 stages are required.

2. The time needed to accomplish the separation is less than 8 hr.

3. The operation may be run at atmospheric pressure.

4. Disproportionation is appreciable in the temperature range of 420–460°C at atmospheric pressure.

5. Adequate agitation of the solid reaction mixture is necessary since good heat transfer is required.

6. Evidence was obtained of the formation of several complex compounds in the reaction between the feed zirconium tetrachloride and the reducing bed. The thermal stability of these complexes differed. The most stable complex, which had an apparent stoichiometry of 2 moles of zirconium dichloride to 1 mole of zirconium tetrachloride, was relatively stable up to 420°C and appeared to be selective for zirconium tetrachloride. An unstable intermediate complex apparently forms in the temperature range of 330–370°C by nonselective addition of both hafnium tetrachloride and zirconium tetrachloride to the stable complex. This less-stable complex decomposes in the temperature range of 350–430°C, while the "stable" complex undergoes low to moderate disproportionation. This unstable complex apparently is a source of contamination and is responsible for the poorer separations when reactions are run at low temperatures. Formation of the complex is held to a minimum by operating at an absorption temperature of 400°C.

7. High hafnium concentrates (up to 42%) are obtainable in one stage from 2.3% hafnium feed. It is possible to obtain pure hafnium concentrates by this process.

8. A method for preparation of zirconium dichloride from zirconium sponge powder (325 mesh or finer) and zirconium tetrachloride was developed in which at least 90% of the metal reacted. The reaction conditions are 450°C for about 160 hr in a sealed system.

9. Iron may be removed effectively from zirconium tetrachloride by the zirconium dichloride bed. The iron is reduced and remains in the bed. Other polyvalent metals may also be removed from zirconium tetrachloride in this fashion.

10. Volatile nonreducible metallic halides such as aluminum chloride may be separated into the hafnium-rich fractions.

11. The hafnium content of the reducing bed does not rise appreciably during successive runs.

12. The reducing bed ($ZrCl_2$) has a long life provided the feed is relatively free of reducible impurities. This bed is free flowing and has a bulk density of 0.99 g/ml.

13. Molten tin is a suitable heat-transfer medium for the laboratory process.

14. Liquid tin valves activated by inert gas pressure were designed and found to operate satisfactorily in bench-scale equipment.

Acknowledgement

The work reported herein was performed at the Research Department of the U.S. Industrial Chemicals Company in Cincinnati. The authors wish to thank Dr. Virgil Hansley for his advice and guidance, Mr. B. Saffer, Mr. K. Sieve, and Mr. S. Stregevsky for their assistance with technical manipulations, Mr. Ivan Newnham for his inspiration and cooperation, and Mr. D. Hinshaw for the spectrographic analysis. Mr. Robert Maddox contributed the flow diagrams.

References

1. H. Etherington, in *The Metallurgy of Zirconium,* 1st ed., B. Lustman and F. Kerze, Jr., Eds., McGraw-Hill, New York, 1955, p. 13.
2. J. N. Wanklyn and P. J. Jones, *J. Nucl. Mater.,* **6,** 291–329 (1962).
3. G. L. Miller, *Zirconium,* 2nd ed., Academic Press, New York, 1957, pp. 140–142.
4. G. L. Miller, *Zirconium,* 2nd ed., Academic Press, New York, 1957, p. 143.
5. H. Etherington, in *The Metallurgy of Zirconium,* 1st ed., B. Lustman and F. Kerze, Jr., Eds., McGraw-Hill, New York, 1955, p. 15.
6. I. E. Newnham, *J. Am. Chem. Soc.,* **79,** 5415 (1957).
7. F. Steidler, *Microchemie,* **2,** 98 (1924); through J. W. Mellor, *Inorganic and Theoretical Chemistry,* Longmans, Green, New York, 1952, p. 170.
8. E. M. Larsen, *J. Chem. Educ.,* **28,** 329 (1951).
9. J. M. Herndon, in *The Metallurgy of Zirconium,* 1st ed., B. Lustman and F. Kerze, Jr., Eds., McGraw-Hill, New York, 1955, p. 113.
10. G. L. Miller, *Zirconium,* 2nd ed., Academic Press, New York, 1957, chap. 4.
11. J. M. Googin, *U.S. At. Energy Comm.,* Rept. Y-B65-103 (1956).
12. W. M. Leaders, *U.S. At. Energy Comm.,* Rept. Y-480 (1949).
13. L. G. Overholser, C. J. Barton, and W. R. Grimes, *U.S. At. Energy Comm.,* Rept. Y-477 (1949).
14. W. Fischer and W. Chalybaeus, *Z. Anorg. Allgem. Chem.,* **255,** 79 (1947).
15. W. J. Kroll and W. W. Stephens, *Ind. Eng. Chem.,* **42,** 395 (1950).
16. H. A. Wilhelm, *U.S. At. Energy Comm.,* Rept. ISC-144 (1951).
17. H. W. Chandler, U.S. Pat. 3,276,862 (1966).
18. R. C. Young, *J. Am. Chem. Soc.,* **53,** 2148 (1931).
19. W. C. Schumb and C. K. Morehouse, *J. Am. Chem. Soc.,* **69,** 2696 (1947).
20. O. Ruff and R. Wallstein, *Z. Anorg. Allgem. Chem.,* **128,** 96 (1923).
21. J. H. deBoer and J. D. Fast, *Z. Anorg. Allgem. Chem.,* **187,** 177 (1930).
22. J. D. Fast, *Z. Anorg. Allgem. Chem.,* **239,** 145 (1938).
23. E. M. Larsen and J. J. Leddy, *J. Am. Chem. Soc.,* **78,** 5983 (1956).

24. I. E. Newnham, U.S. Pat. 2,791,485 (1957).
25. P. Prakash and C. V. Sundaram, United Nations International Conference of the Peaceful Uses of Atomic Energy, 2d Geneva, 1958, through *Nucl. Sci. Abstr.*, **13**, 842 (1959).
26. L. F. Dahl, T. Chiang, P. W. Seabaugh, and E. M. Larsen, *Inorg. Chem.*, **3**, 1236 (1964).
27. J. A. Watts, *Inorg. Chem.*, **5**, 281 (1966).
28. H. L. Schlafer and H. W. Wille, *Z. Anorg. Allgem. Chem.*, **327**, 253 (1964); through J. A. Watts, *Inorg. Chem.*, **5**, 281 (1966).
29. B. Swaroop and S. N. Flengas, *Can. J. Phys.*, **42**, 1886 (1964); through J. A. Watts, *Inorg. Chem.*, **5**, 281 (1966).
30. S. J. Gregg, *Surface Chemistry of Solids*, 1st ed., Reinhold, New York, 1951, chap. XIV.
31. W. B. Blumenthal, *Ind. Eng. Chem.*, **46**, 528 (1954).
32. S. Schott and V. L. Hansley, U.S. Pat. 2,880,084 (1959).
33. D. M. Mortimore and L. A. Noble, *Anal. Chem.*, **25**, 296 (1953).
34. L. D. Hinshaw and F. D. Miller, paper presented at Pittsburgh Conference on Analytical and Applied Spectroscopy, March 2–6, 1959, paper No. 140.
35. I. E. Newnham, U.S. Pat. 2,953,433 (1960).
36. I. E. Newnham, U.S. Pat. 2,961,293 (1960).
37. H. Greenberg, U.S. Industrial Chem. Co., 1957, private communication.
38. H. Greenberg and H. R. Lubowitz, U.S. Pat. 3,053,620 (1962).
39. D. W. Kuhn, A. D. Ryan, and A. A. Polko, *U.S. At. Energy Comm.*, Rept. Y-552 (1950).
40. L. J. Howell, R. C. Sommer, and H. H. Kellogg, *Trans. Met. Soc. AIME*, **209**, 193 (1957).
41. F. D. Rossini, D. D. Wayman, W. H. Evans, S. Levine, and I. Jaffee, *Selected Values of Thermodynamic Properties, Natl. Bur. Std. Circ. (U.S.)*, **500**, 718, 720 (1952).
42. P. Gross, C. Hayman, and D. L. Levi, *Trans. Faraday Soc.*, **53**, 1285 (1957).
43. J. A. N. Friend, A. T. W. Colley, and R. S. Hayes, *J. Chem. Soc.*, **1930**, 494.
44. Deville and Troost, *Compt. Rend.*, **45**, 821 (1857); through J. A. N. Friend, A. T. W. Colley and R. S. Hayes, *J. Chem. Soc.*, **1930**, 494.

Ultrafiltration

ALAN S. MICHAELS

President, Amicon Corporation
Lexington, Massachusetts

I. DEFINITIONS AND BASIC CONCEPTS

"Ultrafiltration" is a process of separation whereby a solution containing a solute of molecular dimensions significantly greater than those of the solvent is depleted of solute by being forced under a hydraulic-pressure gradient to flow through a suitable membrane. "Reverse osmosis," ultrafiltration, and ordinary filtration differ superficially only in the size scale of the particles which are separated; differentiation among the three is, in large measure, arbitrary. It is, however, convenient to reserve the term "reverse osmosis" for membrane separations involving solutes whose molecular dimensions are within one order of magnitude of those of the solvent, and to use "ultrafiltration" to describe separations involving solutes of molecular

297

dimensions greater than ten solvent molecular diameters and below the limit of resolution of the optical microscope (ca. $0.5\,\mu$). Ultrafiltration thus encompasses all membrane-moderated, pressure-activated separations involving solutions of modest-molecular-weight (ca. 500 and up) solutes, macromolecules, and colloids. At present, ultrafiltration processes are largely confined to aqueous media, and most of what follows relates to aqueous systems. There are, however, no fundamental reasons why ultrafiltration cannot be performed with non-aqueous solvents (utilizing, of course, solvent-resistant membranes); as a matter of fact, there are numerous commercially important petroleum- and petrochemical purifications which can and ultimately will be performed by ultrafiltration with suitably constituted membranes.

Ultrafiltration can be utilized to accomplish one or more of the following:

1. Concentration of solute by removal of solvent.

2. Purification of solvent by removal of solute.

3. Separation of solute A from solute B by ultrafiltration through a membrane permeable to A but not to B (or vice versa).

4. Analysis of a complex solution for specific solutes to which the membrane is permeable.

While each of these procedures can be performed by alternative means (e.g., evaporation, dialysis, ultracentrifugation, chemical precipitation, etc.), ultrafiltration is usually the method of choice from the point of view of speed, efficiency, and cost—particularly when thermally unstable or biologically active materials are involved, or when large volumes of dilute solutions are to be processed.

The basic ultrafiltration process is extremely simple, as illustrated in Figure 1. The solution to be ultrafiltered is confined under pressure (utilizing either compressed gas or a liquid pump) in a cell, in contact with an appropriate ultrafiltration membrane supported on a porous plate. The contents of the cell must be subjected to moderate agita-

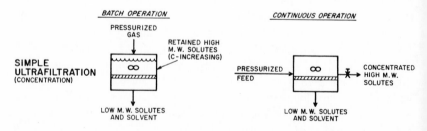

Fig. 1. Basic ultrafiltration process.

tion to avoid accumulation of retained solute on the membrane surface with attendant "blinding" of the membrane. Ultrafiltrate is continuously produced and collected until the retained solute concentration in the cell solution reaches the desired value. With a high-quality ultrafiltration membrane and adequate agitation, the solute concentration in the "retentate" can often be allowed to rise as high as 10–30 wt % of the solution before the ultrafiltration rate begins to decrease significantly. Since most ultrafiltrations are carried out on quite dilute (e.g., 0.01–1 wt % solids) solutions, concentration factors as large as 1000, or solvent recoveries as high as 99%, can be accomplished without difficulty. For the same reason, virtually complete separation of retained solutes from unretained solutes can often be accomplished by direct ultrafiltration.

II. ULTRAFILTRATION MEMBRANES

The heart of successful and useful ultrafiltration is obviously the ultrafiltration membrane itself. To serve its purposes, an ultrafiltration membrane must possess the following characteristics:

1. High hydraulic permeability to solvent (water), that is, the membrane must be capable of transmitting water at high rates per unit membrane area, under modest hydraulic pressures. As a rough rule of thumb, permeabilities in excess of 1 liter/hr-ft^2 at 100 lb/in.^{2} pressure difference are adequate for many laboratory and industrial ultrafiltration operations.

2. Sharp "retention-cutoff" characteristics; that is, the membrane must be capable of retaining *completely* all solutes of molecular weight *above* some specified value, and of passing *completely* all solutes of molecular weight below some second specified value not much different from the first. "Sharp" and "diffuse" cutoff membranes are illustrated in Figure 2.

3. Good mechanical durability, chemical and thermal stability.

4. High flow stability, that is, minimum dependence of hydraulic permeability upon solute type or concentration.

5. High-fouling resistance, that is, little tendency for irreversible reduction in hydraulic permeability due to intrusion of retained solutes into the membrane substance.

6. Excellent manufacturing reproducibility of flow and retention characteristics.

Only over the past few years have advances in membrane technology and polymer physics made it possible to produce membrane structures meeting most of these requirements. Indeed, it is only due

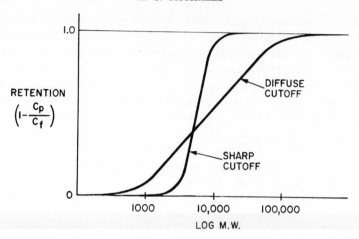

Fig. 2. Retention characteristics of representative ultrafiltration membranes.

to the advent of these new structures that ultrafiltration has become an attractive laboratory- and industrial-separation/purification process. A brief discussion of the structure of ultrafiltration membranes and of the transport mechanisms by which they are believed to function is thus necessary to explain the similarities and differences between ultrafiltration and other separation processes, and to point out its capabilities and limitations.

One can make a rational differentiation between two types of ultra-filtration membranes: the *microporous ultrafilter* and the *diffusive ultrafilter*. The microporous ultrafilter fits the picture normally accorded a "filter" in the traditional sense: it is a rigid, highly voided structure containing interconnected, extremely small random pores—average size perhaps of the order of 500–5000 Å. Through such a structure, solvent (water) flows essentially viscously under a hydraulic-pressure gradient, flow rate proportional to pressure difference. Dissolved solutes, to the extent that their (hydrated) molecular dimensions are smaller than the *smallest* pores within the structure, will pass through little impeded by the matrix; larger-sized molecules, on the other hand, will become trapped within (or upon the external surface of) the membrane, and will thereby be retained. If the *texture* of the porous structure is very uniform—that is, the distribution of pore dimensions within any one small volume element of the matrix is identical to that within any other such element—then all molecules smaller than the minimum pore dimension will permeate the membrane, all larger than the largest pore dimension will be retained external to the membrane, and all solutes of intermediate size will be partially retained within or upon the membrane to a degree which increases with size. The "cutoff level" of the membrane will thus depend upon its mean pore size; the "sharpness of cutoff," on the other hand, will depend upon the *breadth of the pore-size distribution*. Such a microporous filter may be compared to a series of sieves, each of which contains a random array of holes of somewhat differing sizes, stacked randomly upon one another. The probability that a particle of a given size will find its way through the stack depends on the probability of finding a path through the stack which does not require passage through a hole smaller than the particle. A little thought will show that, the fewer sieves in the stack (or the thinner the membrane), the greater is the probability for passage of *any* particle smaller than the *largest* hole. Hence, microporous membranes tend to show higher cutoff levels and sharper cutoffs as their thickness decreases.

It will also be evident that microporous ultrafilters are inherently susceptible to internal plugging or fouling by solute molecules whose dimensions lie within the pore-size distribution of the filter. Fouling will be most pronounced with solutes whose dimensions lie in the lower third of the pore-size distribution, since these will have the least difficulty entering the structure, but the greatest likelihood of lodging in pore constrictions. To minimize this internal fouling problem, one is thus forced to employ, for a specific solute, a microporous ultrafilter whose *mean* pore size is far below that of the dimensions of the solute particle to be retained.

In addition to the solvent-permeability reduction which always accompanies fouling, there is a corresponding change in the rejection spectrum of the membrane: since the lower end of the pore-size distribution is inactivated in the plugging process, most of the residual flow is forced to take place in the larger pores. Consequently, the plugged filter is often less retentive and solute discriminatory than it was before plugging. If pore obstruction becomes complete, however, an effectively much finer pore-size filter may result, and a far more retentive structure formed in the process.

In contrast, the *diffusive ultrafilter* is an ostensibly homogeneous gel membrane through which both solvent (water) and solutes are transported by *molecular diffusion*, under the action of a *concentration* or *activity* gradient. In such a structure, solute and solvent migration occur via random thermal (Brownian) movements of molecules within and between the (similarly oscillating) chain segments comprising the polymer network. Under these circumstances, molecular transport is a *thermally activated* process, that is, the migration of a molecule from one location to another within the matrix requires the concentration of an abnormal amount of kinetic energy in that molecule and its surroundings. The amount of energy required depends, among other things, upon the dimensions of the diffusing molecule, the density of the polymer matrix surrounding the molecule itself, forces of interaction between the molecule and the matrix, and the constraints upon free motion of polymer segments comprising the matrix. As a rule, the more highly expanded or hydrated the gel matrix, the stronger the specific binding affinity between polymer and solvent, and the more flexible are the chains making up the matrix, the higher will be the solvent permeability. Conversely, the "tighter" the gel matrix, the weaker the affinity between solute and polymer, and the more rigid the polymer chains, the lower will be the solute permeability. Thus, membranes prepared from highly hydrophilic polymers which swell to a limited extent in water have the potential for serving as useful diffusive aqueous ultrafiltration membranes.

Despite the rather fundamental difference in water-transport mechanisms in microporous and diffusive ultrafilters, the latter show the same proportionality of solvent flow to pressure as the former. The explanation lies in the fact that, for dilute solutions and relatively low pressures, solvent activity is nearly directly proportional to hydrostatic pressure. Hence no inferences about solvent-flow mechanisms in membranes can be drawn from pressure–flow measurements alone. Where differences do appear, however, is in the *temperature dependence* of solvent permeability. For microporous filters the temperature variation is attributable solely to the change of fluidity of the solvent with temperature, whereas for diffusive ultrafilters, the variation reflects the *activation energy* for diffusion of solvent within the membrane. Usually, the latter is greater than the former; hence, diffusive ultrafilters normally display greater changes in solvent permeability with temperature than do microporous ultrafilters.

Solute transport through a diffusive ultrafilter is, to a first approximation, governed by Fick's laws of diffusion within the membrane. The solute flux is thus proportional to the solute concentration gradient in the membrane, and the solute diffusivity. Since the surfaces of the membrane are in essential solution equilibrium with the solutions with which they are in contact, the gradient of solute concentration is determined solely by the solute concentrations in the feed and ultrafiltrate, and is virtually *pressure independent*. This means that, as the pressure difference across a diffusive ultrafilter is increased (causing a proportionate increase in solvent flow), the solute flux is but little altered—leading to the curious result that the *retention efficiency* (fraction of solute retained) of the membrane increases as flow rate increases. In theory, therefore, the retention efficiency of a diffusive ultrafilter for *any* solute could be raised to a level approaching 100% if the operating pressure were raised high enough; in actuality, however, it is found that solute transport through such membranes is not completely independent of the rate of solvent transport, but increases somewhat with the solvent flux. Consequently, the retention efficiency from a specific solute, while it increases with solvent-flow rate, approaches a finite limiting value below 100%, which is not increased by further elevation of the hydraulic-pressure gradient.

If two solutes of differing molecular dimensions are compared with respect to their transport rates through a diffusive ultrafilter, under equivalent concentration gradients, the rates are roughly proportional to the solute *diffusion coefficients* within the membrane. These diffusion coefficients are found to be exceedingly sensitive to molecular size and shape, varying roughly exponentially with the square of the molecular diameter. As a consequence, two solute molecules differing by as little as 10% in linear dimension may possess diffusion coefficients in a given membrane which differ by an order of magnitude. Thus, although a diffusive ultrafilter is not truly "impenetrable" to a specific solute or solute mixture, it is nonetheless highly effective at selectively passing solvent relative to solute, and at selectively passing one solute species with respect to another of slightly larger molecular size. For this reason, diffusive ultrafilters are peculiarly suited for the efficient separation of relatively small-molecule species from dilute solutions, and for the differential separation of closely related molecular species.

Since a diffusive ultrafilter contains no "pores" in the conventional sense, and since the concentration within the membrane of any solute retained by the membrane is low and time independent, such a filter is not "plugged" by retained solute; that is, there is no decline in solvent permeability with time at constant pressure. This property is particularly important for continuous concentration–separation operations, where constant flow and retention properties are essential to stable, long-term performance.

A. Anisotropic, Diffusive Ultrafiltration Membranes

Rough calculation of the diffusional rate of transport of water through an idealized hydrogel under a pressure gradient reveals that,

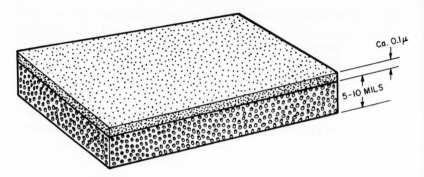

Fig. 3. Schematic of an anisotropic, diffusive ultrafiltration membrane.

for a membrane a few thousandths of an inch in thickness, subjected to a pressure difference of the order of 100 lb/in.², a water-flow rate of less than 100 ml/ft²-hr can be expected. This is too low to be of much practical interest for ultrafiltration. To be attractive for laboratory or industrial use, therefore, a diffusive ultrafilter with an effective thickness in the *micron* or *submicron* range would have to be fabricated. Since present-day polymer film-fabrication technology does not permit manufacture of films thinner than about 0.5 mils (13 μ) without pinhole defects, conventional plastic films fall far short of meeting the requirements of ultrafiltration membranes.

In the past few years, novel polymer-fabrication procedures have been developed for producing anisotropic membrane structures of a character represented schematically in Figure 3. These structures consist of an extremely thin (0.1–10 μ) layer of homogeneous polymer supported upon a much thicker (20 μ–1 mm) layer of microporous, open-celled sponge. An electron micrograph of the cross section of such a membrane is shown in Figure 4. If desired, these membranes can be further supported upon a fibrous sheet (e.g., paper) to provide greater strength and durability. These membranes are always used with the thin-film or "skin" side exposed to the high-pressure solution. The support provided to the "skin" by the spongy substructure is adequate to prevent film rupture even at high pressures, and the hydraulic resistance to liquid flow through the spongy layer is small compared with the resistance of the skin. As a consequence, the filtration characteristics of this composite structure are essentially those of the microscopically thin surface layer which constitutes the major barrier. Surprisingly, membranes of this quite-sophisticated structure can be manufactured in quantity with excellent reproducibility of properties and with few skin-layer defects.

Virtually all diffusive ultrafiltration membranes now in use are of this anisotropic type. Control of solute retention and water permeability is effected by proper choice of polymeric material and adjustment of the degree of consolidation of the outermost gel layer.

In contrast, most *microporous* ultrafilters now available are isotropic structures, whose flow and retention properties are independent of flow direction. Such microporous membranes are produced in the form of sheets of one to ten thousandths of an inch in thickness; their specific hydraulic permeabilities (even for the smallest mean pore-size structure (ca. 50 mμ available) are sufficiently high to yield water-flow rates far in excess of 1 liter/hr-ft^2 at 100 lb/in.^{2}. These membranes are, however, virtually nonretentive for solutes of molecular weight under about one million. While yet finer pore-texture microporous filters can be prepared, their hydraulic permeabilities are too low to obtain practical flow rates with liquids.

Recently, success has been realized in producing anisotropic microporous membrane structures whose texture is shown schematically in Figure 5. In such a membrane, the pore size (and local porosity) increases rapidly from one face to the other. Since, in viscous flow, the hydraulic flow resistance varies inversely with the square of the mean pore diameter, the overall permeability of the anisotropic struc-

Fig. 4. Cross-section electron micrograph of an anisotropic diffusive ultrafiltration membrane (Diaflo UM-1) (courtesy Amicon Corporation, Cambridge, Massachusetts).

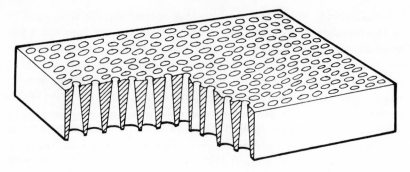

Fig. 5.   Schematic of a "tapered-pore" microporous membrane.

ture is substantially higher than that of an isotropic membrane of equivalent mean pore size.   Moreover, the *retention* capability of the anisotropic membrane is governed by the pore texture of the finest pore layer; hence, the retention cutoff point is far lower for an anisotropic membrane than for an isotropic membrane of comparable mean pore size.   An additional important advantage of the anisotropic homogeneous membrane is its superior resistance to plugging by retained solutes (provided it is used with the fine-textured side in contact with the feed solution); this is merely a consequence of the fact that a particle barely large enough to penetrate the topmost layer of the membrane cannot become trapped within the membrane because of the larger pore sizes in the substructure.   Hence, anisotropic microporous membranes are more permeable, more retentive, and more plug resistant than their isotropic counterparts.   Such membranes are just beginning to appear on the market now and are expected to become an increasingly important factor in laboratory and industrial ultrafiltration operations.

III. TRANSPORT KINETICS IN ULTRAFILTRATION

A. "Diffusive" Ultrafiltration

If solute and solvent transport take place by molecular diffusion within an essentially homogeneous, isotropic membrane, the transmembrane fluxes at steady state can be approximated by the following relations (6,7):

$$J_w = K_w \frac{(\Delta p - \Delta \pi)}{t} = \bar{c}_w' \bar{D}_w \frac{\bar{V}_w}{RT} \frac{(\Delta p - \Delta \pi)}{t} \tag{1}$$

where J_w is the solvent (water) flux: K_w, the "specific permeability" of the membrane; $\bar{c}_w'$, the mean water content of the membrane; $\bar{D}_w$, the average diffusivity of water in the membrane; $\bar{V}_w$, the molar volume of (bulk) water; Δp, the hydrostatic' pressure difference across the membrane; $\Delta\pi$, the solute osmotic-pressure difference across the membrane; and, t, the membrane thickness. As a rule, one expects (and finds) K_w (or $\bar{c}_w'$ and $\bar{D}_w$) to be relatively independent of pressure or solute concentration (for dilute solutions). [According to Schlögl, Katzir, Staverman, and others, eq. (1) should be modified by the insertion of a "reflection coefficient," σ, as a multiplier of $\Delta\pi$, although for sufficiently dilute solutions (where $\Delta\pi \ll \Delta p$), this correction is of little importance.] The temperature dependence of J_w, at constant $\Delta p - \Delta\pi$, can be used to determine the enthalpy of solvent sorption by the membrane and the activation energy for diffusion of solvent within the membrane.

For *solute*, the transmission flux is given by (4,6,7):

$$J_s = -D_s \frac{dc_s'}{dt} + k_s' \frac{J_w c_s'}{\bar{c}_w'} \tag{2}$$

The first term on the right-hand side of eq. (2) represents the "normal" diffusion flux of solute through the membrane, where D_s is the local diffusion coefficient of solute, and c_s' is the local solute concentration in the membrane. The second term accounts for so-called "coupling" of solute- and solvent transport, and is presumed to arise from frictional "drag" of moving solvent molecules on molecules of solute within the membrane. This term is, in essence, a "convective flux" of solute driven by the net flux of solvent, where k_s' is a dimensionless coupling coefficient of magnitude between zero and unity. In the opinion of some, the concept and existence of coupled flows of solute and solvent is inconsistent with a diffusive flow mechanism since diffusion implies concentration-biased random, thermal motion of individual molecules and such a process could not be envisioned to exert a mechanical drag in the usual sense on solute molecules. However, a high net diffusive flux of solvent through a membrane would undoubtedly result in momentum interchange between solvent and solute molecules due to collision, as well as the more familiar momentum interchange between solvent and polymer and solute and polymer. Such momentum interchange would cause, at best, a small (and generally undetectable) reduction in the high net diffusive flux of solvent, but could give rise to a significant increase in the net diffusive flux of solute. This is thought to be the origin of "coupled diffusion."

If it is assumed that solution equilibrium is established at both boundaries of the membrane, that the solute diffusivity is essentially concentration independent, and that the distribution coefficient of solute between membrane and solution is constant $(c_s' = k_s c_s)$, eq. (2) can be integrated over the thickness of the membrane to yield

$$J_s = \frac{k_s k_s' J_w}{\bar{c}_w'} \left[\frac{c_{1s} - c_{2s} \exp\left(-k_s' J_w t / c_w' D_s\right)}{1 - \exp\left(-k_s' J_w t / \bar{c}_w' D_s\right)} \right] \tag{3}$$

where c_{1s} and c_{2s} are the solute concentrations in the upstream and downstream solutions, respectively. As k_s' or J_w approaches zero, eq. (3) reduces to:

$$J_s = k_s D_s[(c_{1s} - c_{2s})/t] \tag{3a}$$

which is the simple uncoupled diffusion equation. If the exponential argument in eq. (3) is small enough to permit expansion to the first three terms only (exp $x = 1 + x + x^2/2$), and recognizing that, by mass conservation, $J_s = J_w c_{2s}'$ eq. (3) becomes:

$$J_w c_{2s} = J_s = \frac{k_s D_s}{1 - k_s k_s'/\bar{c}_w'} \frac{c_{1s} - c_{2s}}{t} \left(1 + \frac{k_s' t}{2\bar{c}_w' D_s} J_w\right) \tag{3b}$$

Rearrangement of eq. (3b) yields

$$\frac{c_{1s}}{c_{2s}} = 1 + \frac{J_w t}{k_s D_s} \frac{1 - k_s k_s'/\bar{c}_w'}{1 + k_s' J_w t/2c_w' D_s} \tag{4}$$

Hence, a plot of $J_s/(c_{1s} - c_{2s})$ vs. J_w (or Δp) yields a function whose value approaches $k_s D_s/t(1 - k_s k_s'/\bar{c}_w')$ as $J_w \to 0$, while a plot of (c_{1s}/c_{2s}) vs. J_w should yield (at small values of J_w) a straight line of slope $t/k_s D_s(1 - k_s k_s'/\bar{c}_w)$. These two parameters are sufficient to characterize the complete solute flux–concentration–pressure relationship, if the solute transport process is indeed diffusive. Again, the *temperature* dependence of the solute flux can be used to estimate the enthalpy of solute dissolution in the membrane, the solute-diffusion activation energy, and the activation energy associated with the "coupled solute–solvent complex" through the term k_s'.

Equation (4) can be rearranged to relate the "solute-rejection efficiency" of the membrane, R, $(R = 1 - c_{2s}/c_{1s})$ to the water flow through the membrane, for small values of J_w:

$$R = \alpha t J_w/(1 + \alpha t J_w) \tag{4a}$$

where

$$\alpha = (1/k_s D_s)(1 - k_s k_s'/\bar{c}_w)$$

which illustrates the often-observed increase in rejection efficiency with increasing flow rate during ultrafiltration. It is interesting to note, however, that if the thickness of a diffusive ultrafiltration membrane is reduced, its solute-rejection efficiency will be reduced unless the solvent flux through the membrane is proportionately increased. Comparison of eq· (1) with eq. (4a) discloses that the solute-rejection efficiency is uniquely determined by the driving potential across the membrane (i.e., $\Delta P - \Delta \pi$), irrespective of its thickness or solvent flow rate; this emphasizes the thermodynamic–molecular-kinetic aspect of ultrafiltration, wherein separation efficiency is directly related to the energy expended in carrying out the transport process.

At sufficiently high pressure differentials across the membrane, eq. (3) reduces to

$$J_s/J_w = c_{2s} = k_s k_s' c_{1s}/\bar{c}_w \tag{3c}$$

whereupon

$$\lim_{\Delta p \to \infty} R = 1 - k_s k_s'/\bar{c}_w \tag{3d}$$

Hence, the limiting rejection efficiency for a particular membrane and solute is determined by (1) the equilibrium distribution coefficient of the solute between membrane and solution, and (2) the solute–solvent coupling or cross coefficient within the membrane. As a rule, the coupling coefficient k_s' (like the diffusion coefficient D_s) decreases extremely rapidly with increasing molecular size of solute, thereby accounting for the frequently remarkable solute discrimination of diffusive ultrafiltration membranes.

As might be expected, minor changes in gel-network texture (as reflected, for example, by gel water content) of ultrafiltration membranes give rise to major changes in solute-rejection properties. As the network is consolidated the water permeability decreases (because of the decrease in D_w and c_w'), but the permeability to larger solute molecules decreases far more rapidly. As a consequence, the cutoff limit on solute leakage drops dramatically. Figure 6 presents water-flow- and selected solute-rejection data for a series of polyelectrolyte-complex ultrafiltration membranes (Diaflo) whose gel–water contents were varied over narrow limits, and illustrates that but small increases in network "tightness" cause a major increase in retentivity for solutes of lower molecular weights.

Representative water-permeability and solute-retentivity data for several types of commercially available diffusive ultrafiltration–reverse-osmosis membranes are shown in Table I. As will be seen,

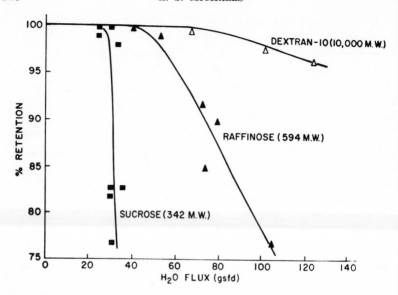

Fig. 6. Variation of solute-retention with water-permeability (or gel water content) of a series of anisotropic polyion complex ultrafiltration membranes.

the anisotropic, skin-type membranes display the highest water permeabilities consistent with high solute retentivity.

B. Microporous Ultrafiltration

In a sheet of random, isotropic microporous medium, where viscous (Poiseuille) flow obtains, the pure solvent (water) flux can be represented by the following relation:

$$J_w = (K_w'/\eta_w)(\Delta p/t) \cong \epsilon(\bar{r}^2/20\eta_w)(\Delta p/t) \qquad (5)$$

where K_w' is the hydraulic permeability, η_w is the viscosity of the solvent, $\bar{r}$ is the "hydraulic mean" pore radius, and ϵ is the porosity of the medium. The constant of 20 which appears in the relationship has been empirically determined to apply to random beds of packed powders, and takes account of anomalies of pore shape, nonuniformity of pore cross section, and sinuosities of flow paths in the medium.

A porous medium can be characterized, experimentally or mathematically, by a pore-size distribution function, $f(r)$, defined by:

$$f(r) = \frac{1}{\epsilon} \frac{d\epsilon}{dr} \qquad (6)$$

TABLE I
Comparative Characteristics of Commercially Available Ultrafiltration Membranes

Description	Manufacturer	Type	Water permeability at 100 psi, ml/cm²-min (gal/ft²-day)	NaCl (58)	Urea (60)	Sucrose (342)	Raffinose (594)	PE glycol (6,000)	Dextran (10,000)	Bovine serum albumin (60,000)	Phenyl alanine (165) pH6	pH10
Gel cellophane	DuPont	Homogeneous (?)	0.004 (1.5)	0	0	15	25	—	—	100	—	—
PEM membrane	Union Carbide	Cellulosic	0.02 (8.5)	0	0	0	0	0	0	100	—	—
P membrane	Gelman; Schleicher & Schuell	Isotropic microporous cellulosic; Homogeneous cellulosic	0.08 (35)	0	0	0	0	0	0	100	—	—
Loeb-type CA-type A	General Atomics	Anisotropic, diffusive cellulose acetate	0.001 (0.4)	~30	~0	95	~100	—	—	100	—	—
CA-type B	General Atomics	Anisotropic, diffusive cellulose acetate	0.007 (2.7)	~5	~0	50		—	—	100	—	—
CA-type C	General Atomics	Anisotropic, diffusive cellulose acetate	0.003 (1.4)	~10	~0	80	~100	—	—	100	—	—
Diaflo UM-1	Amicon Corp.	Anisotropic, diffusive	0.5 (175)	0	0	0	0	80	100	100	0	0
Diaflo UM-2	Amicon Corp.	Anisotropic, diffusive	0.2 (60)	0	0	50	90	100	100	100	0	0
Diaflo UM-3[a]	Amicon Corp.	Anisotropic, diffusive	0.1 (25)	5	0	90	95	100	100	100	15	95
Diaflo XM-50	Amicon Corp.	Anisotropic, microporous	1.0 (350)	0	0	0	0	0	0	100	0	0
VF	Millipore	Isotropic, microporous	3.7 (1,000)	75 Å pore size; retains ca. 200,000 MW solutes								
HA	Millipore	Isotropic, microporous	500 (15,000)	4,500 Å pore size; retains ca. 0.5 µ particles								

[a] Ion exchange membrane.

where $f(r)$ is the fraction of the total pore space in the medium whose radial dimensions lie between r and $r + dr$.

If the medium were assumed to be comprised of a bundle of parallel capillaries fitting the distribution function given by eq. (9), then the fraction of the total permeability, K_w', attributable to capillaries of radius lying between r and $r + dr$ would be approximated by:

$$dK_w' = \frac{r^2}{20} d\epsilon = \frac{\epsilon r^2 f(r)\ dr}{20} \tag{7}$$

Integrating between the limits of pore dimension, $r_{min} \to r_{max}$:

$$K_w' = \frac{\epsilon}{20} \int_{r\ min}^{r\ max} r^2 f(r)\ dr = \frac{r^{-2}\epsilon}{20} \tag{8}$$

Equation (11) defines the hydraulic-mean pore radius—because of the square dependency of flow on pore radius, the mean radius is heavily weighted toward the upper end of the pore-size distribution. This means that, in any microporous medium, by far the majority of the fluid flows through the larger pores—far more than the population density of large pores might suggest. As a consequence, the solute retentivity of a microporous membrane is invariably lower than might be inferred from the mean pore size, since most flow takes place through pores far larger than the mean pore diameter.

The retentivity of a membrane for a solute of effective molecular radius r' can be approximated by determining the fraction of the fluid flowing through pores of radius less than r'. If it is assumed that retention is complete for pores $<r'$, and zero for pores $>r'$, then, from eq. (11):

$$\frac{c_{2s}}{c_{1s}} = \int_{r'}^{r\ max} r^2 f(r)\ dr / \bar{r}^2 \tag{9}$$

This relationship indicates that the retentivity of a microporous membrane for a particular solute is independent of membrane porosity, water permeability (per se), or operating pressure, but solely governed by the pore-size distribution. It is also of interest to note that the retention selectivity of such a membrane toward a pair of solutes differing but slightly in molecular radius (say between r_A and r_B) is given by:

$$\frac{c_{2A}}{c_{2B}} \frac{c_{1B}}{c_{1A}} = \frac{\displaystyle\int_{r_A}^{r\ max} r^2 f(r)\ dr}{\displaystyle\int_{r_B}^{r\ max} r^2 f(r)\ dr} \tag{10}$$

where $r_B = r_A + \Delta r$ or

$$\frac{c_{2A}}{c_{2B}} \cdot \frac{c_{1B}}{c_{1A}} = 1 + \frac{\int_{r_A}^{r_B} r^2 f(r)\, dr}{\int_{r_B}^{r\,\text{max}} r^2 f(r)\, dr} = 1 + \frac{r_A{}^2 f(r_A)\, \Delta r}{\int_{r_B}^{r\,\text{max}} r^2 f(r)\, dr} \qquad (11)$$

Since the numerator of the fraction on the right-hand side of eq. (13a) is (except for the case where $r_B = r$ max—that is, the membrane is virtually completely retentive for species B) much smaller than the denominator, the retention selectivity for B relative to A will be small. Paradoxically, therefore, a sieve-type membrane is basically less discriminating between solutes of slightly differing molecular dimensions than is a diffusive ultrafilter.

When the solvent contains a solute which is partly retained by the membrane, the solvent flux is altered in a manner reflected by:

$$J_w = (K_w'/t)\alpha(\Delta p - \Delta\pi) + [K_w'(1 - \alpha)/t]\beta(\Delta p) \qquad (12)$$

where α is the fraction of the pure solvent flux which passes through pores of dimension *smaller* than those of the retained solute molecule, and β is a "resistance factor" (of magnitude less than unity) which accounts for the increased resistance to solvent flow offered by solute within those pores of the membrane large enough to admit solute. β approaches unity as solute concentration approaches zero. If $\Delta\pi = 0$, combining eqs. (5) and (6) yields:

$$J_w/J_{w0} = \alpha + \beta(1 - \alpha) \qquad (13)$$

Hence, plots of the ratio of solvent flux in the presence of solute to that of pure solvent, versus solute concentration and solute molecular size or shape, will permit estimation of the pore size and pore-size distribution of the membrane, and of the drag resistance of solute upon pore flow of solvent. In contrast, the presence of solute has no significant effect upon solvent flow in diffusive membranes, except through its effect on the osmotic pressure of the solution.

Solute flux through a microporous membrane can, consistent with eq. (12), be represented by :

$$J_s = [K_w'(1 - \alpha)\,\Delta p/t](c_{1s}) \qquad (14)$$

That is, solute is transported purely convectively with solvent through those pores of the membrane large enough to admit solute molecules. Combining eqs. (8) and (6) yields:

$$c_{1s}/c_{2s} = 1 + [\alpha/\beta(1 - \alpha)] \qquad (15)$$

Equation (9) indicates that the retentivity of a microporous membrane will be independent of pressure or solvent flux, but dependent solely on the membrane pore structure *and* upstream solute concentration. This is in marked contrast to the situation with a diffusive membrane, where retentivity is pressure dependent, and solute concentration, independent.

Moreover, with a microporous membrane, solute flux (at constant pressure) should [eq. (14)] vary with temperature at the same rate as solvent flux, since only solvent fluidity should display significant temperature dependence. This also means that solute retentivity [eq. (15)] should be virtually *temperature independent*, unlike the situation with a diffusive membrane.

Examination of eqs. (9)–(15) reveals that, in order for a microporous ultrafilter to be able to effectively retain a solute species of a given molecular size, and yet not become internally clogged by the molecules it retains, either (*1*) the *mean* pore size (if the distribution is normal) must be far below the solute molecular dimensions, or (*2*) the pore-size distribution must be exceedingly narrow and peaked sharply around a pore dimension somewhat below that of the solute molecule. The former constraint requires that the solvent flow primarily through pores far smaller than are needed to keep out the solute (with consequent low solvent permeability). The latter is virtually impossible to achieve by conventional microporous filter-manufacturing methods; however, a close approach to this ideal distribution has recently been achieved by the General Electric Company with its Nucleopore Filters, made by chemical etching of radiation-damaged plastic films. The nature of the film-treatment process, however, limits the fraction of the film area which can be perforated; consequently, flow rates through these filters are often considerably lower than are desired for preparative separations.

For the most part, because of their limited discriminatory power for solutes, their low permeabilities when tailored for retention of molecules of ca. 1000 Å diameter and smaller, and their pronounced clogging tendency, microporous ultrafilters are used for detection, analysis, and removal of trace quantities of solutes or dispensoids in fluids, rather than for preparative separations or concentration.

IV. CONCENTRATION POLARIZATION AND BLINDING IN ULTRAFILTRATION

If a solution is driven under a hydraulic-pressure gradient through a membrane which in whole or in part retains the solute, since the convective flux of solute toward the membrane with the solvent must initially exceed the rate at which solute passes through the membrane, the solute must accumulate in the solution contacting the upstream face of the membrane. A steady state will ultimately be reached wherein any solute convectively transported to the membrane must be removed by leakage through the membrane, by molecular diffusion away from the membrane into the bulk of the upstream fluid, or by turbulent mixing with the upstream fluid. The consequence of these interacting transfer processes is the development of a layer of solution in contact with the upstream membrane surface, which is more concentrated in solute than the feed liquid. This is so called "concentration polarization"; it is illustrated schematically in Figure 7.

Since the membrane responds in its transport behavior only to the immediate boundary conditions it "sees," it naturally displays the solvent- and solute-transport rates characteristic of this layer of more concentrated upstream solution. If the membrane is not completely solute impermeable, the consequence of this polarization is increased solute leakage, and an *anomalously low rejection efficiency*. Further, if the solute is present in sufficient concentration to exert significant osmotic pressure, the polarization reduces the effective driving force for solvent transport and thus *reduces water flow*. It is obvious that, in a completely unstirred system, it would be physically impossible to effect any solute–solvent separation, even with a membrane which displays inherently high solute retentivity. Hence, it is essential to conduct ultrafiltration under conditions where retained solute can be removed diffusively or convectively from the upstream membrane surface. This can be accomplished, for example, by vigorous mechanical stirring of the upstream liquid (see Fig. 8), or by causing the feed

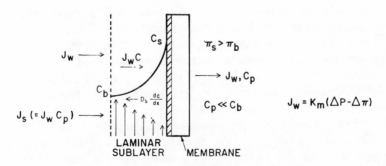

Fig. 7. Schematic diagram of concentration polarization near ultrafiltration membrane with low molecular-weight solutes.

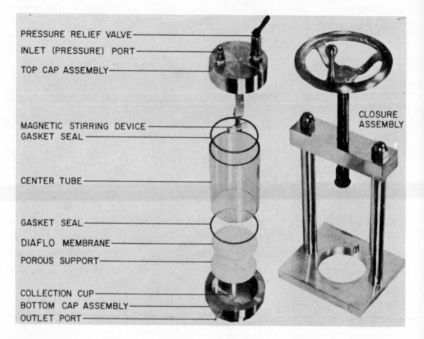

PRESSURE RELIEF VALVE
INLET (PRESSURE) PORT
TOP CAP ASSEMBLY

MAGNETIC STIRRING DEVICE
GASKET SEAL

CENTER TUBE

GASKET SEAL
DIAFLO MEMBRANE
POROUS SUPPORT

COLLECTION CUP
BOTTOM CAP ASSEMBLY
OUTLET PORT

CLOSURE
ASSEMBLY

Fig. 8. A stirred batch-type ultrafiltration cell (courtesy Amicon Corporation, Cambridge, Massachusetts).

liquid to flow under turbulent conditions parallel to the membrane surface, such that only a small fraction of the liquid is allowed to permeate the membrane during a single pass. Also, by causing the liquid to flow across the membrane surface in an extremely thin film, concentration polarization can also be minimized by molecular diffusion away from the membrane, even when flow is laminar.

The tendency toward concentration polarization during ultrafiltration increases rapidly with the membrane permeation flux, with solute molecular weight (due to decreased diffusivity in solution), and with solution viscosity. In many cases (particularly with macromolecular solutes), concentration polarization is the limiting factor governing ultrafiltration rate and separation efficiency, and heroic measures such as ultrasonic agitation of the upstream fluid, extremely high fluid velocities across the membrane, or a mechanical "wiping" of the upstream surface, may be required to combat it. For viscous solutions, moderate elevation of the temperature (despite its usual adverse effect on membrane solute retentivity) often brings about dramatic improvement in ultrafiltration rates and efficiency, solely because of improved turbulence and relief of polarization.

With solutions of proteins and other hydrocolloids, concentration polarization is further complicated by another problem peculiar to concentrated solutions of highly hydrophilic materials: such substances tend to form highly viscous or truly gelatinous mixtures when their concentration in water exceeds 5–15 wt %. As a consequence, accumulation of these materials on an ultrafiltration membrane surface leads to a slimy, coherent film which is not only difficult to detach and redisperse in the liquid by simple fluid shear, but is also rather impermeable to water. This situation is shown schematically in Figure 9. The result is a rapid and marked reduction in ultrafiltration rate, which cannot be restored except by interrupting the process and physically removing the adherent film from the membrane. This "blinding" or "surface fouling" of the membrane is most likely to occur when ultrafiltering rather concentrated (2 wt % or more) hydrocolloid solutions, under conditions where upstream fluid agitation is poor. If agitation of the upstream liquid is adequate, and if the fluid shear stress at membrane surface is higher than the yield stress of the gel film which might form, blinding can usually be prevented; if conditions conducive to slime formation develop even for a short time, however, consolidated gel material may deposit on the membrane, and restoration of high-flow conditions may not dislodge it. Attempts to compensate for blinding by increasing the operating pressure are usually fruitless, since the slime layer tends to consolidate further

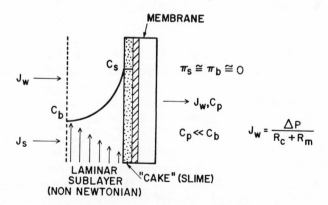

Fig. 9. Schematic diagram of hydrocolloid "cake" formation on an ultrafiltration membrane with high molecular-weight solutes.

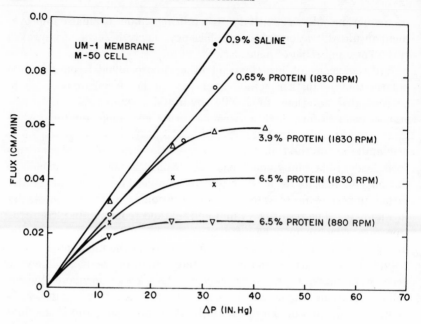

Fig. 10. Effects of stirring rate, pressure, and solute concentration on ultra-filtration rates of serum albumin solutions through Diaflo ultrafiltration membranes.

under these conditions.   Frequently, one observes no increase in flow—and sometimes an actual decrease—as pressure is raised.   Some typical data showing the effects of agitation (stirring rate), pressure, and solute concentration upon the ultrafiltration of serum albumin solutions are shown in Figure 10.

Blinding can often be alleviated or minimized by taking advantage of the fact that, during active ultrafiltration, the slime layer or cake is held onto the membrane surface by the pressure differential across it.   Periodic interruption of flow through the membrane (by, for example, periodically closing off the ultrafiltrate line) may allow the cake to be sloughed off into the flowing upstream fluid.   Such periodic flow interruption will, of course, be beneficial to overall throughput if the time required for cake accumulation is long compared to the time required to dislodge it; if this is not the case, then no gain will result from the cycling process.   As a rule, flow cycling of this type is advantageous with relatively dilute feed solutions (less than ca. 1% solids), but not with more concentrated solutions.

Obviously, the chemical constitution and texture of the membrane surface plays an important role in blinding tendency, inasmuch as these properties influence the adhesion of the cake to the membrane. Very smooth membranes, and those whose surfaces carry the same sign-of-charge as the solute being ultrafiltered are least prone to this type of contamination.

In general, the specific water permeability of most diffusive ultrafiltration membranes is independent of the nature of the solutes present in the solution being processed, and of solute concentration below levels where substantial changes in solution fluidity are observed. These generalizations appear to apply to solutes which are virtually completely retained by the membrane (90% retention or more) or virtually completely passed (10% retention or less). However, for solutes to which the membrane is moderately permeable (10–90% retention), one frequently observes a marked *reduction* in water permeability in the presence of such solutes; this reduction is reversible, in that if the membrane is subsequently permeated by pure water, its permeability rapidly returns to its normal value. The reasons for this behavior are obscure, although it appears to be due to a dynamic accumulation of solute within the membrane during ultrafiltration, with consequent increase in resistance to water transport. This phenomenon can give rise to operational difficulties in ultrafiltration of solutions containing a continuous spectrum of molecules of differing molecular weight (e.g., synthetic polymer solutions), under which conditions anomalously low filtration rates may be encountered.

In summary, successful ultrafiltration requires the maintenance of adequate feed-fluid agitation upstream of the membrane to minimize concentration polarization and maximize filtration rates and separation efficiencies. Blinding or caking is a common occurrence with concentrated hydrocolloid solutions, which can be controlled by proper agitation or periodic flow interruption. Hence, properly designed and operated ultrafiltration equipment is no less important to successful separation or purification by this process than is selection of the proper membrane.

V. MEMBRANE DURABILITY

Most of today's ultrafiltration membranes are rather sophisticated structures which are unable to withstand the same kinds of physical or chemical abuse to which conventional filtration media are frequently subjected. For example, anisotropic membranes owe their properties to a submicron-thick skin layer, which, if perforated or abraded, will no longer be solute retentive. Hence, excessive handling or deliberate abrasion of the skin layer will destroy the membrane. Gentle rubbing with soft materials to remove adhering deposits or rinsing in running water will usually not cause damage.

Most ultrafiltration membranes are comprised of hydrophilic polymers and often cannot be desiccated without irreversible changes in properties or complete destruction. Many undergo structural changes at elevated temperatures which markedly alter their transport characteristics. Some (e.g., those made from cellulose esters) are sensitive to hydrolytic degradation in acidic or basic media and are susceptible to enzymatic deterioration or microorganism attack. Many are swollen and structurally altered by organic solvents (particularly alcohols, ketones, or phenols), concentrated electrolyte solutions, or detergent solutions. Few can withstand steam sterilization; where sterility is required, treatment with formaldehyde, aqueous ethanol, a dilute aqueous bactericide solution (e.g., Zephiran, hexylresorcinol), or radiation is frequently recommended.

Many ultrafiltration membranes have an operating-pressure limit above which permeability begins to decline rapidly, or mechanical breakdown occurs. Low-permeability membranes can frequently be subjected to pressures as high as several thousand pounds per square inch without failure, while high-permeability membranes may have upper-pressure limits of ca. 300 psi.

VI. LABORATORY APPLICATIONS OF ULTRAFILTRATION AND ULTRAFILTRATION MEMBRANES

As mentioned in the introduction to this chapter, ultrafiltration provides a fast, efficient, and inexpensive means for *concentration, purification,* or *separation* of solutes in solution, and thus serves as a valuable adjunct or alternative to conventional *preparative* and *analytical* procedures utilized by the chemical, medical, and biochemical research community.

A. Concentration

One of the dominant uses of ultrafiltration today is for selective removal of solvent from dilute solutions. The objectives of ultrafiltrative concentration are either (a) to facilitate isolation and quantitative recovery of specific compounds or mixtures or (b) to facilitate detection and analysis of specific compounds whose concentration in the original solution before treatment is too low for accurate determination. Ultrafiltration is a particularly attractive concentration procedure for such compounds as proteins, nucleic acids, polypeptides, enzymes, antibodies and antigens, viruses, and other biocolloids whose chemical structure or biological activity are likely to be altered by common concentrative procedures such as precipitation, lyophilization, evaporation, freeze drying, and the like. Evidence is strong (1) that ultrafiltrative concentration does not alter solute structure or properties; moreover, ultrafiltrative concentration can be conducted rapidly and at low temperature if desired, so that troublesome degradative processes common to labile biologicals can be minimized, and high yields of active material assured.

Concentration by ultrafiltration is also highly attractive for use with exceedingly dilute solutions of any solute or hydrocolloid whose isolation by conventional means would be time consuming, awkward, or inefficient. For example, a liter of a 0.1 wt % solution of polymer in water can be concentrated to 10 ml of a 10%-by-weight solution in a matter of a few hours, utilizing a 3-in. diameter ultrafiltration membrane at modest operating pressure. The resulting concentrate can then be readily desiccated by freeze or vacuum drying, if desired.

Ultrafiltrative concentration is rapidly gaining popularity as an adjunct to gel partition chromatography, for both preparative and analytical purposes. As a rule, efficient chromatographic separation of complex mixtures can be accomplished only under conditions where column effluents are maintained exceedingly dilute. Column fractions are thus frequently difficult to analyze accurately, and solute isolation from such effluents is often formidable. As shown in Figure 11, by the simple expedient of delivering the chromatographic column effluent (continuously, or in intermittent aliquots) under pressure to a stirred ultrafiltration cell fitted with a properly retentive membrane, 90% or more of the solvent can be removed from the solution, leaving behind a concentrate which can easily be analyzed, or subsequently dehydrated to obtain the isolated component.

Ultrafiltrative concentration is now beginning to receive increasing attention as an important *clinical-laboratory* tool for the detection and analysis of trace components of medical significance in body fluids. For example, ultrafiltration of cerebrospinal fluid, urine, lymph, or blood plasma makes it possible to concentrate by several orders of magnitude, traces of macrosolutes (e.g., albumin, Bence-Jones protein, bilirubin, hemoglobin) present in such fluids, and thereby raise their concentrations to levels where accurate quantitative analysis is feasible. Similar procedures are also being practiced by biochemical researchers in the development of new detection meth-

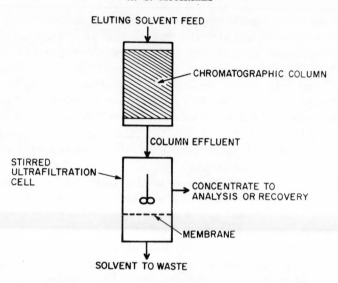

Fig. 11. Diagram of a combined column-chromatography and ultrafiltration fractionation system.

ods for and elucidation of the causes of, metabolic, neoplastic, and viral diseases.

B. Purification

Ultrafiltration is proving to be a very useful technique for isolating, and obtaining in high purity, specific solutes which occur in admixture with other solutes of differing molecular size or chemical constitution. For example, tissue homogenates, culture broths, polymer solutions and latices, and aqueous chemical-reaction media invariably contain simple inorganic electrolytes, low molecular-weight organics, macromolecules, and colloidal or cellular material, any one or more of which may constitute an undesirable impurity whose removal is necessary.

In those cases where the impurity components are of low molecular weight (e.g., salts in admixture with proteins, simple amino acids in admixture with polymers), ultrafiltration can be employed to yield an impurity-depleted retentate and an ultrafiltrate free of the desired product. In such cases, ultrafiltration performs the same functions as conventional membrane dialysis, but at far greater economy of time, space, and labor. Direct ultrafiltration of such mixtures obviously yields a retentate in which the *volume concentration* of impurity is the same as that in the original solution, but in which the ratio of *retained* solute to impurity is increased; e.g., ultrafiltrate removal of 90% of the volume of the original solution will result in a retentate containing only 10% of original impurity, but 100% of the desired product. Obviously, if the retentate is rediluted to the original solution volume with pure solvent, and reultrafiltered to remove 90% of the volume, the second-stage retentate will contain only 1% of the original impurity components. As a rule, with high-flow ultrafiltration membranes, impurity removal by repetitive ultrafiltration and redilution can be accomplished in 1/10 to 1/100 the time required to accomplish the same result by dialysis.

An alternative and convenient technique of ultrafiltrative purification to remove low molecular-weight impurities is continuous "diafiltration," which is schematically illustrated in Figure 12. The solution to be purified is charged to a stirred ultrafiltration cell fitted with a suitable membrane, and pure solvent is fed to the cell under pressure. This procedure results in the delivery of an ultrafiltrate which contains the same concentration of impurity as exists within the cell volume at any instant, and which is produced at a rate equal to the rate of solvent addition. Under these circumstances, the concentration of product in the solution within the cell remains constant, while that of impurity decreases. The concentration of impurity remaining in the cell liquid at any time during diafiltration is given by the simple relationship:

$$C_i/C_{i0} = \exp\left(-V_s/V_0\right)$$

where C_i is the impurity concentration at any time C_{i0}, the impurity concentration in the original solution, V_s, the total volume of solvent delivered to the cell (or ultrafiltrate collected), and V_0, the initial solution volume charged to the cell. Thus, for example, injection of two volumes of solvent per volume of solution charged to the cell will reduce the impurity concentration to about 13% of its initial value.

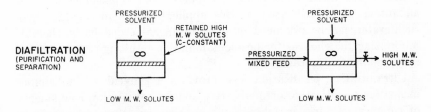

Fig. 12. Flow-diagram for diafiltration.

In those cases where the desired product is of low molecular weight, and membrane permeable, whereas the impurity is of high molecular weight and membrane impermeable, ultrafiltration will yield the pure product directly in the ultrafiltrate, at the same concentration as it occurs in the feed solution. Under such circumstances, direct ultrafiltration is usually preferable to diafiltration, since the latter process—while assuring higher yields of product—produces an unnecessarily dilute product solution.

Ultrafiltrative purification by removal of membrane-retentive impurities is enjoying increasing popularity in the laboratory for isolating and identifying low molecular-weight products formed as a consequence of biological processes (e.g., fermentation processes or cell metabolic processes) or of chemical conversions involving colloidal or macromolecular reactants (e.g., enzymatic hydrolysis, chemical oxidation or degradation of polymers, formation of soluble products by interaction of colloidally dispersed chemical reactants). In many instances, the ultrafiltration cell is actually utilized as the reaction vessel, whereupon the kinetics of the chemical conversion process can be studied by measuring the concentration of product species in the ultrafiltrate as a fraction of time. Moreover, continuous conversion-purification processes can be conducted in ultrafiltration equipment conveniently via a variant of the diafiltration procedure: one delivers continuously to the ultrafiltration cell fresh reactant mixture under pressure, and removes ultrafiltrate at the same rate. By adjusting the cell volume and feed rate to values sufficient to provide proper "residence time" for the reactants within the cell, one can achieve essentially steady-state conditions within the system so that conversion to product is quantitative, and reactant concentrations within the cell remain constant.

Ultrafiltration is also used for the preparation of particle-, microorganism-, virus-, and macrosolute-free water for a variety of laboratory preparative procedures. Such water is particularly useful for the preparation of sterile biologicals, and of "mote-free," optically homogeneous water for use in turbidimetry and nephelometry. The advantage of ultrafiltration in this application lies in the fact that such water can be produced quickly on an as-needed basis, thereby minimizing the chances for postcontamination resulting from unnecessary handling or storage.

Ultrafiltrative purification also provides a convenient and simple means for quantitative or qualitative analysis of specific components of complex mixtures in solution via analytical methods which would otherwise be useless because of component interferences. For example, ultrafiltration of blood through a microsolute-permeable membrane yields an ultrafiltrate which can be quantitatively analyzed for glucose by simple chemical-colorimetric means without interference by proteins, polysaccharides, or other constituents that would otherwise be included in the glucose titer. Similarly, a mixture containing simple amino acids in the presence of polypeptides, of monomer in the presence of polymer, can be quickly ultrafiltered to yield a filtrate analyzable for the membrane-permeable species via methods which would not discriminate between monomeric and polymeric forms of the same (or similar) substances. Also, mixtures which contain macromolecular color bodies and which thus cannot be analyzed for microsolute species by simple colorimetric methods, can be ultrafiltered to yield colorless and transparent filtrates readily amenable to such analysis.

C. Separation

Ultrafiltration is now gaining recognition as a new technique for the separation (for analysis or isolation) of molecular species which differ from one another in size, shape, or ionic constitution. This is accomplished either by direct ultrafiltration of a solution of the desired mixture through a suitably discriminatory membrane (wherein the retained solute is concentrated in the retentate, and the permeable solute is obtained at its initial concentration in the ultrafiltrate), or by diafiltration (wherein the retained solute is obtained in its initial concentration in the retentate, and the permeable solute in diluted form in the ultrafiltrate).

At the present time, ultrafiltration membranes are available with retentivity characteristics that allow fractionation of micro- and macrosolutes into rather broad molecular-weight ranges, but which do not afford very sharp differentiation between molecules of slightly different dimensions. As a rule, highly retentive membranes (i.e., those which are capable of retaining relatively small solute molecules) display much shaper cutoff characteristics than less-retentive membranes. For example, a membrane displaying 50% retentivity for a species of molecular weight 500 will be virtually completely retentive for a MW 1000 molecule, and completely permeable to a MW 250 molecule; whereas a membrane displaying 50% retentivity for a MW 50,000 molecule might be expected to show ca. 80% retentivity for MW 100,000, and ca. 20% retentivity for MW 25,000.

Thus, ultrafiltrative fractionation or separation is most satisfactorily utilized today in the processing of mixtures which contain classes of solutes which encompass rather divergent molecular-size regimes. In the case of solutes of biological origin, fortunately, macromolecular species tend to fall into quite narrow molecular weight-distribution categories, each widely separated from one another. Normal blood plasma, for example, contains albumin (MW ca. 69,000), globulins (MW ca. 100,000–150,000), and macroglobulins (MW ca. 1,000,000). Hence, an albumin-retentive membrane will allow separation of normal from lower molecular-weight abnormal proteins in plasma, whereas an albumin-permeable membrane will permit separation of albumin from the globulins. Consecutive ultrafiltration of such mixtures through a series of increasing-retentivity membranes (the ultrafiltrate from each membrane being supplied as feed to the next in the series) thus can perform the function of a "molecular-sieve stack," in which the individual retentates contain predominantly the species retained by that particular membrane. This procedure has been used successfully to perform rather sophisticated protein fractionations (2).

Molecular weight "group fractionation" by ultrafiltration is proving to be an important and useful timesaver in the column-chromatographic fractionation of biological and synthetic macromolecules. In general, the column loading and elution conditions, as well as the choice of column adsorbent, are governed by the molecular-weight range of the macrosolutes to be separated. If the mixture to be fractionated encompasses a very broad molecular-weight range, its direct fractionation by gel-permeation chromatography is often tedious, since the mixture must be passed through several different columns in sequence in order to properly separate the individual components. If, however, the initial mixture is first separated by ultrafiltration into two or more rough cuts, and the resulting filtrates or retentates subjected to separate column-chromatographic fractionation, not only are the final fractionations greatly expedited, but their accuracy is enhanced by virtue of the fact that "loading solutions" delivered to the columns can be maintained at quite high concentrations. For preparative chromatographic fractionation, it is evident that the application of *separative* ultrafiltration prior to column processing, and *concentrative* ultrafiltration of the column effluent, can result in substantially more rapid and cleaner fractionation of components and recovery of fractions in higher yield.

Separative ultrafiltration is also successfully being used to narrow the (normally quite broad) molecular-weight distribution of synthetic water-soluble polymers, by selective removal of either the high or low molecular-weight components. For example, Figure 13 shows the molecular-weight distributions (determined by column chromatography) of fractions of a sample of commercial poly(vinyl pyrollidone) obtained by passing the polymer solution through an ultrafiltration membrane retentive for ca. 50,000 MW species and higher. The efficiency of the fractionation is reflected not only in the distribution curves, but also in the intrinsic viscosities of the individual fractions. Ultrafiltrative fractionation of this sort, when supplemented with simple viscometric and solids-concentration measurements on the filtrate and retentate, can rapidly provide important insight into the character of the molecular-weight distribution, which would otherwise require much more complicated analytical procedures. Thus, if the retentate from such fractionation is found to contain a very small quantity of exceedingly high molecular-weight (high intrinsic-viscosity) material, while the filtrate contains a much larger amount of quite low molecular-weight material, the original distribution must have been badly skewed. The utility of ultrafiltration in polymer molecular-weight and structure-characterization studies is thus evident. Similarly, the procedure can be used as a rapid means for preparing significant quantities of polymer from which the high or low ends of the distribution have been quite effectively eliminated.

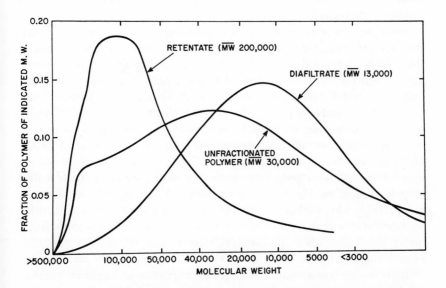

Fig. 13. Molecular-weight distribution curves for a crude sample of poly-(vinyl pyrrolidone), and fractions obtained by ultrafiltration of the polymer (in aqueous solution) through a 50,000 MW cutoff Diaflo XM-4B membrane. (Data courtesy of H. Schreiner, Linde Division, Union Carbide Corp.)

Rather unusual separations can often be accomplished by ultrafiltration using membranes which display significant ion-exchange capacity. Due to electrostatic effects associated with Donnan coion exclusion from such structures, membranes of this type are (as long as the total ionic strength of solution being filtered is low) much more retentive toward ionized solutes than toward un-ionized solutes of equal or larger molecular dimensions. Due to this phenomenon, it is thus possible to separate mixtures of simple electrolytes (sodium chloride, magnesium sulfate, etc.) from nonelectrolytes such as sugars and alcohols. A particularly interesting and useful application of this effect is found in the ultrafiltrative separation of ampholytes such as amino acids. At its isoelectric point, an amino acid is effectively uncharged and is highly permeable through an ionic membrane; above or below the isoelectric point, however, the compound is ionized and virtually membrane impermeable. Thus, if an amino acid mixture is adjusted to a pH corresponding to the isoelectric point of only one of the components and ultrafiltered, only that isoelectric species will be found in the filtrate. Obviously, by stepwise adjustment of solution pH during ultrafiltration, it is possible, by ultrafiltrate analysis, to determine the amounts of the various amino acid components present in the mixture, provided the isoelectric point of each of the expected components is known.

Separative ultrafiltration is also proving to be a very powerful tool for studying macrosolute-associative processes in aqueous solution, including ion binding by proteins, lipid protein interactions, and micellization. A solution containing both a microsolute (e.g., calcium lactate) and a macrosolute (e.g., casein), when ultrafiltrated through a membrane permeable only to the microsolute, will yield a filtrate containing only that portion of the microsolute which is not molecularly or ionically associated with the macrosolute. Similarly, a solution of an association colloid such as a surface-active agent will yield an ultrafiltrate containing only unassociated molecules or (depending on the membrane selection) the smaller association complexes. This makes it possible to determine association equilibrium constants or binding equilibration constants easily and rapidly, under conditions where the analytical procedure cannot influence the equilibrium. This property of ultrafiltration membranes can also be utilized to determine quantitatively the proportion of a specific macrosolute present in a complex mixture if a microsolute exists (e.g., a dye) which reacts specifically with that compound: one adds the reactant to the mixture in modest excess of the coreactant present, ultrafilters the mixture, and analyzes the ultrafiltrate for the added reagent. By difference, the amount consumed in reaction is readily determined.

At the present time, the utilization of ultrafiltration for preparative and analytical separations is far from being developed to the level of sophistication of which the process appears to be capable. Today, the technique is mainly being utilized to perform "class separations" which allow the employment of simple, nondiscriminatory analytical procedures for specific compounds or groups of compounds where previously only much more complicated discriminatory analyses were effective. Ultimately, however, one can expect to witness the development of a broad spectrum of sharply discriminatory ultrafiltration membranes which, when used in proper sequence with complex mixtures, will allow precise analysis and clean separation of specific compounds, with much less expenditure of laboratory time and money, and requiring far less manipulative and interpretative skill, then current separation–fractionation techniques demand.

VII. INDUSTRIAL APPLICATIONS

There are numerous important industrial concentration, purification, and separation problems which have been shown, or promise, to be amenable to solution by ultrafiltration, with considerable improvement in efficiency and economy over alternative processing methods. The most significant of these are described briefly below.

A. Waste Treatment and Potable-Water Production

The majority of industrial and municipal wastes, and polluted water supplies resulting therefrom, contain as their major objectionable components inorganic and organic colloids, biodegradable organic macrosolutes, microorganisms, viruses, and other high molecular-weight substances contributing taste, odor, or color. The dominant fraction of these pollutants is sufficiently large in molecular dimensions to be virtually completely retained by suitable ultrafiltration membranes; moreover, the concentration of such substances in even the most objectionable waste streams seldom exceeds 1 wt %. This means that, by single-stage ultrafiltration, of the order of 90% of the water in such streams can be recovered in reusable and usually potable form as ultrafiltrate. The residual concentrated retentate can then be chemically treated at reasonable cost, or evaporated and the organic residue used as fuel; in the case of many industrial wastes, the concentrated retentate frequently contains significant by-product values, which can be subsequently processed, recovered, and sold.

At present, considerable progress is being made toward application of ultrafiltration to the recovery of potable or reusable water from sewage sludges produced in "activated-sludge" secondary waste-treatment processes (8). Sludges containing upwards of 10% total suspended and dissolved solids can be ultrafiltered at reasonable rates (2–10 gallons per day per square foot of membrane area) at low pressures (30–100 psi) to yield essentially sterile water containing less than 100 ppm of oxidizable organic matter (COD). Similarly, many turbid or otherwise contaminated natural water supplies can be ultrafiltered cheaply and with high water recovery to yield clear, taste-free, color-free and odor-free water.

Pulp and paper-mill waste streams (particularly spent bleach liquors and diluted pulp-digestor streams) have been shown to be capable of contaminant removal by ultrafiltration (10). The decontaminated water obtained in the process is generally satisfactory for reuse by the mill, thereby greatly reducing fresh-water consumption and stream-pollution loads; moreover, the concentrated retentates from the process can be economically reduced by evaporation and burned, or useful chemicals (lignins, sugars) recovered.

Effluents from food-processing plants (particularly dairies, meatpacking houses, canneries) are also being evaluated for decontamination by ultrafiltration. While, again, the major objective of such treatment is pollution abatement, the concentrates obtained from such processing are frequently rich in nutritive components (proteins, sugars, etc.) which can be recovered and used as fertilizer, animal feed, and in some cases, food for human consumption.

The desalination or demineralization of water by reverse osmosis, utilizing electrolyte retentive membranes, has received widespread attention over the past five years, and has reached an advanced stage of development through the intensive support of the Office of Saline Water (U.S. Department of Interior), the State of California, and numerous foreign government agencies. At present, brackish waters (5000 ppm of salts or less) can be successfully demineralized to potable (less than 500 ppm) levels at costs comparable with other desalting processes. Several modest-sized reverse-osmosis pilot plants are now in operation, and larger-sized plants (ca. 1 million gallons/day capacity) are in design or under construction.

B. Pharmaceutical and Medicinal Manufacture

Ultrafiltration is proving to be a valuable and economic processing aid for the separation, purification, and recovery of a wide variety of pharmaceutical and biological products.   The ultrafiltration of fermentation broths, for example, allows the continuous separation of fermentation products (which are in the main, compounds of low-to-moderate molecular weight, e.g., antibiotics, hormones, vitamins, polypeptides, and, amino acids) from culture organisms and macromolecules, circumventing the need for centrifugation, ordinary filtration with filter aids, selective solvent extraction, and the like.   The product-containing filtrate, which is usually quite dilute in product and contaminated with electrolytes and low molecular-weight metabolic residues, can then be ultrafiltered through a product-retentive membrane, yielding a highly concentrated and purified retentate from which product recovery (by evaporation, freeze-drying, or the like) is simple and inexpensive.

Ultrafiltration is also a very attractive process for the concentration and purification of macromolecular biologicals, such as vaccines, enzymes, blood proteins, and nucleic acids.   Homogenates, sera, or cultures containing these substances are first centrifuged or filtered to remove cells or colloidal debris, and then ultrafiltered or diafiltered through a macromolecule-retentive membrane.   In this fashion, microsolutes and/or water are removed and a purified product concentrate is obtained as the retentate.   Large volumes of very dilute solutions can be processed very rapidly in this fashion to yield concentrates of high biological activity.

Ultrafiltration is promising to become an important tool for blood fractionation and processing, due to the advent of protein-discriminatory membranes.   At present, it is possible by ultrafiltration to separate plasma into globulin and albumin fractions, thereby allowing the isolation and manufacture of globulins by a far simpler process than is now employed.   Under development at present is a globulin-permeable membrane; ultrafiltration of blood plasma with this membrane may permit the production of virus-free plasma—a most important objective, long sought by the medical community.

C. Food and Beverage Processing

Ultrafiltration is emerging as a promising new technique for clarification, sterilization, and concentration of liquid foodstuffs and beverages, and for the separation of specific components from liquid foods. For example, bacteria, cell residues, and other colloidal matter present in beverages, normally marketed as clear liquids (e.g., apple juice, beer, wine, etc.) which give rise to turbidity or opalescence, can be ultrafiltered through high molecular-weight solute-retentive membranes to yield clear, sterile filtrates which contain all the flavor and nutrient components present in the initial liquid.

Utilizing microsolute-retentive ultrafiltration membranes, a variety of liquid foodstuffs can be concentrated without loss of flavor or nutritional value; in this operation, the primary or sole ultrafiltrate component is water (9). This concentration or dehydration process has decided advantages over alternative dehydration procedures, such as vacuum, spray, or freeze-drying, in that essential oils and other volatiles are conserved, taste changes due to thermal alteration of components are avoided, and equipment and operating costs are reduced. Orange juice, tomato juice, coffee, and milk are representative of the types of food products amenable to ultrafiltrative concentration (5).

Preliminary field tests have been performed on ultrafiltration systems for the purification of raw beet-sugar juice prior to evaporation and crystallization. Filtration through sucrose-permeable, macrosolute-retentive membranes yields an ultrafiltrate essentially free of colloid, polysaccharides, color bodies, and other objectionable components which are normally removed by chemical and sedimentative processes ("defecation"); ultrafiltration can perform these functions with considerable cost saving for the sugar refinery.

Ultrafiltration has also become of interest for the isolation and recovery of food products and industrial chemicals produced by fermentation, or chemical conversion, of natural products, e.g., dextrose from starch, alcohol from grain, amino acids from packing-house wastes. By continuous ultrafiltration of such reaction mixtures through a membrane permeable to the desired end product and recycling the concentrated retentate to the reactor, it is possible to produce a purified solution of the desired product as ultrafiltrate, and maintain the optimum microorganism population (or enzyme concentration) in the reactor. In this fashion, considerably higher average production rates can be maintained, and subsequent product purification operations greatly simplified.

The processing of dairy products is promising to become one of the most important outlets for ultrafiltration in the food industry. Salt- and sugar-free milk-protein concentrates can easily be made by ultrafiltering whole or skim milk through casein–lactalbumin retentive membranes. Cheese-whey can be deacidulated, desalted, desugared, and concentrated similarly to obtain edible protein concentrates; the ultrafiltrate from such processing can be refiltered through a lactose-retentive membrane, to yield a purified lactose concentrate from which the sugar can be readily crystallized and recovered.

One of the most attractive features of ultrafiltration for food processing is the low capital and operating cost of the procedure. For large-scale systems (e.g., 10,000 gallons/day and higher), costs are estimated to be in the vicinity of fifty cents to two dollars per thousand gallons of liquid treated—considerably lower than most other alternative processes.

VIII. MEDICAL APPLICATIONS: THE DIAFILTRATION ARTIFICIAL KIDNEY

A novel—and potentially important—application for ultrafiltration is now under experimentation (3) for use in the treatment of acute or chronic renal failure, as an alternative to "hemodialysis." This involves the continuous diafiltration of blood from a living human being, to remove blood-borne toxins and impurities, and to return purified blood to the patient. By use of an ultrafiltration membrane which is retentive only for albumin and higher molecular-weight blood components, one can ultrafilter blood and obtain a filtrate which contains all microsolutes in the same concentrations as they are present in the protein and cell-free serum. By simultaneous infusion into the blood of sterile solution containing (in the proper concentration) only the necessary components (salts, sugar, etc.), one can maintain the gross composition of the blood unaltered, but remove essentially all the unwanted substances. This purification process can, in all probability, be carried out in a much more compact device than hemodialyzers now in use, and will allow elimination of undesirable blood solutes which are difficult to remove by simple dialysis.

References

1. W. F. Blatt, M. P. Feinberg, H. B. Hopfenberg, and C. A. Saravis, "Protein Solutions: Concentration by a Rapid Method," in *Science*, **150**, 224 (1965).
2. W. F. Blatt, S. Robinson, A. Robbins, and C. A. Saravis, "An Ultrafiltration Membrane for the Resolution and Purification of Bovine α-Lactalbumin," in *Anal. Biochem.*, **18**, 81 (1967).
3. L. W. Henderson, A. Besarab, A. Michaels, and L. W. Bluemle, Jr., "Blood Purification by Ultrafiltration and Fluid Replacement," in *Trans. Am. Soc. Artificial Internal Organs*, **13**, 216 (1967).
4. J. S. Johnson, L. Dresnex, and K. A. Kraus, in *Principles of Desalination*, K. S. Spiegler, Ed., Academic Press, New York, 1966, Chap. 8.
5. R. L. Merson and A. I. Morgan, Jr., "Reverse Osmosis for Food Concentration," paper presented at Office of Saline Water (U.S. Dept. of Interior), Research Conference on Reverse Osmosis, San Diego, Calif., February 13–15, 1967.
6. V. Merten, Ed., *Reverse Osmosis*, MIT Press, Cambridge, Mass., 1967.
7. A. S. Michaels, H. J. Bixler, and R. M. Hodges, "Kinetics of Water and

Salt Transport in Cellulose Acetate Reverse Osmosis Desalination Membranes," in *J. Colloid Sci.*, **20,** No. 9, 1054, 1965.

8. R. W. Okey and P. L. Stavenger, "Reverse Osmosis Applications in Industrial Waste Treatment," paper presented at Symposium on Membrane Processes for Industry, Southern Research Institute, Birmingham, Ala., May 1966.

9. S. Sourirajan, "Reverse Osmosis Separation and Concentration of Sucrose in Aqueous Solutions Using Porous Cellulose Acetate Membranes," in *Ind. Eng. Chem. Process Design Develop.*, **6,** 154 (1967).

10. A. J. Wiley, A. C. F. Ammerlaan, and G. A. Dubey, "Application of Reverse Osmosis to Processing of Spent Liquors from the Pulp and Paper Industry," Office of Saline Water, Research Conference on Reverse Osmosis, San Diego, Calif., February 13–15, 1967.

Gas and Vapor Separations by Means of Membranes

KARL KAMMERMEYER

University of Iowa
Iowa City, Iowa

I. INTRODUCTION

A discussion of advances in a given field of endeavor implies that there is a time basis which should serve as a start. The brief treatise on "Barrier Separations" in *Technique of Organic Chemistry* (1) seems to be an acceptable basis. This was the first time that the subject was covered on a review basis. So looking back, it should be helpful in deciding what material could be considered as advances in the field.

As is often the case, new developments in other areas exert an influence upon progress in specific endeavors. Thus, the enormous growth in computer science and the perfection of gas-chromatographic procedures contributed materially to developments in barrier studies. Computerization of stagewise operations has lightened the burden of repetitive calculations, particularly as the more involved gas-diffusional operations usually require trial-and-error solutions. Gas chromatography, fortuitously, has become the preferred analytical tool as a result of its versatility and relative simplicity.

A. Delineation of Coverage

The operations which could possibly be included under gas and vapor separation with membranes are so manifold that a thorough coverage would require a very extensive treatise. Much of the fringe-area material, thus, will not be discussed in detail, but will be mentioned in appropriate context.

The material, then, will include new concepts, references to permeability data for both polymeric membranes and microporous media, effects of operating variables upon permeability and flow mechanism, and computer use in iterative calculations for cell performance.

B. Measurements and Analysis

The separative behavior of barriers involves measurements of permeabilities of pure components and, of course, also the carrying out of separation experiments with mixtures. A knowledge of permeabilities of pure components often suffices to predict separations which can be attained with barriers. However, when pressure and temperature variations result in appreciable adsorption or solubility effects, separation experiments will be needed.

Analytically speaking, the greatest benefit in the last 10–15 years has been derived from the development of the gas chromatograph. Small sample size, speed and ease of operation, satisfactory accuracy, and moderate investment in equipment, have made these instruments the preferred analytical tool.

Specialty devices which deserve to be mentioned are some of the new water vapor-sensing units which have come on the market in recent years. These include the water vapor-transfer unit developed by Honeywell (2) and the St. Regis Paper Co. (3), the Alnor Dew-pointer (4), the DuPont Moisture Analyzer (5), the Vapor Corporation Hygrometer (6), and others. All these types of equipment are based on principles set forth in Volume 1 of *Humidity and Moisture* (7), a reference text which would be of great help to anyone interested in water-vapor transfer.

A somewhat novel method for determining permeabilities of organic compounds has been developed by Riemschneider and Riedel (8). The method is useful for organics which are soluble in water, such as the lower aliphatic carboxylic acids, and thus change the electrical conductivity when absorbed.

II. COMPUTATIONAL METHODS

The methods used to compute the performance of a single barrier separation can readily be adapted to digital-computer programming. Weller and Steiner's (9,10) original equations and Nalor and Backer's (11) modification represent model interpretations which are based on fairly stringent simplifications. The most recent analysis of the use of these methods in the interpretation of experimental results was carried out by Breuer (12). Breuer also attempted to improve the earlier models, basing his analysis upon the necessary existence of molecular interdiffusion in the gas phases above and below the barrier. Progressive diffusion of the faster-moving component through the barrier creates a concentration gradient on the surface of the barrier in the direction of gas flow along the barrier. Breuer was able to allow for interdiffusion, by first setting up a plug-type flow equation to establish an approximate concentration profile and then adjusting this profile by a finite-difference computer calculation. This procedure gave in most instances a better agreement with experimental data than was possible with the other methods (9–11). As would be expected, the procedure becomes more involved, and it would be difficult to use without a computer. At the present time it does represent, however, as good a model as has been devised. The involvement of a concentration gradient along the bulk-flow direction over the barrier leads, of necessity, to a geometry dependence of the method.

A typical experimental concentration profile of the slower diffusing component is illustrated in Figure 1, which also illustrates schematically what goes on in barrier separation. At any point along the barrier, the slope of the concentration curve represents the driving

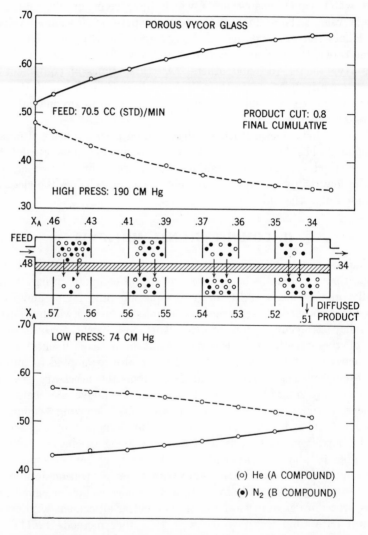

Fig. 1.   Diagram of separation barrier and concentration profiles (12,41).

TABLE I

Computation Guide for Weller-Steiner, Case I (19). (See Appendix for nomenclature.)

Assumption: No concentration gradient on either high- or low-pressure side of barrier, i.e., complete mixing.

$$y_A = \frac{P_H(1 - x_A{}^f) + (\alpha^* - 1)[P_H F + P_L(1 - F)] + \alpha^* P_H x_A{}^f}{2(\alpha^* - 1)P_H F + P_L(1 - F)}$$
$$- \frac{\{[P_H(1 - x_A{}^f) + (\alpha^* - 1)[P_H F + P_L(1 - F)] + \alpha^* P_H x_A{}^f\}^2 - 4(\alpha^* - 1)[P_H F + P_L(1 - F)]\alpha^* P_H x_A{}^f\}^{1/2}}{2(\alpha^* - 1)P_H F + P_L(1 - F)}$$

The above equation is of the form: $y_A = -b \pm (b^2 - 4ac)^{1/2}/2a$ and can be solved directly for y_A. *Note:* Only the minus root is of interest.

Method of Use: Given: $x_A{}^f$, P_H, P_L, α^*, F solve for y_A obtain $x_A{}^0$ by material balance.

TABLE II
Computation Guide for Weller-Steiner, Case II (9,10,19).
(See Appendix for nomenclature.)

Assumption: Laminar flow (no mixing) on both sides of barrier.

$$\ln \frac{n_B{}^0}{n_B{}^f} = R \ln \frac{t^f - B/A}{t^0 - B/A} + S \ln \frac{t^f - \alpha^* + c}{t^0 - \alpha^* + c} + T \ln \frac{t^f - c}{t^0 - c}, \text{ or}$$

$$x_B{}^0 = \frac{x_B{}^f}{1 - F} \left(\frac{t^f - B/A}{t^0 - B/A}\right)^R \left(\frac{t^f - \alpha^* + c}{t^0 - \alpha^* + c}\right)^S \left(\frac{t^f - c}{t^0 - c}\right)_T$$

A, B, C, R, S, T, t^f, t^0 are functions of α^*, P_H, P_L, $x_B{}^f$

Method of Use: Given $x_B{}^f$, P_H, P_L, α^*, F; solve for $x_B{}^0$ by trial and error. Obtain y_B by material balance.

For computer program write W-S-II as follows:

$$x_B{}^0 = f_1(x_B{}^f) \left[\frac{f_2(x_B{}^f)}{f_2(x_B{}^0)}\right]^R \left[\frac{f_3(x_B{}^f)}{f_3(x_B{}^0)}\right]^S \left[\frac{f_4(x_B{}^f)}{f_4(x_B{}^0)}\right]^T$$

force for diffusion of the slower diffusing component, that is N_2, in a backward direction. This condition forms the basis of the Breuer method (12).

Unless one is familiar with the various computational models, it is a time-consuming matter to use them properly. To alleviate this problem Breuer prepared tabular summaries and computer flow sheets. These should be of considerable help in deciding which method one might want to use and how to set up the problem for a solution. The information is contained in Tables I–IV and Figures 2–4. The selection of the method depends on the degree of accuracy which is desired. The Weller-Steiner, Case I method (9,10) is the simplest one to use and gives a reasonably good answer. However, in most cases, the enrichment calculated by this method will be lower

TABLE III
Computation Guide for Naylor-Backer Method (11).
(See Appendix for nomenclature.)

$$y_A = (x_A{}^0)^{-1/\epsilon} \left(\frac{1 - F}{F}\right)\left\{(1 - x_A{}^0)^\sigma \left[\frac{F y_A + (1 - F)x_A{}^0}{1 - \{F y_A + (1 - F)x_A{}^0\}}\right]^\sigma - (x_A{}^0)^\sigma\right\}$$

$\epsilon = \alpha - 1$; $\sigma = \alpha/(\alpha - 1)$; $\alpha = $ function of α^*, P_H, P_L, x_A

Method of Use: Given: $x_A{}^0$, P_H, P_L, F, α^*; solve for y_A by trial and error and obtain $x_A{}^f$ by material balance.

TABLE IV
Computation Guide for Breuer Method (12).
(See Appendix for nomenclature.)

$$\frac{dx_B}{dl} = \frac{1}{2n}\left[K_A\,\Delta P_A - K_B\,\Delta P_B + (1 - 2x_B)\,\frac{dn}{dl} + 2K_{DH}\,\frac{d^2x_B}{dl^2} \right];$$

$$\frac{dn}{dl} = -K_A\,\Delta P_A - K_B\,\Delta P_B$$

$$y_B = \frac{1}{(n^J - n)}\left[x_B{}^J n^J - x_B n + K_{DH}\,\frac{dx_B}{dl} + K_{DL}\,\frac{dy_B}{dl} \right]$$

Method of Use: Given P_H, P_L, L, K_A, K_B, K_{DH}, K_{DL}, n^J, $x_B{}^J$; solve first for case of plug flow, i.e., undirectional flow, no mixing, using "crude Euler stepping procedure." Then correct this solution by appropriate numerical technique. The referenced method diverges when the x_B vs. l curve is significantly concave downward. Experimental data show that, when the diffusion-mixing solution diverges, the plug-flow solution gives as good a prediction as was obtained for cases where the diffusion-mixing solution converged.

than the one actually obtainable. Obviously, the computer solution for this method is simply that of a quadratic equation.

Because the gaseous-diffusion type of separation is, with few exceptions, a partial separation in any individual cell, the subject of multistage cascade procedure is of importance. The most recent treatise of cascade operation is by Hwang (13), where the general case of cascade operation is treated, and three specific applications are considered in detail. Appropriate equations for operating lines are supplied for each situation. *Theory of Isotope Separation in Columns* by

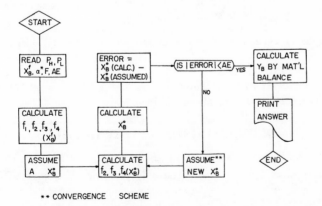

** CONVERGENCE SCHEME

Fig. 2. Computer flow diagram for Weller-Steiner, Case II (9,10,12,41).

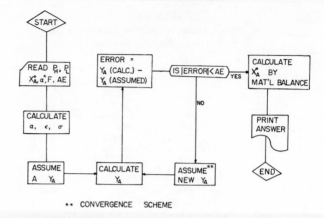

Fig. 3. Computer flow diagram for Naylor-Backer Method (11,12,41).

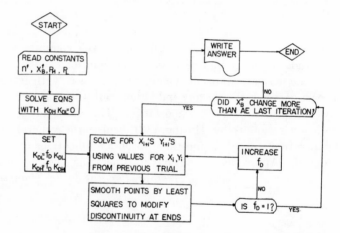

Fig. 4. Computer flow diagram for Breuer Method (12,41).

Rozen (14) should also be consulted, as it represents a comprehensive and authoritative treatise of the subject.

III. MECHANISMS OF SEPARATION

Probably the most significant advancement in the interpretation of barrier separation resulted from the growing awareness of the importance of surface flow in microporous media. It must be remembered that Graham's law and Knudsen flow had been the basic concepts, which used molecular weight as a controlling factor. Some of the early investigations must have given what appeared as inconsistent results and led to unsatisfactory attempts at interpretation (15). The records of the U.S. Atomic Energy Commission probably contain the evidence when the occurrence of surface flow was first recognized as a factor in reducing the separative effectiveness of the gaseous-diffusion process. In retrospect, it is obvious that the system of heavy vapor components (UF$_6$) would be particularly subject to the occurrence of surface flow. In addition, isotopic compounds are unlikely to exhibit noticeable differences in surface-flow properties. This means that surface flow of isotopes normally will inhibit the separation.

The acceleration in permeation studies of polymers surely must have contributed to the realization that the diffusive solubility flow in polymers has, in many respects, a great deal in common with surface flow in microporous media. The polymer studies quickly showed that molecular weight *per se*, of a gas or vapor, had only limited influence, or none at all, on permeation rate. There is good reason to believe that the polymer experience conditioned investigators to look for factors other than molecular weight in order to clearly explain separations in microporous media flow.

IV. SURFACE FLOW IN MICROPOROUS MEDIA

A. Surface Flow—Pure Components

The basic importance of surface flow in separation with microporous media suggests that some of the more important investigations be considered. Thus, studies covering surface flow of pure components are important, even though such information is by itself inadequate to permit definite conclusions on the behavior of mixtures.

Most representative of the progress which has been made in recent years are the papers by Oishi, Barrer, Metzner, Hwang, and their respective co-workers. Some of these publications, of course, also contain information on separation studies, and this material will be discussed in its proper context.

In many respects, a review involves a somewhat arbitrary choice of prominent publications. If these publications, however, are representative of the current activities and also take into account past performances, it is a justified procedure to limit discussion to them. The bibliographies of such papers also present a wealth of reference material, and it would be redundant to repeat it.

The classical text of Barrer (16) takes into account the surface-flow phenomenon and discusses it in some detail as "surface diffusion." Its possible effect upon separative flow is implied. Carman's (17) later treatise does report the definite effect of surface diffusion upon separation by quoting some of Haul's (18) data on the separation of oxygen isotopes. It appears that Huckins (19) may have published the first striking evidence of the potentially pronounced effect of surface flow when he presented permeability data for propane and butane in microporous Vycor glass. Hagerbaumer (20) definitely recognized that surface flow could be used to accomplish separation under appropriate operating conditions, and Wyrick (21) demonstrated that propane and CO_2, having essentially the same molecular weight, can be effectively separated on the basis of their surface-flow properties.

B. Surface Flow—Pressure Effect

Gilliland (22) and his co-workers were among the early investigators who attempted to develop a theoretical interpretation of surface flow. They used the spreading pressure concept to write a force balance and arrived at an equation for the surface flow as a function of amount absorbed and the gas-phase pressure. However, their data for the more highly adsorbed compounds, i.e., hydrocarbons, did not show a pressure effect as the experimental pressure range was too limited.

An important treatise on the effect of pressure upon surface flow (or surface migration) is that of Higashi, Ito, and Oishi (23). These authors presented a theoretical interpretation on the basis of a statistical site-hopping analysis, which, in essence, is a variation and at the same time an improvement of Hill's (24) random-walk surface-migration model. It is apparent that their theoretical equation does correlate available experimental data to a satisfactory degree. Their own permeability data for microporous Vycor glass are given as a function of pressure, but only up to about 30 cm Hg average pressure. The correlations of literature data are accomplished by a surface diffusion coefficient which is a function of surface coverage.

Data for surface flow of adsorbed gases and vapors are critically reviewed and analyzed by Barrer (25) for "microporous ceramic-type materials, including Vycor porous glass, compacted non-porous and porous carbon blacks, compressed alumina–silica cracking catalyst, or Linde silica plugs." The experimental evidence shows that surface-diffusion coefficients are independent of surface concentration at low-adsorptive surface coverage, that is, at relatively low pressures, but that these coefficients become concentration (or pressure) dependent at higher surface coverage.

A more recent paper by Aylmore and Barrer (26) presents permeability coefficients for a carbon plug as a function of pressure. In agreement with earlier data on Vycor (27), there is no effect of pressure on the permeability of gases such as N_2, Ar, Kr, and CO_2 up to pressures of 40 cm Hg. This relatively low pressure, of course, means that only dilute adsorbed films were encountered. The significance of this work is that it indicates a generalized behavior for widely different microporous media. The authors' interpretation of the constancy of permeability is "that the ratio of gas phase to surface flow remains constant throughout the medium, i.e. there is no interconversion of fluxes."

Also, of importance in the area of surface-flow interpretation is a mechanism proposed by Weaver and Metzner (28). By setting up equations which relate force parameters in molecular and surface interactions, they derive expressions to predict variation in adsorbate flux with pressure. In essence, their analysis is analogous to an earlier development by Kaser (29), who used a force-field analysis to interpret temperature effects. Weaver and Metzner's analysis certainly represents a good start on the pressure-effect interpretation. However, it is applied to a very limited amount of experimental data, that is, only to the propane data of Oishi (23) and to one set of isobutane data. The actual mechanism seems rather more complicated, so that considerable further efforts will be required for a comprehensive elucidation.

A discussion of the surface-flow phenomenon would be incomplete without mentioning the continued investigations of Flood (30) and Smith (31), dealing, respectively, with flow through microporous carbon and the effect of surface diffusion in catalytic reactions.

The most important models which have been proposed to interpret surface flow are summarized in Table V.

TABLE V

Investigators	Basic mechanism
Oishi (23)	Site hopping; statistical analysis
Metzner (28)	Force parameter effects
Barrer (25)	Activated site hopping
Flood (30)	Hydrodynamic flow with additive surface-flow term. Pore size large enough to get laminar flow
Gilliland (22)	Spreading pressure and force balance
Hwang (35)	Site hopping with local equilibrium of gas and adsorbed phase

C. Flow of Condensed Vapors

Several reports in the literature indicate that the permeability of condensable vapors in the high surface-coverage region reaches a maximum when it is plotted against pressure, as shown in Figure 5 (32). No quantitative explanation has been proposed, but a qualitative interpretation can be fashioned by dividing the pressure range into three regions and applying the appropriate equation to each flow mechanism.

In the lower-pressure region, the ordinary gas phase and surface-flow equations (23,33,34) are applied, giving a continuous increase of permeability with pressure. The intermediate-pressure region yields some pores plugged with the condensed liquid; therefore the permeability drops rapidly with the increase in pressure (32). The ordinary viscous-flow equation is applied to this region, giving a rapid increase of the flow resistance due to the buildup of condensate. Finally, in the higher-pressure region, the entire pore matrix is filled with the condensed liquid. The flow mechanism is again the viscous

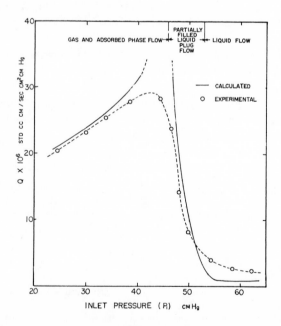

Fig. 5.   Effect of pressure on permeability of n-butane; microporous Vycor glass system (32).

liquid flow, but the permeability does not change much with the pressure.

It should be noted that capillary force plays no role in the total driving force across the porous medium, since both sides of the medium are at the same temperature and can be considered to have the same average pore radius.

D. Surface Flow—Temperature Effect

Until quite recently, the temperature ranges covered by the various pertinent investigations were too narrow to give an indication of any unusual temperature effect. Only when the experiments by Hwang (35,36) were published did it become evident that surface flow was affected by temperature in a pronounced and rather unexpected manner. The situation at present is not completely resolved, as experiments at elevated temperatures, that is, above 300°C, are needed to verify the conjectural behavior.

Because of its potential importance, especially also in conjunction with the pressure effect, some of these new findings should be summarized.

1. HELIUM FLOW

Most of the reported work in surface diffusion is based on the concept that the helium permeability can be used to estimate the surface flow from the total observed flow. The underlying reasons are:

a. the amount of helium adsorption at ordinary condition is very small,

b. the permeability of helium is independent of the average pressure, and

c. the product of the permeability and the square root of temperature shows little variation with temperature.

Therefore, the helium flow appears to obey the Knudsen equation quite well. But the above reasons are not sufficient proof that all of the helium flow is Knudsen flow. The correct way is to treat helium as any other gas in regard to the possible presence of surface flow. This can be done by studying the temperature effect on the total permeability.

2. SURFACE DIFFUSION

If Fick's law is used for the surface diffusion, the flow rate is a product of the surface diffusivity and the surface concentration gradi-

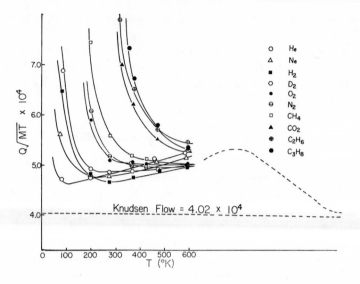

Fig. 6. Effect of temperature on adjusted permeabilities of glass and vapors; microporous Vycor glass system (35,36).

ent. Assuming the site-hopping model as the mechanism of the surface migration, the surface diffusivity can be calculated, as the adsorption isotherm provides the relationship between the surface concentration and the gas-phase pressure. The final equation is as follows:

$$Q(MT)^{1/2} = A + BT \exp{(\Delta/T)} \tag{1}$$

where Q is the total permeability, M is molecular weight, T is absolute temperature, A is a constant corresponding to the gas-phase flow, and B and Δ are constants which determine the surface flow. The second term in eq. (1) represents the contribution of surface transport. This equation then permits the determination of the surface diffusion in relation to the total flow, and also predicts the permeability of a new gas for a given microporous medium. It can be used in the study of separation factors for a binary mixture as a function of temperature. Figure 6 shows how well eq. (1) fits the experimental data for various gases over a wide temperature range. This analysis indicates that the surface diffusion of helium in microporous Vycor glass may be 13–25% of the total flow. If this is a correct picture, every study in this field should be reexamined.

3. HIGH-TEMPERATURE LIMIT

Equation (1) is good only for a moderate temperature range (less than 600°K). In ref. 35 a discussion is given for the limit of extreme high temperature based on the more general equation. The main point is that the surface flow would increase slightly with temperature until it reaches a maximum; then it must decrease again to asymptotically reach zero surface flow at a higher temperature level. This means the total flow approaches the Knudsen flow.

E. Surface Flow—Generalized Relationship

In an earlier publication (27) an attempt was made to correlate the permeabilities of various gases and vapors flowing through microporous Vycor glass. It was possible to obtain a reasonably good relationship for the permeabilities as a function of either boiling point or critical temperature. The permeabilities which were plotted in this manner all represent values extrapolated to zero pressure. Also, the plot had to be made for a fixed temperature.

An up-to-date version of the correlation of total flow, expressed as $Q(M)^{1/2}$, with the critical pressure, is shown in Figure 7. The permeability data are for the most part taken from Hwang's studies (35,36).

The new curve follows, in principle, the behavior of the earlier

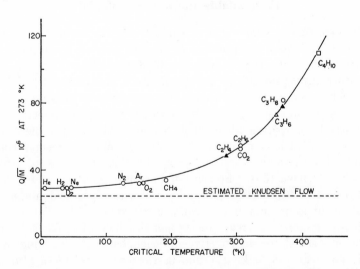

Fig. 7. Generalized permeability correlation, microporous Vycor glass system. (O) Hwang (35,36); (□) Higashi (23); (△) Gilliland (22); (▲) Huckins (19).

curve, but it represents conditions at 0°C. New compounds entered
on the curve are D_2, Ne, and C_4H_{10}. The last point is from Oishi's
(37) studies and extends the earlier curve beyond the propane point.

The most significant development, however, is the recognition of the
existence of surface flow of helium. This places the basic molecular
flow line well below the previously accepted basis. As indicated in
Figure 7, the base line is considered to be at an estimated value, but
it is located reasonably close to what should be a true value. Because
the plot is for the quantity $Q(M)^{1/2}$, the base line represents the Knudsen
flow for all compounds. Consequently, the difference between the
base line and the curve gives an estimate of the surface flow.

The curve should be used with some degree of caution. It is too
much to expect that such a relatively simple correlation would be
exact. If the boiling points of the compounds were used as a plotting
parameter instead of the critical temperature as in the earlier curve
(27), the picture would again be similar.

The solid matrix acts, in principle, as a chemically inert component. It exerts an influence on the surface flow only because it is
a capillary system and thus leads to adsorption. Consequently, the
behavior of porous Vycor glass can be expected to be representative
of that of any microporous medium. This means that similar curves
should be obtainable for any other microporous solid matrix.

F. Available Microporous Media

The investigator who wants to study flow and separation in microporous media faces the problem of finding suitable materials. The
barrier used by the U.S. Atomic Energy Commission is not available
for either investigations or industrial use. The barriers which are
presumably being used by the Soviet and French gaseous diffusion
installations likewise represent a politically sensitive subject. What
little has been written about such barriers appears in a news item
(38) and in various publications resulting from the Second United
Nations International Conference on the Peaceful Uses of Atomic
Energy in Geneva (39). French attempts at making barriers included
treatment of a 40:60 gold–silver alloy foil with nitric acid, and anodic
oxidation of aluminum foil in sulfuric or oxalic acid. These procedures are designed to create the proper size holes, i.e., 50–300 Å pore
diameter, needed for an effective diffusion barrier. Sintering of ceramic powders and making Teflon barriers on metal mesh also were
investigated. There is no information on Soviet work.

Thus, the investigator has to look more or less to what one must

TABLE VI
Available Microporous Media (41)[a]

Medium	Material	Safe[b] temperature limit	Pore diameter
Porous Vycor glass, Corning Glass Works	Mainly silica glass	300–400°C	About 50 Å average; fairly sharp distribution
Flotronic Membrane, Selas Corp. of America	Silver metal	~300°C	Maximum pore size; five sizes 0.2–5.0 μ
Porous tungsten, Hughes Aircraft Co.	Tungsten metal	High	Smallest presently available 1.2 μ
Porous porcelain, Selas Corp. of America	Porcelain	~1300°C	Smallest maximum pore diam. 0.3 μ

[a] A number of new porous materials, particularly porous metals, are available down to about 2-μ pore size. To be usable at near-atmospheric pressures, a pore size of less than 0.1 μ is desirable.

[b] No structural change.

still call experimental materials. Of these, microporous Vycor glass is the only material readily available which possesses a porous matrix in the proper size range. This material and other materials which approach, but do not reach the Vycor glass suitability, are listed in Table VI. Some investigators have had good success in using compressed powder sections. Suitable materials are graphite, alumina, and silica powders (24,30,31,40).

V. DIFFUSIVE SOLUBILITY FLOW IN POLYMERS

The flow behavior of gases and vapors through polymer films was covered extensively at a recent NATO Advanced Study Institute (41), which was concerned with the use of polymer films in all types of separations. The status of diffusive solubility flow in polymers can be characterized by stating that flow rates, or permeabilities, for both gases and vapors, can readily be taken as those of the pure components when determined under the proper operating conditions of pressure and temperature.

A. Gas and Vapor Permeation

Briefly, permeabilities of gases such as H_2, He, N_2, O_2, and CO_2 are not affected by pressure up to levels of about 100 psi. The tem-

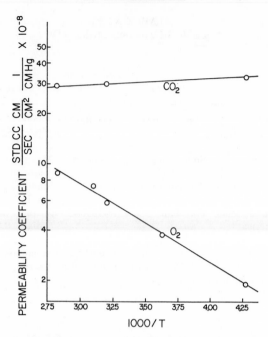

Fig. 8. Permeabilities versus temperature; silicone rubber membrane (G.E. 632A) system (42).

perature variation can usually be expressed by the straight-line plot of the logarithm of permeability against the absolute temperature (commonly $1/T$), so that increasing temperature yields higher permeabilities. The possible existence of breaks in the curves due to polymer-phase changes must be kept in mind. Also, there are some polymer–gas systems which show a different behavior; such anomalous behavior can be utilized to advantage in selecting the most favorable operating conditions, as shown in Figure 8 for a silicone rubber (42).

The most striking illustration of diffusive solubility flow of vapors in polymers is still the graph presented by Sobolev et al. (43), which was reported as early as 1957. The effect of both pressure and temperature is shown in Figure 9 for the permeation of methyl bromide through a film of polyethylene. The permeability is a somewhat involved function of the two parameters. At sufficiently low total pressure, in this case at about 100 mm Hg, the temperature effect is the same as that normally encountered with gases and polymers, so that an increasing temperature results in a continuous increase in permeability. The phenomena become more interesting at higher pressure levels. There, the lower temperature favors increased solubility in the polymer, as does the higher pressure. When temperature is increased (moving to the left on the abscissa scale), the solubility decreases at first and reaches a minimum which is dictated by the interplay of temperature and pressure effects. Finally, the continued increase in temperature leads to a progressive loosening of the polymer matrix with correspondingly increased vapor flow through the solid phase.

Several of the more recent compilations of permeability values are of interest as they contain both new data and valuable lists of references. A thorough coverage of silicone rubber data is given by Robb (44), and many new data are reported by Barrer and Chio (45). A comprehensive collection of available data from the packaging viewpoint was published by Lebovitz (46). The latter publication contains 119 references. It, therefore, represents an authoritative and comprehensive coverage of the permeability literature and it would be redundant to repeat this wealth of information. Another compilation of recent permeability data is contained in a treatise by Stern (47), where design calculations for a number of systems are presented. Also, the article on "Permeability and Chemical Resistance"

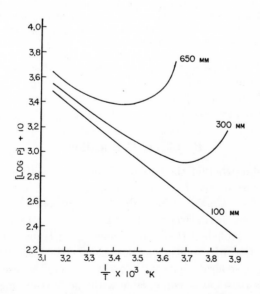

Fig. 9. Permeability of methyl bromide in polyethylene (43).

by Rogers (48), in *Engineering Design for Plastics,* should be mentioned as a desirable reference text.

Interestingly, there have been a number of studies published which deal with the influence of structural factors upon gas and vapor flow in polymer films. To mention a few, one should cite Michaels and Hausslein (49) investigation of the effect of elastic factors upon sorption and transport properties of polyethylene, the work of Rogers (50) on membranes with a gradient inhomogencity, the study of Ash, Barrer, and Palmer (51) of diffusion in multiple laminates, and a similar subject by the Szwarc school of investigators on permeability valves (52). The problems encountered in using a silicon rubber membrane for a model of a blood oxygenator were studied in detail by Buckles (53). This work includes variations in filler content of polymers.

1. PROMISING FUTURE DEVELOPMENT

A fascinating new development may be in the making, as indicated by a recent technical news item (54). Listed among the 1966 IR 100 winners of most significant new products is an "immobilized liquid membrane" for separating CO_2 from gas mixtures, developed by Robb and Ward of the General Electric Company. The material is described as follows:

> Essentially a porous cellulose acetate membrane impregnated with an aqueous solution the material is characterized by chemical reactions at the gas-liquid interface and within the membrane, and the reactions can be catalyzed to achieve still higher separation factors.

It is stated that these membrane types exhibit separation factors "hundreds to thousands of times higher than purely diffusive membranes." This intriguing situation is certain to stimulate intense interest. Additional information was published recently (54a).

B. Liquid Permeation

It is debatable whether the so-called liquid-permeation phenomenon should be discussed within the context of this treatise. In a manner of speaking, the operation where the phase in contact with one side (or both sides) of the barrier is a liquid represents a hybrid operation. If it could be established that the mechanism of transport is uniquely different from that of vapor or gas permeation, then the process should be treated as a separate entity. To this writer, the use of a liquid phase which is reasonably compatible with the barrier substance leads to a situation where the barrier properties are modified. Then, one cannot refer any more to the original barrier material when permeation results are cited.

Further substantiating evidence can be deduced from the findings of Stannett and Yasuda (55) with "conditioned films" in studies of cyclohexane and benzene permeation through polyethylene and of acetone and acetonitrile through vulcanized rubber. Their results actually deserve a more detailed treatment, but for the sake of conciseness, it must suffice to quote their statement concerning polyethylene: "Once the films have been conditioned, the liquid points become identical with the saturated vapor points as would be anticipated on thermodynamic grounds." The "points" refer to a plot of permeability versus the relative vapor pressure.

The specific meaning of the statement is that the permeability obtained with liquid on one side of the film was the same as that obtained when a saturated vapor was in contact with the film. This same behavior was established for water vapor and polyethylene by Varsanyi (56), who also developed useful equations for calculating permeabilities as a function of the respective humidity driving forces.

A completely different situation is the case where a liquid phase in no way permeates into or reacts with the polymer barrier, and where the liquid phase is used as a solvent for a particular gas. In this situation one can postulate that the gas or vapor-phase resistances on the barrier faces should be essentially nil. The studies of Yasuda and Stone (57) on permeation of oxygen dissolved in water illustrate this condition. A comparison of "gas-to-gas" permeabilities of oxygen with those of "O_2 in water-to-O_2 in water" shows that the O_2 permeabilities from and to the water phase were appreciably higher. It is significant that the same behavior was exhibited by membranes of widely different natures, to wit, silicone rubber, polyethylene, Teflon, and Teflon FEP, all of which are rather inert with respect to liquid water. Consequently, it is safe to assume that this phenomenon can be expected to occur with any system of similar relationships.

If the crude assumption is made that the surface resistances are essentially zero when the liquid is in direct contact with both surfaces of the barrier, one can make an approximate calculation for the surface resistances in gas-to-gas permeation. Yasuda's data would indicate that the two gas-phase resistances (one on each side of the barrier in gas-to-gas operation) amounted to some 75% of the total resistance in transfer through Teflon, about 82% in silicone rubber, and perhaps as much as 96% with polyethylene and Teflon FEP. These estimates must be treated with caution because of the great and perhaps unjustified assumption made.

VI. SEPARATIONS

The activities in the use of barriers to separate gas and vapor mixtures will be discussed under two main categories, that is, microporous barriers and nonporous media. Within each category most of the available material will concern experimental-scale work, as large-scale applications are, so far, quite limited.

A. Microporous Media

1. GENERAL BEHAVIOR

Some relatively new phenomena of a generalized behavior deserve discussion. When flow rates of a variety of gases and vapors are plotted against temperature, they present the type of illustration shown in Figure 10, that is, a transfer coefficient chart. The microporous medium used for the experiments was porous Vycor glass (58).

What makes this graph particularly significant is that every intersection of any two curves represents a nonseparative composition, that is an azeotrope. [Strictly speaking, the term azeotrope implies a boiling mixture. The term has, however, been used to signify a nonseparative condition in its broad sense (59).] It is reasonable

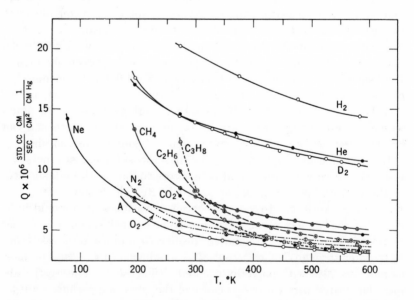

Fig. 10. Standard permeability of gases and vapors versus temperature, °K (58).

to state that this extensive existence of gas-diffusional azeotropes has never been recognized. Certainly, individual occurrences have been reported (35) from the same thesis investigation, but only the "azeotrope graph" of Figure 10 highlights the actual situation.

So, while gaseous diffusion can usually be counted upon to be useful in separating distillation azeotropes (constant boiling mixtures, CBMs), the gaseous-diffusion process will be plagued with its own "gaseous-diffusion azeotropes," GDAs, in analogy to the CBMs of distillation. However, while the composition of distillation azeotropes can be varied only by changing the distillation pressure (excluding the use of extractive distillation), a GDA can be manipulated independently by a change in either pressure or temperature.

2. SPECIFIC INVESTIGATIONS

The literature which describes actual separation experiments is rather limited. Thus, publications by Oishi et al., Barrer, Breuer, and Tock would represent the available material.

a. Mainly Pressure Effect

Oishi and his co-workers reported on the separation of binary gas mixtures (37) and on isotope separations (60). All of the experiments were carried out at 35°C. The single gas mixture tested was the system n-butane–propane. The isotopic compositions included naturally occurring mixtures of the isotopes of nitrogen, methane, propane, and n-butane. In every case the pronounced effect of the pressure ratio (low pressure/high pressure) is demonstrated, so that the separation factor tended to approach the theoretical $(M_2/M_1)^{1/2}$ value at zero downstream pressure (M_1 = molecular weight of lighter component).

The authors found that they could predict quite well what the separation efficiency should be, especially for N_2 and CH_4, as the respective isotope mixtures separated in accordance with Knudsen-flow behavior. With the C_3H_8 isotopes and the C_4H_{10} isotopes, however, a certain amount of surface flow was encountered, and the authors state that the ratio of the surface flows was essentially the same as that of the Knudsen flows of the isotopes. At the time that the Oishi work was published, the great potential effect of temperature on the surface flow, and in turn on the separation factor, had not been recognized. It would seem that the equality of the respective separation effects was a coincidental occurrence at the specific operating temperature of 35°C. One would not expect this coincidence at other operating temperatures.

The very recent investigation by Breuer (12), which led to the new computational model described in a previous section, involved numerous separation experiments. Breuer's data cover 138 runs on a microporous flat-plate Vycor glass membrane for the systems He–N$_2$, and He–CO$_2$. The glass barrier was 20 cm long, 6 cm wide, and had a thickness of 0.135 cm. The data all conform to previous findings, but they cover a greater range of operating parameters. The most significant contribution was the determination of concentration gradients as illustrated in Figure 1. This concentration profile is a typical one, but its slopes will vary, of course, with the operating conditions. It is especially sensitive to the flow rate and the "cut," that is, the fraction of the feed stream which is forced to permeate the barrier. One important conclusion is that diffusion cells, in general, should be designed long and narrow, rather than short and wide, so as to minimize the effective concentration gradient which becomes responsible for back diffusion.

b. Pressure Effect—Temperature Variation

A series of separation experiments with microporous media was reviewed by Barrer (25). The earlier data of Ash, Barrer, and Pope (61), covering mixtures of H$_2$–SO$_2$, Ne–CO$_2$, N$_2$–CO$_2$, and N$_2$–Ar, were interpreted on the basis of surface flow and possible blockage of the microporous carbon plugs to an extent that the less adsorbable component was almost prevented from flowing through the barrier. The experiments were carried out at such low temperatures, from -10 down to $-180°C$, that extensive adsorption (and possibly some capillary condensation) was attained. The result was a high degree of separation in line with what one would have expected from the behavior of the propane–CO$_2$ system which was reported by Wyrick (21). The Barrer studies confirm decisively that surface flow can have a completely controlling influence upon separation by gaseous diffusion. In particular, the system H$_2$–SO$_2$ presents an extreme example of a rather nonsorbable and a very sorbable component, and thus shows what could be accomplished under unusually favorable conditions.

c. Temperature Effect

In the previously cited paper by Hwang (35) concerning the temperature effect upon surface flow, there is a graph which shows the striking effect of temperature upon the separation factor for the system O$_2$–CO$_2$. (Fig. 6 of ref. 35.) The experimental data were taken from the research of Tock (58), and an improved version of the results

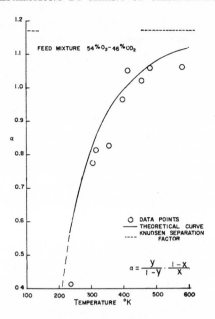

Fig. 11. Separation factor α versus temperature O_2–CO_2 system with microporous Vycor glass (58).

is shown in Figure 11. When the appropriate pressure correction was applied, as discussed in the next paragraph, the experimental points can be expressed very well by means of the curve which is calculated from pure gas-flow data, and the two curves shown in the earlier figure of ref. 35 are not needed.

The α's for experimental points are calculated from:

$$\alpha_{exp} = [y/(1 - y)]/[x_w/(1 - x_w)]$$

where y is the mole fraction of the more permeable component in the permeated stream and x_w is the mole fraction of the same component in the unpermeated stream. Both y and x_w are measured experimentally. The α's for the theoretical curve are obtained using the ratio of Hwang's permeability coefficients and Weller-Steiner (W.S.), Case I (9,10) to correct for back pressure.

W.S. Case I:

$$\frac{y}{1 - y} = \frac{\alpha^*_{1,2}(x_w - P'y)}{(1 - x_w) - P'(1 - y)}$$

where $\alpha^*_{1,2} = Q_1/Q_2, P' = \bar{P}$ downstream pressure/$\bar{P}$ upstream pressure, and Q_1 and Q_2 are permeabilities of the pure gases; thus,

$$\alpha_{\text{theo}} = \frac{y}{1 - y} \frac{1 - x_w}{x_w} = \frac{\alpha^*_{1,2}(x_w - P'y)}{(1 - x_w) - P'(1 - y)} \frac{1 - x_w}{x_w}$$

In the W.S. Case I equation, when $P' \to 0$, i.e., permeation into a vacuum, α^* becomes equal to α, or, the ratio of the permeability coefficients, Q_1/Q_2, equals α, the experimental separation factor.

It is evident that the variation in operating temperature is a powerful tool, as well as a rather simple one to manipulate, which permits selecting favorable process conditions. Analogously, one would utilize the similar conditions pictured in Figures 8 and 9, for separations with polymeric barriers.

3. LIQUID PERMEATION

When one of the phases is a liquid in direct contact with a barrier surface, the process is described as liquid permeation; this, however, would seem to be a misnomer.

Most of the work in the area has been done with polymeric membranes, and a more detailed discussion is presented in a later section. The areas of reverse osmosis, dialysis, electrodialysis, etc., are not of concern in this treatise.

For the sake of completeness, it should be mentioned that separations in the true liquid phase can be accomplished with microporous media. Such separations were reported by Hagerbaumer (62) for a number of solvent mixtures. Additional experiments and a theoretical interpretation were published by Cleland (63).

4. LARGE-SCALE OPERATIONS

While considerable progress has been achieved in the area of experimental investigations, there has not yet been any successful large-scale application other than the original AEC gaseous-diffusion plants. The reason, of course, is that the economics of the process is so demanding that unusually favorable process conditions must be realized. This has simply not yet been accomplished.

However, a rather unexpected possibility is cited in a news item of January 1967 (64,65). There are indications that the gaseous diffusion plants of the U.S. Atomic Energy Commission may be considered for release to private industry. This would constitute an unprecedented and perhaps courageous step, because it might make the highly developed barrier technology available to technologists all over the world. The obvious conclusion is that anyone could enrich uranium with relative ease, with a concomitant spread of nuclear weapons capabilities.

B. Nonporous Media

1. GENERAL BEHAVIOR

The behavior of individual gases and vapors, which was discussed under diffusive solubility flow, indicates that mixtures in separative flow through polymers should exhibit the same phenomena as is experienced with microporous media. The information which is available for polymers, however, is mostly concerned with temperature variation. Very little has been published on pressure effects. The Sobolev data (43) have already been discussed (see Fig. 9). Reference to an earlier publication (ref. 66, Fig. 3) shows that anomalous temperature effects can occur, which would lead to the occurrence of nonseparative mixtures. Of course, pressure effects become prominent with vapors which are both more condensable and more soluble in the respective polymers. Consequently, nonseparative mixtures will also be encountered (refs. 67 and 68, Fig. 1).

2. SPECIFIC INVESTIGATIONS—EXPERIMENTAL

a. Small Scale

A rather interesting study with polymeric barriers was carried out by Primrose (69). He used H_2–CO_2 mixtures with a fiber-glass barrier (8 mil Dexiglass), a fiber-glass filter (18 mil Ultra Efficient Filter Media), and films of uncoated cellophane, cellulose acetate, and silicone rubber. The scale of operation must be considered experimental, as the barriers consisted of 6-in. circular sections. Adsorption studies were conducted with the respective polymer–gas systems and the adsorption (perhaps better labeled solution) values were utilized in calculating theoretical separations. Thus, by using the Weller-Steiner Case I equation and an adsorptive correction from Russell's (70,71) computational procedure, Primrose was able to obtain a 5% agreement between data and theory. In principle, the studies confirm similar work of previous investigators.

The intriguing problem of removing the carbon dioxide from the air of a space cabin led Major and Tock (72) to develop a cross-flow gaseous-diffusion cell with multilayers of silicone rubber films as the barrier material. Previous studies had shown that silicone rubber was well suited for CO_2 permeation (73). In order to permit a compact cell design, it was necessary to devise the anchoring of the silicone rubber on a porous paper substrate without permitting appreciable penetration of the polymer into the porous substrate. The permeability values of the gases, that is, CO_2, O_2, and N_2 were as expected and so were the separation results. The objective of building a large area into a given volume was accomplished to a satisfactory degree, and cell efficiencies of 96% or better were attained routinely. Design calculations for this process were reported by Stern (47) who developed estimates of power consumption and weight penalty for the power.

A spectacular demonstration of the use of very thin silicone rubber membranes was carried out by Robb (74) when he kept a small animal alive in a silicone membrane enclosure under water. This situation represents the use of counterdiffusion of CO_2 to, and O_2 from, the surrounding water.

Silicone rubber may be a very suitable barrier for many organic compounds, as born out by some data reported by Robb (44). Also, while CO_2 seems to be somewhat unique in its high permeation ratio through silicone rubber, there are other compounds which behave similarly, as for instance, diborane. Birdwhistell (75) obtained a patent for the separation of this compound from H_2 and N_2.

b. Medium Scale

Some experiments, on a scale which is a good deal larger than usual, were carried out by Breuer (12) with silicone rubber. The barrier was 70 in. long, 4 in. wide, and had a thickness of 30 mils (0.030 in.). Eighteen separation experiments were carried out with the system N_2–CO_2. As in the case of the microporous Vycor glass barrier, the separation results were in line with previous findings. Also, the profiles were of the same nature, which was to be expected. In the experiments which used a high cut, the faster diffusing component, which is the CO_2, became so depleted near the cell outlet on the upstream side, that considerable difficulty was encountered in obtaining accurate stream analyses.

c. Design Calculations

The most comprehensive design analyses of several gaseous systems have been carried out by Stern (47). The following examples are treated in detail: (1) recovery of helium from natural gas, (2) separation of oxygen from air, and (3) carbon dioxide control in breathing atmospheres.

The *helium recovery* by means of a Teflon FEP barrier appears to be the most promising of the processes. The technological requirements are formidable, as the barrier septum has to operate at a pressure drop of some 900 psi. The optimum operating conditions developed from the analysis are summarized in Table VII (47).

TABLE VII

Recovery of Helium from Natural Gas by Permeation
through Teflon FEP[a] Membranes

Optimum Operating Conditions for Large Plants

Feed gas: 0.45 He, 17.06 H_2, 76.43 CH_4, 6.06 hydrocarbons
Feed gas compressed to 1000 psig
0.001-in.-thick membrane
Three permeation stages
80°C operating temperature
7.3-psia permeator back pressure
60% recovery of helium in natural gas
Interstage gas compressed to 950 psig
72% helium in product gas

[a] Teflon FEP: a copolymer of tetrafluoroethylene and hexafluoropropylene.

Stern's quotation concerning the economic status of the process is as follows:

The described process is the most efficient permeation technique now available, but is still not quite competitive with the more advanced cryogenic processes for the large-scale recovery of helium. However, the development of more permeable membranes, along with more sophisticated permeation cycles, could greatly increase the efficiency of the membrane process.

Oxygen separation from air is analyzed with the use of an ethyl cellulose film (1 mil thick). In the single-stage process, Stern computes enrichment of oxygen to 32.6%, and that a 91.1% O_2 concentration should be attainable with five stages. On the basis of economics, however, the process could not be competitive on a large scale with conventional technology.

Pertinent comments on the analysis of the CO_2 removal process have already been made [see Major and Tock's diffusion-cell study (72)]. Present thinking in the space program is still concerned with short-duration or, at most, medium-length excursions. For these situations there are more highly developed conventional processes which have the advantage of lower power requirements. However, the inherent advantage of being a single-phase operation could make the gaseous-diffusion process an attractive choice for space excursions of long duration.

3. LIQUID PERMEATION

The considerations of the permeation mechanism for a liquid phase in direct contact with the barrier, presented in an earlier section. should serve to explain separation of mixtures.

When the liquid phase can penetrate with relative ease into the polymer matrix, it becomes impossible to determine just how far the liquid front will approach the off-side of the barrier. It is then conceivable that the thickness over which vapor separation is effected is extremely small—the product is taken off as a vapor and not as a liquid. This could lead to an extremely short permeation path, perhaps $\frac{1}{1000}$ mil, and the result would be a tremendously increased permeability, if the permeability coefficient were to be expressed on the basis of the original film thickness, or perhaps even on a swollen-film thickness. This concept was suggested by Binning (76) when he first presented his pioneering studies in this field. An excellent analysis by Long (77) of the possible mechanism indicates that the two-zone model comes close to a proper interpretation.

4. HELIUM SEPARATION THROUGH QUARTZ

In general, one is not likely to think of a glass or glasslike composition as being a suitable barrier for separations. Obviously, the structure is so "tight" that whatever flow takes place is quite low. Thus, if such a material were to be used as a barrier, it must show some particular advantages. This happens to be the case with the helium–quartz system.

Actually, the fact that some gases may diffuse through glass has been known for a long time. However, one must credit Norton's (78) publications of recent findings with creating a definite interest in the quartzlike and glassy materials for helium separation. The advantage which makes the process possible is that the separation of helium from other gases approaches what one might call the ideal process, that is, essentially a "go, no-go" operation or an almost complete one-step separation.

The discussion of the process by McAfee (79) is so comprehensive that only a few items should be mentioned here. To get an idea of the flow rates, a round numbers comparison is of interest. Quoted in standard permeabilities (see ordinate of Fig. 8), the flow of helium through silica is about 2×10^{-9}, through microporous Vycor glass, 15×10^{-6}, and through polyethylene, roughly 1×10^{-9}. However, in order to quote realistic values, the silica figure is for a temperature of 700°C, while the other two values are for room temperature. This points up one of the difficult processing requirements for glasses. The temperature must be quite high, otherwise the permeation rates are simply too low for serious consideration. Some years ago, a considerable amount of effort went into possible designs for using quartz tubing in heat-exchange arrangement. The design called for an operating temperature of 900°C and high pressure differentials. There is reason to believe that a successful design of workable header construction could not be accomplished and thus the effort was shelved (80).

5. LARGE-SCALE OPERATION—HYDROGEN PURIFICATION

The only truly large-scale operation which has been developed, using, albeit, a metallic barrier, is the purification of hydrogen. The process is accomplished in an essentially one-step separation when hydrogen diffuses through a palladium foil with the virtual exclusion of other gaseous components.

The literature on hydrogen diffusion through palladium and other metals is exhaustive but pertinent in the present instance only as to the most recent references. Because the palladium barrier constitutes the heart of the process, the International Nickel Company's bulletin on palladium (81) dealing with the properties and applications of the metal is of definite interest. The publication lists 129 references, and one of them, an *Engelhard Industries Technical Bulletin* (82) contains a very recent series of scientific papers on palladium and hydrogen, including the latest information on permeation rates (83) of hydrogen and deuterium through a 75% Pd–25% Ag alloy.

The most recent report on experimental work is by deRosset (84). The study was carried out with a 0.8 mil palladium membrane at gas-flow rates of more than 250 scf/hr-sq ft of gas mixture. One mixture contained $64.3H_2$ and $35.7N_2$, and another was composed of $65.9H_2$, $33.9CH_4$, and $0.2N_2$. Permeability measurements with essentially pure H_2 also were obtained. A strict comparison of permeabilities of metals and other media is not possible as the diffusion process in metals is proportional to a fractional power of the pressure, generally the one-half power. However, an order-of-magnitude estimate yields a value of 5.10^{-5} at 450°C scf/hr-sq ft (see Fig. 8). Most separation experiments were carried out at a pressure drop of 400 psi. It was found that the presence of up to 35% nitrogen or methane did not interfere with the hydrogen diffusion. DeRosset's data showed that the diffusion rate was directly proportional to pressure drop at low pressures, and at higher pressures it became linear with the 0.8 power of pressure.

After a number of news items (85,86) and a review by Stormont (87) had indicated that commercial use of hydrogen diffusion was under serious consideration, a publication in March 1965 by McBride and McKinley (88) lifted the curtain on the actual installations in use by the Union Carbide Corp. Operating conditions require feed pressures of about 500 psi and temperatures 300–400°C or higher. The potentially deleterious effect of gas impurities, such as CH_4, C_2H_4, CO, and H_2S, especially at higher temperatures, imposes a practical upper temperature limit. Economic evaluations are presented and show that profitable operation is attainable with plants that produce about 4 million cubic feet of high-purity hydrogen per day (greater than 99.99% H_2). Sulfur compounds and CO must be removed from the feed stock to as large an extent as possible, as they are particularly prone to poisoning the barrier in forming sulfides and carbides.

Another equally interesting account of a commercial installation by Meinhold (89) appeared later in the year. It gave a brief description of a unit operated by Humble Oil and Refining Co. In this instance, the main purpose is the removal of hydrogen from a gas mixture to prepare a relatively pure carbon monoxide synthesis stream. Operating conditions are stated to be 700°F (370°C) and about 450 psi pressure drop, with a daily throughput of nearly 150,000 scf. The process is conducted in 24 diffusion cells requiring a space of only 6.5 × 13 × 6 ft. The membrane is made of a Pd–Ag alloy which may be the reason that the quoted operating temperature can be used with the carbon monoxide (81,83).

In a publication dealing with hydrogen-upgrading processes, Alexis (90) compares palladium diffusion with cryogenics and the so-called heatless adsorption. Most of the analysis deals with the latter process, that is, selective adsorption of impurities where the main gas stream passes through the adsorber with little loss in pressure. There is not sufficient information presented to permit an actual economic comparison, but it is implied that high gas-flow rates, also high plant capacity, would favor the cryogenic and adsorption processes. Generalized statements by McBride and McKinley give some useful guidelines regarding the possible choice of the diffusion process; to quote ref. 88:

Favorable Factors	Unfavorable Factors
High Feed Pressure	Low Feed Pressure
High Feed Concentration	Low Feed Concentration
Absence of Contaminants	Contaminants Present
High Purity Required	High Purity Not Required
Low Use Pressure	High Use Pressure

The status of commercial-scale hydrogen-diffusion plants of the Union Carbide Corp. is as follows (91):

Nine installations have been placed into service during the last three years. Only one of these plants is not now in use because external conditions made its operation uneconomical. The eight plants are producing 25 million cubic feet of pure hydrogen per day, and the largest installation produces 7.2 million cubic feet. Some of the feed gas streams contain as little as 25% hydrogen. The combined capacity of the operating units is somewhat less than had originally been projected. It was found advisable to lower the operating temperature and the pressure differential in order to extend the life of the barrier. Experience indicates that barrier life and reliability of the units will continue to be satisfactory at the less-severe operating conditions. The hydrogen produced by the units is used in ammonia synthesis and in organic chemicals manufacture, as well as for market outlets of commercial hydrogen.

VII. FINALE

So, how does the picture look today? A process was born under wartime pressure and secrecy, bursting on the scene as a marvel of technology and engineering. Some 25 years later, however, it is still struggling for a foothold in large-scale use. What does this mean? Commercialization requires favorable economics. Evidently, this has been attained thus far only with the hydrogen-diffusion process through a palladium barrier. This process is the sought-for one-step separation (92). It is undoubtedly this very tempting possibility which has kept interest alive.

Single-step, complete separations will still be hard to come by. So, efforts will be directed toward reduction of required stages. Stern's (47) analyses show what can be expected. Needed improvements are twofold: greater selectivity of barriers to reduce stage numbers and higher permeabilities to reduce area requirements.

Selectivity improvement with *microporous* barriers would seem to be dependent on surface-flow manipulation. This is a system-dependent variable, which is just beginning to be recognized as a promising operating variable. With *polymeric* barriers, the greatest possibility of selectivity improvement is probably polymer tailoring for specific systems. This also is then a system-dependent affair, but with perhaps a greater degree of freedom than surface-flow variation.

Improvement in permeability, which means increasing rate of flow per unit area of barrier can be approached by perfecting barrier structure and reducing barrier thickness. For *porous* media, this means producing a microstructure of appropriate pore diameter and high porosity. If this is attained, the matter of barrier thickness will not be so critical. With *polymeric* media, also metallic foils, considerable progress should be possible in reducing barrier thickness. Raising permeabilities is going to be a rather more difficult problem as it would call for a more open polymer structure than has been effected so far.

In the final analysis, the gaseous-diffusion process has been, and still is, a process with tantalizing implications. It is a foregone conclusion that it will continue to engage the interest of research workers and engineers alike, and that the success of hydrogen purification will provide a tremendous stimulus to search for new uses.

VIII. APPENDIX

A. Digital Computer-Convergence Scheme

1. DIGITAL COMPUTER PROGRAM

In trial-and-error solutions mentioned previously, we have $x_C = f_1(x_T)$

$$x_T = x \text{ trial}$$

$$x_C = x \text{ calculated from } x_T$$

We are interested in the case where $x_C = x_T$ (allowed error)

First plot a few points of $f_1(x_T)$ vs. x. This will give the general nature of the function $f_1(x_T)$. It will also indicate the sign of the error as a function of whether the assumed trial x is greater than or less than the desired solution.

2. METHOD OF SOLUTION

1. Select an x_T which is known to be less than the desired solution.
2. Solve for x_C; $x_C < x_T$, therefore, error $= x_C - x_T < 0$
3. Add increment Δ to x_T; x_T (new) $= x_T$ (old) $+ \Delta$
4. Solve for x_C. Check sign of error; if negative, add another Δ; if positive, return to original x_T and add $\Delta/2$.
5. Repeat steps 3 and 4 until desired degree of accuracy is obtained.

IX. LIST OF SYMBOLS

A. Nomenclature

x Mole fraction, high-pressure side of cell
y Mole fraction, low-pressure side of cell
P_H High pressure
P_L Low pressure
P_r P_L/P_H
Q Permeability coefficient
α^* Ratio of permeabilities Q_A/Q_B
n Moles gas in a stream (moles/unit time)
F "Cut" fraction of feed-stream permeating barrier
AE   Allowable error (desired accuracy) in iterative solutions

Subscripts:
A Designates gas component with higher permeability
B Designates gas component with lower permeability

Superscripts:
f Feed stream
o Unpermeated stream

B. Definition of Functions

Weller-Steiner I
$$a = (\alpha^* - 1)[P_H F - P_L(1 - F)]$$
$$c = \alpha^* P_H x_A{}^f$$
$$b = - [P_H(1 - x_A{}^f) + a + c]$$

Weller-Steiner II
$$A = \tfrac{1}{2}[(1 - \alpha^*)P_L/P_H + \alpha^*]$$
$$C = -\tfrac{1}{2}[(1 - \alpha^*)P_L/P_H - 1]$$
$$B = -AC + \alpha^*/2$$
$$S = [\alpha^*(A - 1) + C]/[(2A - 1)(\alpha/2 - C)]$$
$$T = 1/(1 - A - B/C)$$
$$R = 1/(2A - 1)$$
$$t = -Ai + [A^2 i^2 + 2Bi + C^2]^{1/2}$$
$$i = n_A/n_B$$

Naylor-Backer
$$\alpha = \frac{\alpha^* + 1}{2} - \frac{P(\alpha^* - 1)}{2x_A{}^o} - \frac{1}{2x_A{}^o}$$
$$+ \left[\left(\frac{\alpha^* - 1}{2}\right)^2 + \frac{(\alpha^* - 1) - P(\alpha^{*2} - 1)}{2x_A{}^o} + \frac{P(\alpha^* - 1) + 1^2}{2x_A{}^o} \right]^{1/2}$$

Breuer
$$K_A = Q_A w/t; \; K_B = Q_B w/t$$
$$\Delta P_A = P_H x_A - P_L y_A; \; \Delta P_B = P_H x_B - P_L y_B$$
$$K_{DH} = D_{AB} S_H; \; K_{DL} = D_{AB} S_L$$
where DH = diffusion, high-pressure side
DL = diffusion, low-pressure side
D_{AB} = diffusion coefficient for binary mixture of A and B
t = barrier thickness
w = barrier width
S_H = cross-sectional area for diffusion, high-pressure side
S_L = cross-sectional area for diffusion, low-pressure side
l = barrier length; L = total length
h = interval spacing in finite-difference numerical method
f_D = fraction used to multiply diffusion constant

General References

N. N. Li, R. B. Long, and E. J. Henley, "Membrane Separation Process," in *Ind. Eng. Chem.*, **57**, No. 3, 18 (1965).

Charles Gelman, "Membrane Technology. Part I. Historical Development and Applications," in *Anal. Chem.*, **37**, No. 6, 29A (1965); II. H. Z. Friedlander and R. N. Rickles, "Theory and Development," in *Anal. Chem.*, **37**, No. **6**, 27A (1965).

R. N. Rickles and H. Z. Friedlander, "Membrane Separation Processes," in *Chem. Eng.*, **73**, 111 (Feb. 28, 1966), 121 (March 28, 1966), 163 (April 25, 1966), 217 (June 20, 1966).

R. N. Rickles, "Molecular Transport in Membranes," in *Ind. Eng. Chem.*, **58**, No. 6, 18 (1966).

K. S. Spiegler, "Diffusion of Gases across Porous Media," in *Ind. Eng. Chem. Fundamentals*, **5**, No. 4, 529 (1966).

References

1. K. Kammermeyer, "Barrier Separations," in *Separation and Purification* (*Technique of Organic Chemistry*, Vol. 3, Part 1), Arnold Weissberger, Ed., Interscience, New York, 1956, pp. 37–64.
2. E. J. Amdur, *Instr. Control Systems*, **39**, No. 8, 97 (1966).
3. O. H. Ranger and M. J. Gluckman, *Mod. Packaging*, **37**, No. 11, 153, 202 (1964).
4. Alnor Instrument Co., Product Bulletins 72-11 and 2051.
5. DuPont 510 Moisture Analyzer, Instrument Products Division Bulletin, E. I. DuPont Co.
6. R. C. Bridgeman, *Instr. Control Systems*, **39**, No. 1, 95 (1966).
7. R. E. Ruskin, Ed., *Principles and Methods of Measuring Humidity in Gases* (*Humidity and Moisture*, Vol. 1), Reinhold, New York, 1965.
8. R. Riemschneider and E. Riedel, *Kuntstoffe—Plastics*, **56**, No. 5, 355 (1966).
9. S. Weller and W. A. Steiner, *J. Appl. Phys.*, **21**, No. 4, 279 (1950).
10. S. Weller and W. A. Steiner, *Chem. Eng. Progr.*, **46**, No. 11, 585 (1950).
11. R. W. Naylor and P. O. Backer, *A.I.Ch.E. J.*, **1**, No. 1, 95 (1955).
12. M. E. Breuer, Ph.D. Thesis, Univ. of Iowa (1966); *Separation Science*, **2**, No. 3, 319 (1967).
13. S-T. Hwang and K. Kammermeyer, *Can. J. Chem. Eng.*, **43**, No. 1, 36 (1965).
14. A. M. Rozen, "Theory of Isotope Separations in Columns (U.S.S.R.)"; translation available from Office of Tech. Serv. (Joint Publications Research Service: 11213).
15. K. Kammermeyer and H. T. Ward, *Ind. Eng. Chem.*, **33**, No. 4, 474 (1941).
16. R. M. Barrer, *Diffusion in and Through Solids*, Cambridge University Press, London, 1941.
17. P. C. Carman, *Flow of Gases Through Porous Media*, Academic Press, New York, 1956.
18. R. A. W. Haul, *Naturwissenchaften*, **41**, 255 (1954).
19. H. E. Huckins, Jr., and K. Kammermeyer, *Chem. Eng. Progr.*, **49**, 180, 294, 517 (1953).
20. D. H. Hagerbaumer and K. Kammermeyer, *Chem. Eng. Progr. Symp. Ser.*, **50**, No. 10, 560 (1954).
21. K. Kammermeyer and D. D. Wyrick, *Ind. Eng. Chem.*, **50**, No. 9, 1309 (1958).
22. E. R. Gilliland, R. F. Baddour, and J. L. Russell, *A.I.Ch.E. J.*, **4**, No. 1, 90 (1958).
23. K. Higashi, H. Ito, and J. Oishi, *J. At. Energy Soc., Japan*, **5**, No. 10, 846 (1963).
24. T. L. Hill, *J. Chem. Phys.*, **25**, 730 (1956).

25. R. M. Barrer, *Proc. Brit. Ceram. Soc.*, **5**, 21 (Dec. 1965); *A.I.Ch.E.–Inst. Chem. Engrs. (London) Symp. Ser.*, **1**, 112 (1965).
26. L. A. G. Aylmore and R. M. Barrer, *Proc. Roy. Soc. (London) Ser. A*, **290**, No. 1423, 477 (1966).
27. K. Kammermeyer and L. D. Rutz, *Chem. Eng. Progr. Symp. Ser.*, **55**, No. 24, 163 (1959).
28. J. A. Weaver and A. B. Metzner, *A.I.Ch.E. J.*, **12**, No. 4, 655 (1966).
29. J. D. Kaser, Ph.D. Thesis, Univ. of Iowa (June 1963).
30. R. F. Bartholemew and E. A. Flood, *Can. J. Chem.*, **43**, 1968 (1965).
31. R. J. Arrowsmith and J. M. Smith, *Ind. Eng. Chem. Fundamentals*, **5**, No. 3, 327 (1966).
32. J. F. Haman, Ph.D. Thesis, Univ. of Iowa (1965).
33. J. R. McIntosh, Ph.D. Thesis, Univ. of Iowa (1966).
34. G. P. Perkinson, Ph.D. Thesis, Mass. Inst. Tech. (1965).
35. S-T. Hwang and K. Kammermeyer, *Can. J. Chem. Eng.*, **44**, No. 2, 82 (1966).
36. S-T. Hwang and K. Kammermeyer, *Separation Science*, **1**, No. 5, 629 (1966).
37. K. Higashi, H. Ito, and J. Oishi, *J. Nucl. Sci. Technol. (Tokyo)*, **1**, No. 8, 298 (1964).
38. "Europe Looks at U-235," *Chem. Eng. News*, **36**, No. 40, 68 (1958).
39. *Proc. 2nd Intern. Conf. Peaceful Uses At. Energy*, Geneva, 380 (1958).
40. J. R. Dacey, *Advan. Chem. Ser.*, **33**, 172 (1961).
41. Paper presented at NATO Advanced Study Institute on Synthetic Polymer Membrane Systems, Ravello, Italy (Sept. 1966).
42. R. W. Tock, M.S. Thesis, Univ. of Iowa (1964).
43. I. Sobolev, J. A. Meyer, V. Stannett, and M. Szwarc, *Ind. Eng. Chem.*, **49**, No. 3, 441 (1957).
44. W. L. Robb, General Electric Co., Report No. 65-C-031 (Oct. 1966).
45. R. M. Barrer and H. T. Chio, *J. Polymer Sci. C*, **10**, 11 (1965).
46. A. Lebovitz, *Mod. Plastics*, **43**, No. 7, 139, 144, 194 (1966).
47. S. A. Stern, "Industrial Applications of Membrane Processes: The Separation of Gas Mixtures," presented at *Proc. Symp. Membrane Processes Ind.*, **1966**, 196.
48. C. E. Rogers, "Permeability and Chemical Resistance," in *Engineering Design for Plastics*, Reinhold, New York, 1964.
49. A. S. Michaels and R. W. Hausslein, *J. Polymer Sci. C*, **10**, 61 (1965); see also A. S. Michaels, ref. 47, pp. 157–195.
50. C. E. Rogers, *J. Polymer Sci. C*, **10**, 93 (1965).
51. R. Ash, R. M. Barrer, and D. G. Palmer, *Brit. J. Appl. Phys.*, **16**, 873 (1965).
52. C. E. Rogers, V. Stannett, and M. Szwarc, *Ind. Eng. Chem.*, **49**, No. 11, 1933 (1957).
53. R. G. Buckles, Ph.D. Thesis, Mass. Inst. Tech. (1966).
54. W. L. Robb and W. J. Ward, III, *Ind. Res.*, **8**, No. 12, 81 (Dec. 1966); see also ref. 41.
54a. W. J. Ward, III, and W. L. Robb, *Science*, **156**, No. 3781, 1481 (1967).
55. V. Stannett and H. Yasuda, *J. Polymer Sci. B*, **1**, 289 (1963).
56. I. Varsanyi, *Rep. Central Food Inst., Budapest*, **3**, 15 (1965).
57. H. Yasuda and Wm. Stone, Jr., *J. Polymer Sci. A-1*, **4**, 1314 (1966).
58. R. W. Tock, Ph.D. Thesis, Univ. of Iowa (1967).
59. J. A. Seiner, *Ind. Eng. Chem. Fundamentals*, **4**, No. 4, 477 (1965).

60. K. Higashi, A. Oya, and J. Oishi, *J. Nucl. Sci. Tech.*, **3**, No. 2, 51 (1966).
61. R. Ash, R. M. Barrer, and C. G. Pope, *Proc. Roy. Soc. (London), Ser. A*, **271**, 1, 19 (1963).
62. K. Kammermeyer and D. H. Hagerbaumer, *A.I.Ch.E. J.*, **1**, No. 2, 215 (1955).
63. R. L. Cleland, *Trans. Faraday Soc.*, **62**, No. 2, 336 (1966).
64. *Technol. Week,* **20**, No. 2, 26 (Jan. 9, 1967).
65. *Chem. Eng. News,* **45**, No. 2, 17 (Jan. 9, 1967).
66. K. Kammermeyer, *Chem. Eng. Progr. Symp. Ser.*, **55**, No. 24, 115 (1959).
67. N. N. Li and E. J. Henley, *A.I.Ch.E. J.*, **10**, No. 5, 666 (1964).
68. N. N. Li, R. B. Long, and E. J. Henley, "Membrane Separation Processes," in *Ind. Eng. Chem.*, **57**, No. 3, 18 (1965).
69. R. A. Primrose, Ph.D. Thesis, Virginia Polytechnic Inst. (1965).
70. J. L. Russell, Sc.D. Thesis, Mass. Inst. Tech. (1955).
71. E. R. Gilliland, R. F. Baddour, and J. L. Russell, *A.I.Ch.E. J.*, **4**, No. 1, 90 (1958).
72. C. J. Major and R. W. Tock, *Atmosphere in Space Cabins and Closed Environments,* Appleton-Century Crofts, New York (1966), pp. 120–144.
73. K. Kammermeyer, *Ind. Eng. Chem.*, **49**, No. 10, 1685 (1957); U.S. Pat. 2,966,235 (Dec. 27, 1960).
74. W. L. Robb, *Chem. En.*, **71**, No. 24, 94 (1964); General Electric Co., Res. Lab. Bull., Winter 1964–65, pp. 7–8.
75. R. K. Birdwhistell, U.S. Pat. 2,862,575 (Dec. 2, 1958).
76. R. C. Binning, R. J. Lee, J. F. Jennings, and E. C. Martin, *Ind. Eng. Chem.*, **53**, 45 (1961).
77. R. B. Long, *Ind. and Eng. Chem. Fundamentals,* **4**, No. 4, 445 (1965).
78. F. J. Norton, *Gen. Elec. Rev.* (September 1952).
78a. F. J. Norton, *J. Am. Ceram. Soc.*, **36**, No. 3, 90 (1953).
78b. F. J. Norton and A. U. Seybolt, *Trans. AIME,* **230**, 595 (1964).
79. K. B. McAfee, "Diffusion Separation," in *Kirk–Othmer Encyclopedia of Chemical Technology,* 2nd Suppl., A. Standen, Ed., Interscience, New York, 1960, pp. 297–315; also *Bell Lab. Record,* **39** (Oct. 1961).
80. Personal communication.
81. "Palladium, the Metal, Its Properties and Applications," The International Nickel Co., Inc. (25 M 10-66).
82. *Engelhard Ind. Tech. Bull.,* **7**, No. 1/2 (June/Sept. 1966).
83. L. R. Rubin, *Engelhard Ind. Tech. Bull.,* **7**, No. 1/2, 55 (June/Sept. 1966), **2**, No. 1 (June 1961).
84. A. J. deRosset, *Ind. Eng. Chem.*, **52**, No. 6, 525 (1960).
85. *Chem. Week,* **96**, No. 7, 28, 49, 50, 52 (Feb. 13, 1965).
86. *Chem. Eng.*, **72**, No. 5, 36,38 (March 1, 1965).
87. D. H. Stormont, *Oil Gas J.*, **63**, No. 10, 125 (March 8, 1965).
88. R. B. McBride and D. L. McKinley, *Chem. Eng. Progr.*, **61**, No. 3, 81 (1965).
89. T. F. Meinhold, *Chem. Process.,* **12**, 66 (Sept. 1965).
90. R. W. Alexis, paper presented at Symposium on Adsorption Processes, Detroit, Mich., Dec. 4–8 (1966), *Am. Inst. Chem. Engrs.,* preprint 34F.
91. Personal communication by Dr. D. L. McKinley.
92. K. Kammermeyer and J. O. Osburn, *Separation Science,* **1**, No. 1, 7 (1966).

Author Index

Numbers in parentheses are reference numbers and show that an author's work is referred to although his name is not mentioned in the text. Numbers in *italics* indicate the pages on which the full references appear.

A

Abramson, H. A., 202(56), *242*
Ackers, G. K., 216(107), *243*
Adamson, A. W., 19(23), 34(23), *55*
Alameri, E., 227, *245*
Albertsson, P. A., 230, *245*
Alder, P. J., 25(29), 30(29), *56*
Alderweireldt, F., 90(3), *131*
Alexis, R. W., 366, *372*
Allison, A. C., 232(168), *245*
Alnor Instrument Co., 337(4), *370*
Ambard, L., 210(84), *243*
Amdur, E. J., 337(2), *370*
Ammerlaan, A. C. F., 330(10), *334*
Anastassiadis, P. A., 193(16), *241*
Anderson, N. G., 206, 209(82), *242, 243*
Annicolas, D., 193(15), 212(93), *241, 243*
Archibald, W. J., 205, 206(60), *242*
Arrowsmith, R. J., 345(31), 351(31), *371*
Ash, R., 354, 358, *371, 372*
Ashworth, J. N., 227(147), *245*
Askonas, B. A., 230(164), *245*
Aubel-Sadron, G., 207, *242*
Avrameas, S., 199, *241*
Aylmore, L. A. G., 344, *371*

B

Baarson, R. E., 4(10), *55*
Bachrach, H. L., 208(78), *243*
Back, N., 140(7), *142*
Backer, P. O., 337, 340(11), 342, *370*
Baddour, R. F., 165(14,15), 166(14), 167(15), *186*, 344(22), 345(22), 349(22), 361(71), *370, 372*

Badgley, D. B., 202(54), *242*
Ball, J. S., 79(41), *82*
Banfield, D. L., 6(14), 25(29), 30(14,29), 52, *55, 56*
Barrer, R. M., 151, *186*, 343–345, 349(25), 353, 354, 357, 358, *370–372*
Barrie, J. A., 148(3), 153(3,8), 154(8), 155(8), 163(3,8), 164(3,8), 165(15), 167(15), *185, 186*
Bartholemew, R. F., 345(30), 351(30), *371*
Barton, C. J., 249(13), *294*
Baur, E. W., 218, *244*
Beck, C. W., 60(25), *81*
Becker, M., 228(154), *245*
Beckman Instruments, 207(73), *242*
Benedict, M., 174, *186*
Bennett, G. F., 59(12), *81*
Bennich, H., 215(103), *243*
Besarab, A., 333(3), *333*
Beyrard, N., 210(85), *243*
Bier, M., 201, *242*
Bikerman, J. J., 19(24), 34(24), 51(51), *56*
Binning, R. C., 364, *372*
Bird, R. B., 35(37), *56*
Birdwhistell, R. K., 362, *372*
Bixler, H. J., 143, 148(2,4), 149(2), 150(2), 152(12), 153(12), 163(2,12), 164(2,12), 165(4,14,15), 166(14), 167(15,16), *185, 186*, 306(7), 307(7), *333*
Blank, F., 201(41), *241*
Blatt, W. F., 210, *243*, 321(1), 326(2), *333*
Block, R. J., 191(12), 192(12), 220(128), *240, 244*

373

Subject Index

A

Acetamide, purification, by normal freezing, 60

Activation energy, for diffusion in membranes, 151

Activity coefficients, in gas chromatography, 86

Activity sequence, for support materials in gas chromatography, 100

Adsorption, equilibrium in, 4
  in foam fractionation, 4
  measured through foam fractionation, 7

Adsorptive bubble separation methods, 3

Adsubble methods, compared, 4
  defined, 3
  foam fractionation, 3
  microflotation, 4
  nomenclature, 3
  nonfoaming types, 4

Agar, in electrophoresis, 192

Agar gels, in electrophoresis, 191

Agarose, in electrophoresis, 191, 192

Agitation, in normal freezing, 59, 60

Albumin, 188, 190, 204, 223, 224

Aluminum, purified by normal freezing, 60

Analysis by ultrafiltration, 298

Apparatus, for foam fractionation in simple mode, 8, 9
  for normal freezing, 76, 77, 79
  for performance in foam fractionators, 14, 15, 16
  for zone refining, 139

Apparatus construction, for preparative gas chromatography, 118

Artificial kidney, 214, 333

Azeotropes, gas-diffusional types, 357

B

Backflash, in gas chromatography infection, 95

Band width, in gas chromatography, 87, 90, 101

Barrier, quartz, 364

Batchwise operation, in foam fractionation, 2, 3

Beer foam, 2

Benzene, purification, by normal freezing, 60

Benzoic acid, purification, by normal freezing, 59

Beverage processing, by ultrafiltration, 331

Biological products, treatment by ultrafiltration, 330

Black spots in foams, 41, 47

Block electrophoresis, 197
  apparatus for, 199
  materials for, 199

Blood fractionation, by ultrafiltration, 331

Boundary layer, film thickness of, 65, 66, 69
  in normal freezing, 64

Boundary phenomena, in normal freezing, 64

Breakage of foams, 51

p-Bromotoluene, purification by normal freezing, 60

Bubble caps, in foam fractionators, 30

Bubble diameters, measurements of, 10

Bubble fractionation, 4, 31, 32

Bubble rupture, in foam fractionation, 10, 49, 50

Bubble separation methods, 3

Bubble size distribution, in foam fractionation, 8, 39, 49

381

C

Capillaries, in foams, 34, 35
Capital outlay, for normal freezing, 58
Carrier gas, in gas chromatography, 106, 109, 110, 114
  rate in gas chromatography, 110
Cascade operation, in gaseous diffusion separation, 341
Cascades, separation, in membrane permeation, 173
Channeling, in foam fractionation, 30
Chelation, in adsorption, 6
Chemical reactivity, in zone refining, 135
Chromatographic columns, filling of, 96, 98
Chromatography, 130
  adsorption, carriers, 220, 221
  ion-exchange, carriers, 222, 223
    in protein separations, 222
  in protein separations, 220
Coacervation methods, 227
Coalescence, of bubbles, in foam fractionation, 8, 9, 10, 48
Coiled columns, in gas chromatography, 96
Collapse of foams, 51
Collection of substances, in preparative gas chromatography, 112
Colligend, adsorption to surfactant layer, 6
  in foam fractionation, 3, 6, 7
Column design, in gas chromatography, 90, 91, 92
Column dimensions, in gas chromatography, choice of, 92
Column efficiency, in foam fractionators, 29
Column electrophoresis, 199
  apparatus, 199
  preparative, 199
Column material, for gas chromatography, 96
Column ovalization, in gas chromatographic columns, 130
Column packing density, in gas chromatography, 98
Column performance, in foam fractionation, 14, 15, 16

Column permeability, in gas chromatography, 98
Column properties, in gas chromatography, 91, 92
Columns, for foam fractionation, 24
Column shape, for gas chromatography, 96
Combined mode of foam fractionation, theory of operation, 23
Combined modes of foam fractionation, 12
Computational methods, for membrane separations, 337
Computational models, for gas membranes, 340
Concentration, effect on recovery in gas chromatography, 113
  by ultrafiltration, 298
Concentration gradient, on membranes for gas separations, 337
Concentration polarization, in membrane permeation, 179
Conditioning, of gas chromatography columns, 99
Constitutional safer cooling, in normal freezing, 65
Continuous electrophoresis, 200
Continuous flow technique, in preparative gas chromatography, 90
Continuous multistage normal freezing unit, 61
Continuous operation, in foam fractionation, 2, 3
Cooling technique, in rotation–convection zone refiner, 138
Countercurrent distribution, 229
Countercurrent methods, for protein separations, 229
Counterions, as colligends, 6, 7
o-Cresol, purification, by normal freezing, 60
Critical micelle concentration, in adsorption, 4
Crystal growth, in normal freezing, 59, 60, 69
Crystallinity, and permeability, in membranes, 164
Crystallization, from melt, 59

D

Decontamination factor, in foam fractionation, 13
Density gradient, in electrophoresis, 190, 194
Desalination, by electrodialysis, 214
Detection, in gas chromatography, 107
Detector-end filled gas chromatography column, 99
Detector filament, life time, 108
Detectors, in gas chromatography, 111
Detergents, removal by foam fractionation, 2
Diafiltration, 323
Dialysis, 203, 212
  for osmotic concentration of proteins, 212
o-Dichlorobenzene, purification, by normal freezing, 60
Diffusion, in membrane permeation, 145
  of swelling penetrants in membrane permeation, 159
Diffusional impedance, in membranes, 165
Diffusion coefficients, in electrophoresis, 190
Diffusion constants, for membranes, correlation of, 149
Diffusive solubility flow, in polymers, 351
Diffusivity, in normal freezing, 66
Directional freezing, 57
Distillation, 58
Distribution coefficient, in zone refining, 135
Distribution coefficients, in normal freezing, 65, 66
Double diffusion precipitation methods, 231
Double layer, electrical, 189

E

Efficiency, in foam fractionators, 29
  ϕ, in normal freezing, 74
  of preparative gas chromatographic column, 86
Elasticity, of films, in foams, 50

Electroconvection methods, 201
Electrodecantation, 201
Electrodialysis, 203, 214, 225
  apparatus for, 215
Electrokinetic potential, 189
Electroosmatic transport, 192
Electroosmosis, 192
Electropherograms, 192
  information of, 194
  scanning of, 193, 194
  staining of, 192, 194
Electrophoresis, 189, 217
  of adsorbed proteins, 202
  apparatus zone types, 195
  block, 197, 198
  chamber, rapid, 195
  of coated particles, 203
  column type, 199
  continuous, 200
  forced flow in, 201
  gel types, 191
  molecular sieve, 218
  polyacrylamide, 219
  in rotating tube, 194
  starch block, 198
  starch gel, 218
  strip types, 191, 192
Electrophoretic fractionation, 202
Electrophoretic methods, 202
Electrophoretic mobility, 189, 190
Enriching mode, of foam fractionation, 12
  theory of operation, 21
Enrichment, in foam fractionation, 16
Equilibrium adsorption, 4
Equilibrium constant, in adsorption, 7
Equilibrium curves, in foam fractionation, 24, 25, 26, 27
External coalescence, in foams, 48, 51

F

Fibrogen, 188
Fickian diffusion, 145
Filling of gas chromatographic columns, 96, 98, 99
Film elasticity, in foams, 50
Film rupture, in foams, 50